Julia Newberry

Scientific Principles for Physical Geographers

Pearson
Education

Scientific Principles for Physical Geographers

Ian Bradbury, John Boyle and Andy Morse

An imprint of Pearson Education

Harlow, England · London · New York · Reading, Massachusetts · San Francisco · Toronto · Don Mills, Ontario · Sydney
Tokyo · Singapore · Hong Kong · Seoul · Taipei · Cape Town · Madrid · Mexico City · Amsterdam · Munich · Paris · Milan

Pearson Education Limited
Edinburgh Gate
Harlow
Essex CM20 2JE
England

and Associated Companies throughout the world

Visit us on the World Wide Web at:
www.pearsoneduc.com

First published 2002

ISBN 0 582 36936 3

British Library Cataloguing-in-Publication Data
A catalogue record for this book can be obtained from the British Library

Library of Congress Cataloging-in-Publication Data
Bradbury, Ian K., 1944–
Scientific principles for physical geographers / Ian Bradbury, John Boyle and Andy Morse.
p. cm.
Includes bibliographical references (p.).
ISBN 0-582-36936-3
1. Physical geography. 2. Physical sciences. I. Boyle, John. II. Morse, Andy. III. Title.

GB54.5.B69 2001
910′02–dc21

2001032733

10 9 8 7 6 5 4 3 2 1
05 04 03 02

Typeset in $9\frac{1}{2}/12\frac{1}{2}$pt Concorde BE by 60
Produced by Pearson Education Asia Pte. Ltd.
Printed in Malaysia, CLP

Brief contents

Contents

Preface

This book attempts to meet the need for an accessible introduction to those scientific principles necessary for mainstream physical geography courses in higher education. Those of us who teach in this area perennially face the problem of dealing with students who, while enthusiastic about the subject area, have received little scientific training or have forgotten much of what they learned. A number of factors are involved here: systematic physical geography appears to be given less prominence in many Advanced Level syllabuses (for many years, Advanced Level has been the standard qualification for entrance to degree programmes in England and Wales); many 'A' level students are not taking the basic sciences along with geography; the switch to an integrated science programme rather than individual science subjects at GCSE level (the standard qualification after five years at secondary school) has tended to reduce scientific content. The overall effect is that increasing numbers of students are entering higher education without the necessary scientific training required to cope with an understanding of environmental processes. We are also mindful that many geography teachers in pre- and post-16 education may not have received a training in the sciences that underpin their discipline.

Our feeling is that the usual ways of addressing the 'science problem' are not very satisfactory. If we emphasize the basic sciences in our introductory lectures we encourage a haemorrhage of potential physical geographers to what are (wrongly) perceived to be the softer options within the discipline. But even if we retain the students, the time allocated to scientific principles has an opportunity cost in terms of the course content. Attempting to explain scientific vocabulary and concepts when they are first mentioned is not very satisfactory because time constraints simply do not permit an effective job to be done. Furthermore, the more scientifically literate students may well find such material superfluous. An alternative approach is to suggest additional reading to bring all students up to a common level. In our experience, however, it is a rare student who shows a ready inclination to explore the literature of a scientific discipline they do not consider their own. In any case the student feels that what they are reading has not been written for their special needs. And they would of course be correct: certainly, the material is unlikely to have been placed in an appropriate context. In general we think there is a dearth of material that is sympathetic to the needs of physical geography students.

This project arose from a perceived need for a text that attempts to explain in as simple terms as possible, and from first principles, the science necessary for geography and other environmental studies. Our belief is that there is a scientific vocabulary and a set of scientific principles which are fundamental to an understanding of the way the physical environment 'works'. (We could probably describe them as 'physical', 'chemical', 'biological' and 'mathematical', but have chosen not to do so.) Furthermore, much of what we deal with can be understood at a descriptive level without compromising the science undesirably. After all, for most of our students most of the time it is a knowledge of the vocabulary and a familiarity with the concepts which are necessary to appreciate environmental processes and problems. We are seeking to provide a greater understanding, and to provide a foundation for further investigations. Given the constraints of most undergraduate teaching programmes in our discipline, our belief is that this book will be used primarily on a 'self-teach' basis. To this end we have adopted a conversational style which we hope will be accessible and sympathetic to those without much scientific background. Where possible, we relate the topics to everyday experience and provide examples of their significance for students of physical geography and the environment in general.

The parts of the book that deal with physics and mathematics are not designed to be a full account of all the topics required by an undergraduate studying physical geography. Rather, it is more a sample of some key areas, with worked examples that are intended to enhance confidence and ability in these areas. Many textbooks which cover certain aspects of geography or environmental science assume familiarity with general concepts in physics and mathematics. However, for many undergraduate readers this is an assumption too far.

If others identify gaps in the range of topics included, we hope they will not remain silent. The chosen topics reflect our own experience and interests, which collectively embrace the life sciences, physics, chemistry and geology. There may well be carping at the superficial (and non-mathematical) treatment of some of the contents. While acknowledging such criticism we would restate that our objective is to engage and stimulate our readers, not to lose them.

We would like to thank Sandra Mather, Head of the Graphics Unit in the Geography Department, University of Liverpool, for overseeing the preparation of illustrations so cheerfully and competently, and her colleague Suzanne Yee for her major contribution in this area. We are also very grateful to Matthew Smith of Pearson Education for his enthusiasm for the project, and for his patience as deadlines came and went, and for steering the book to publication.

1 How to use this book

1.1 Introduction

We think it will be useful for you to spend the few minutes necessary to read this part of the book because it provides a guide to its structure and also to how we think it should be used. It may also reveal a little about the authors, and help to establish a rapport with us. We have written this book primarily for students who are studying higher education courses in physical geography, or in another environmental area, but who have little recent experience or formal training in the sciences. We find that a lot of our students come into either, or both, of these categories, and we know from our colleagues elsewhere that this is a common experience. We firmly believe that your understanding of the physical environment, and environmental issues, will benefit enormously from an appreciation of some basic scientific concepts and an ability to use correctly the relevant vocabulary. You may be one of those many people who believe that even rudimentary scientific principles are in some way obscure and quite beyond your intellectual capabilities. If you are carrying such notions, we want you to discard them right here.

We envisage that most of you will be using the book on a self-study basis rather than to supplement material provided in formal courses: time pressures in the curriculum usually preclude satisfactory coverage of the type of material we present here. We approach the topics from first principles, and, where possible, we link them to everyday experience. We have tried hard to make the material accessible, always emphasizing descriptive explanations. You may well find that the book has been written in a more informal and personal manner than is usual for a text at this level. We have done this deliberately: not only has it helped us to communicate the various concepts but it also serves to remind you that the names on the cover are real people. We believe this approach to be justified given our objectives and the situations in which we envisage the book will be used. However, we hope those who are seeking particular items of information will not find the wording excessive. On the other hand, parts of the book may appear somewhat formidable at first sight, and for that reason introductory sections are included also. We ask you to tackle the various sections, sentence by sentence, and to use the cross-referencing as advised to make up for any deficiencies in understanding principles and vocabulary. Do not be overawed because some entries seem lengthy and may contain chemical or mathematical equations. The first point to make here is that if the entry is a lengthy one it is because we need that amount of time to explain the concepts in as simple a way as possible. A much shorter explanation with concise language and unexplained vocabulary could lose you. Secondly, we have been careful not to introduce concepts that require an appreciation of topics not covered in the book.

1.2 Structure

Because of the diverse array of topics we cover, the book cannot be wholly progressive. Certainly we do not intend it to be read in this way. In some ways it is more like a dictionary: where you enter it depends on what you want to find out. That said, the material is not ordered randomly, nor alphabetically. Rather, we have bunched together the various branches of the sciences early on to provide a foundation for the topics that follow. Later in the book there are topics of special relevance in understanding how the environment works. Although there is more of an 'applied' feel in these topics, we are still principally concerned with the science of the various processes but we link them to environmental process where possible. For cross-referencing to the sections and subsections of this book we have used emboldened numbers. These tell you where to look for an explanation of a term or concept, and in some cases we tell you what you will find in the indicated place.

1.3 Reading

We have provided some references that we think will reinforce and enhance your understanding of the topics introduced. Because of the literally thousands of texts introducing the various basic sciences, this is inevitably something of an arbitrary process. What we have done is select books that we think are particularly appropriate to our readership. Some have been aimed at geographers, some are exceptionally good specialist texts, and some are filed generally under 'popular science'. We are not using 'popular' here in a derogatory sense: the authors of these books are usually eminent in their own field but they also have the ability to explain quite difficult concepts in an engaging and stimulating way.

2 Units and measurement

2.1 Introduction: the SI system

Measurement is the act of determining the extent of something in relation to a standard. Standardization of measurement, while important in everyday affairs of course, is absolutely critical in science. **Units** are the dimensions or quantities on which measurements are based. For many of the phenomena we wish to measure, e.g. distance and weight, different units are in use, which can cause some confusion unless conversion values are known. In the interests of standardization the international scientific community has agreed a set of units, known as **SI units** from the French *Système International d'Unités*. For each unit there is an agreed symbol. Sometimes the symbol is a capital letter, or it begins with one. This indicates the symbol is named after someone. The other symbols consist of one or two lower case letters. The SI units for a range of common phenomena are introduced in this section together with some elaboration about what it is they measure. Some of the other units you may encounter, and the values used to convert them to the appropriate SI units, are also shown.

2.2 Base units

There are seven **base** quantities or units in the SI system, the most fundamental of which are mass, length and time. The seven base units are shown in Table 2.1. In this book we encounter mostly the first, second, third, fourth and sixth of these terms and their derived units. We shall come back to derived units in a moment.

The decimal system is used to express values as multiples or submultiples of the base unit. (The decimal system is based on the number 10.) Multiples and submultiples are usually chosen for ease of use or because of conventions within the particular field of study. It is helpful to make the significant

Table 2.1 Units of measurement.

Unit of	*Unit*	*Symbol*
Length	metre	m
Mass	kilogram	kg
Time	second	s
Temperature interval	kelvin	K
Electric current	ampere	A
Amount of substance	mole	mol
Luminous intensity	candela	cd

figures manageable. For example, we would say that the distance between two places is 120 kilometres rather than 120 000 metres. A set of prefixes is in use to indicate the multiple (or submultiples) being referred to, while power terms – which may be positive or negative – show the particular multiple of ten. (This is a useful application of the use of powers introduced in **26**; if in doubt, check.) As the metre (m) is the standard unit of length, for scientific purposes the distances above should strictly be expressed as 120×10^3 m (120 000 m). The prefixes used, together with the appropriate power terms, are shown in Table 2.2. Note carefully which symbols are capitals and which are in lower case. Note how the table is not arranged in a completely ascending or descending order but with the prefixes paired, e.g. a millionth prefix paired with a million. The symbol for micro is the Greek letter μ, and it denotes a millionth of the unit it prefixes. Mega, on the other hand, denotes a million of the unit it prefixes. You may notice that most of the power terms are multiples of three (10^3, 10^{-3}, 10^6, 10^{-6}, 10^9, 10^{-9}) and so on. This is the preferred way for presenting unit values. However, while the prefixes hecto-, deca-, deci- and centi- are not the preferred SI units, they are frequently used. For example, centimetres are widely employed because of convenience, although the metre is to be preferred for scientific purposes. In addition, there are units which are widely used but which do not appear in the table.

Table 2.2 Prefixes.

Prefix	*Symbol*	*Factor*	*Prefix*	*Symbol*	*Factor*
exa-	E	10^{18}	atto-	a	10^{-18}
peta-	P	10^{15}	femto-	f	10^{-15}
tera-	T	10^{12}	pico-	p	10^{-12}
giga-	G	10^9	nano-	n	10^{-9}
mega-	M	10^6	micro-	μ	10^{-6}
kilo-	K	10^3	milli-	m	10^{-3}
hecto-	H	10^2	centi-	c	10^{-2}
deca-/deka-	da	10^1	deci-	d	10^{-1}

Tonnes (1 tonne = 1000 kilograms), for example, are commonly used, although formally we should use megagrams. Time is another exception. We could say we will meet in 7200 seconds, but 2 hours is rather more conventional. Time merits further comment because it is based on a sexagesimal system (or base 60). There are 60 seconds in a minute and 60 minutes in an hour. It was the Babylonians who invented this system and it was subsequently adopted by the Greeks. It is worth noting that any calculation involving time is normally converted to seconds before carrying out the calculation.

2.3 Derived units

So far we have considered the base units of the SI system. In addition there are **derived units**, so called because they are derived from the base units. In the cases of area and volume, derived units are obtained by multiplying together base units, but in other cases they involve combinations of units. A good example of the latter is density, which is mass per unit volume. Density is used here to show how such units are expressed. We could write in full, 'kilograms per cubic metre' but it is more convenient to use kg/m^3, or $kg\,m^{-3}$. The slash in the former expression simply means 'per', and will be very familiar. It tells us that we divide the total mass (in kg) by the total volume (m^3) to obtain a value for the mass per unit of volume. However, it is the latter expression ($kg\,m^{-3}$) which is preferred for scientific purposes. The minus sign here can also be read as 'per': in effect it means we are dividing. It is an example of how, in a fraction, a positive exponent in the denominator becomes a negative exponent if it is moved to the numerator, and vice versa. That is to say, a/b can be written as $a\,b^{-1}$.

2.4 Measurement of the environment

The above has provided some important background; we are now in a position to look at some key units and consider aspects of their measurement.

2.4.1 Length (distance)

The unit of length is the metre (m). We can easily relate to lengths of a few millimetres, centimetres and kilometres, but objects of a few micrometres in size, as the name suggests, require a microscope to be seen. The dimensions of plant and animal cells are usually expressed in micrometres (μm). It is difficult to conceive of anything as small as a few nanometres (nm) but for some phenomena, for example radiation wavelength (**22**), nanometres are entirely appropriate. Also, we hear much about nanotechnology at present because of our increasing ability to use extremely tiny objects in science and industry.

Other common measures of length are: inches (1 inch = 0.0254 m or 2.54 cm), feet (1 foot = 0.3048 m), yards (1 yard = 0.9144 m), miles (1 mile = 1609.344 m), and nautical miles (1 international nautical mile = 1852 m).

2.4.2 Area

An area measure refers to the space occupied by a surface. It is expressed either as a unit of length (for example metres) followed by a power of two, as in 100 m^2, or simply as a unit of area (for example hectares or acres). The hectare is the metric unit of area (1 hectare = 10 000 m^2). To convert acres to hectares multiply by 0.4047.

The areas of surfaces which conform to certain geometric shapes can be calculated by formulae. For **squares** and **rectangles** it is the multiple of two adjacent sides; for triangles it is 0.5 × base × height; for circles it is pi × radius^2 where pi ≈ 3.14.

2.4.3 Volume

A volume is the amount of space occupied by a body. The standard unit for volume is the **cubic metre** (m^3), but as one cubic metre is rather too large for many objects, cubic centimetres (cm^3) are often used instead.

The volume of a liquid is normally expressed in **litres**: 1 litre = 1000 cm^3. One millilitre (1 ml), therefore, is the same as 1 cm^3. To convert British (imperial) gallons to litres, multiply by 4.546; to convert American gallons to litres multiply by 3.785.

Formulae are used to estimate volumes which conform to certain shapes. For a three-dimensional space in which all angles are 90°, the volume is the product of the three sides. For a **sphere** the volume is 1.333 × pi × radius^3. For a **cylinder** the volume is pi × radius^3 × length of cylinder (pi ≈ 3.14).

2.4.4 Mass and weight

Both **mass** and **weight** refer to a quantity of matter (see **21.3** and **21.6** for some elaboration). We think of weight as the 'heaviness' of a body, but more formally, the weight of a body is defined as the force with which that object is attracted by gravity. Mass, however, refers to the amount of matter in an object. So the mass of a body, for example a piece of equipment or a human being, is a constant, and it is independent of its location. In contrast, the weight of a body depends on its position. For example, a piece of equipment or a human being *weighs* much less on the moon than on the Earth. This is because the gravitational force on the moon is just one-sixth that on Earth. The reason for the large difference in gravitational forces between the Earth and the moon is the difference in their masses, which, as we have just remarked, refers to the amount of matter in a body. Because mass and weight are defined differently, strictly they should be expressed in different units: kilograms for mass and newtons for weight (one newton is defined formally as the force required to give a mass of one kilogram an acceleration of 1 m s^{-2}).

The SI unit of mass is the kilogram (i.e. 1000 grams). Milligrams (mg) and micrograms (μg) are commonly encountered, and so increasingly are nanograms (ng). At the other end of the scale, 1000 kg = 1 tonne.

Still in common use, particularly in the USA, are pounds (2.205 lb = 1 kg) and imperial tons (0.984 ton = 1 tonne).

2.4.5 Heat

As heat is a form of energy it is measured in joules (see **28**). The **temperature** of a body is defined in terms of the rate at which it gains or loses heat, and this in turn is an expression of the kinetic energy of its chemical constituents (**28.2.6**).

The **temperature scale** used for scientific purposes is the **kelvin** scale, named after Lord Kelvin (1824–1907). On this scale the zero point (0 K) is the lowest temperature theoretically attainable, while 273.15 K and 373.15 K are the freezing and boiling points respectively of pure water (at an atmospheric pressure of 1 atmosphere or 1.0135×10^5 pascals – see **2.4.8**). Note that on the kelvin scale K is used on its own, that is without degrees.

The **Celsius** scale, named after the Swedish scientist who proposed it in the mid-eighteenth century, is a very practical one because the freezing and the boiling points of pure water are set at 0 °C and 100 °C, respectively. The **Fahrenheit** scale is still in everyday use but should be avoided for scientific purposes.

Note that C = (F – 32)/1.8 where C and F are degrees Celsius and Fahrenheit, respectively.

2.4.6 Density

The density (D) of a substance or body is its mass (m) per unit volume (V), i.e. $D = m/V$. Different substances have different densities but, in addition, the density of a substance may change according to external conditions. The change in the density of water (see **41.4.3**) in response to a change in temperature is a prime example.

2.4.7 Force

A force is an agency that can bring about movement in a stationary body, cause a body already in motion to alter its velocity or direction, or change the size or shape of a body. The derived SI unit for force is the **newton** (N). It is defined as the force which gives a body of one kilogram mass an acceleration of one metre per second per second; i.e. 1 newton = 1 kg m s^{-2}. (A fuller account is provided in **21**.)

2.4.8 Pressure

Pressure is a force (see **2.4.7**) per unit area acting upon a surface. This relationship can be expressed simply as $P = F/A$ where P = pressure, F = force and A = area. This formalizes a common observation: the greater the area over which a given force is distributed, the lower the pressure. The SI unit for pressure is the **pascal** (Pa); 1 Pa = 1 N m^{-2} (N denotes newtons).

A number of other units of pressure are shown below. Multiplying by the given value converts them to pascals: atmospheres (×101 325); millimetres of mercury, mmHg (×133.32); pounds per square inch (×6883.49); bars (×100 000, i.e. 10^5); dynes cm^{-2} (×0.1).

2.4.9 Energy

Energy, i.e. the capacity to do work (see **28.1**), is expressed in a number of ways, but the SI unit is the **joule** (J). One joule is defined as the work done when a force of one newton moves through one metre in the direction of the force.

Other energy units are calories and ergs. Energy values of foodstuffs are still often expressed in calories. To convert calories to joules, multiply by 4.1860. However, it is important to distinguish calories of common parlance as just used from 'large' calories, written as Calories (note the capital C) or kilogram calories (kcals): 1 Calorie or Kcal = 1000 calories.

The energy content of a food substance can be determined by measuring the heat generated when a sample of known weight is completely combusted. The instrument used for this purpose is called a calorimeter. Another energy unit formerly in common use is the erg. To convert ergs to joules multiply by 10^7.

2.4.10 Power

Power is the *rate* of doing work. It is expressed in units of work per unit time. The derived SI unit of power is the **watt** (W), one watt being equal to one joule per second. (The joule is defined in **2.4.9**.) To convert horsepower to watts multiply by 745.7.

2.5 Dimensions and dimensional analysis

Units are very important. At the simplest level this is illustrated by statements such as: 'the temperature is five degrees'. Clearly, unless we know the units such information is worthless. Dimensional analysis takes this focus on units one step further. The central principle is that equations used to describe physical relations must have the same units on each side. Thus, for an equation such as $F = ma$ (force = mass × acceleration), F must have the same fundamental units as ma. We stress the word 'fundamental', because most of the units we use are not fundamental: they are what we term *derived* units. This is clear from the example above. On the left side of the equation we have force, typically measured in newtons (N). On the right we have mass (kg) times acceleration ($m\,s^{-2}$), which gives a combined unit of $kg\,m\,s^{-2}$. Thus, in terms of our derived units, the two sides of the equation *appear* different. In fact, the derived unit of the newton is actually also $kg\,m\,s^{-2}$ when expressed in fundamental units, so the two sides of the equation do use equivalent units. Formally, we say that both sides of the equation have the same **dimensions**.

While this may appear rather abstract, there are a number of useful aspects to a consideration of the dimensions of an equation:

- We commonly alter or combine equations for convenience. It is easy to make mistakes in doing this, and verifying that our final derived equation has the same fundamental dimensions is very important.
- More commonly, we are using a known equation but the parameters may be measured in different derived units. For example, $F = ma$ does not hold if m is in grams, while F is in newtons and a in $m\,s^{-2}$.
- Finally, we often have an equation for deriving a parameter yet do not know in what units the result is measured. Using dimensional analysis we can derive the units from the equation, as long as we know the units of the measured or given parameters.

In order to consider the dimensions of a unit, we first need to identify the fundamental units. These are mass, length and time, which in the SI system have the units kilogram, metre and second, respectively. In dimensional analysis it is conventional to avoid the units, and to refer to the dimensions using the symbols M (mass), L (distance) and T (time). (It is important not to muddle M (mass) with m (metres).) Consider some examples:

- Density is measured in $kg\,m^{-3}$, hence its dimensions are ML^{-3}.
- Acceleration is measured in $m\,s^{-2}$, hence its dimensions are LT^{-2}.
- Force (see **2.4.7**) is measured in newtons. These can be expressed as $kg\,m\,s^{-2}$, hence the dimensions of force are MLT^{-2}.
- Pressure (**2.4.8**) is measured in pascals (Pa). This can be expressed as newtons per metre squared ($N\,m^{-2}$). From the point above, newtons can be expressed dimensionally as MLT^{-2}. Combining these (i.e. $MLT^{-2} \times L^{-2}$) gives $ML^{-1}T^{-2}$.

One useful case is where several terms of an equation are included within brackets. For example, if:

$$a = b(c + d + e)$$

Then c, d and e must all have the same dimensions.

Worked example

Show by using dimensional analysis that the pressure of a column of water is equal to its density (ρ) × the acceleration due to gravity (g) × the height of the column (h). The dimensions of these are:

$$\rho = ML^{-3} \qquad g = LT^{-2} \qquad h = L$$

The product of these three terms includes:

- One incidence of M. This is simply as M. Thus, the combined expression includes M in its dimensions.

- Three incidences of L. The three incidences must be multiplied: $L^{-3} \times L \times L = L^{-1}$. Thus the combined expression includes L^{-1} in its dimensions.
- One incidence of T. This is as T^{-2}. Thus, the combined expression includes T^{-2} in its dimensions.

Combining the three different dimensions, we are left with $M\,L^{-1}\,T^{-2}$. These are the same dimensions as we derived for pressure in the previous examples.

2.6 Scalars and vectors

You may come across the terms **scalar** and **vector** in some textbooks. Simply put, scalar quantities are those that have a magnitude and no direction. For some variables, such as temperature, mass, pressure, energy and time, direction is not important. So the speed (of a car or the wind) is also a scalar quantity because it does not specify the direction of movement. In contrast, vector quantities have both magnitude *and* direction. Velocity and force are examples of vector quantities.

2.7 Useful symbols

2.7.1 The Greek alphabet

Many scientific and mathematical constants and variables are represented as Greek symbols. It is therefore useful to recognize them and to be familiar with their names (Table 2.3).

Table 2.3 The Greek alphabet.

Letter	*Name*	*Letter*	*Name*
A α	alpha	N ν	nu
B β	beta	Ξ ξ	xi
Γ γ	gamma	O o	omicron
Δ δ	delta	Π π	pi
E ε	epsilon	P ρ	rho
Z ζ	zeta	Σ σ	sigma
H η	eta	T τ	tau
Θ θ	theta	Y υ	upsilon
I ι	iota	Φ ϕ	phi
K κ	kappa	X χ	chi
Λ λ	lambda	Ψ ψ	psi
M μ	mu	Ω ω	omega

Table 2.4 Mathematical symbols.

Symbol	*Description/name*	*Symbol*	*Description/name*
$+$	plus	$>$	is greater than
$-$	minus	$\leq$	is equal to or less than
$\div$ or /	divided by	$\geq$	is equal to or greater than
$\times$ or .	multiplied by	$\pm$	plus or minus
$=$	equals	∞	infinity
$\neq$	not equal to	$\sum$	summation of
$\equiv$	identical to	$\surd$	square root
$\approx$	approximately equal to	δx	small change in x
$<$	is less than	Δx	large change in x

2.7.2 Mathematical symbols

Table 2.4 lists some of the common mathematical symbols you are likely to encounter. For a full list you should see a formulae or science data book.

2.8 A few simple examples

1 *Time*. How would 86.4 megaseconds normally be expressed?

2 *Area and volume*. If we want to measure the area of a room, we measure the width and length of the room. It might be, for example, 3 m by 4 m. This room would be 12 square metres (or 12 m^2), which means you could fit 12 tiles each of 1 square metre (1 m by 1 m) on the floor. If the room is 3 m in depth (height), then the volume of the room would be the area multiplied by the depth or 3 m by 4 m by 3 m, which equals 36 cubic metres (metres by metres by metres, or m^3). This means you could fit 36 separate 1 m by 1 m by 1 m boxes in the room. So the simple base unit for length, i.e. metres, can be used to derive square and cubic measurements.

Confusion can arise with the presentation of squared measurements. You will hear people say ‘twenty metres squared’ when they refer to an area that has dimensions of 20 m by 20 m. They should say ‘twenty square metres’, which is 20 m^2. ‘Twenty metres squared’ is represented as $(20\ m)^2$. Be aware of the difference!

3 Chemical elements

3.1 Definition and symbols

Matter is defined as that which occupies space (see **18.1**). Our discussion of the composition of matter begins with the **chemical elements**. The term 'element' is used for substances which cannot be further divided by chemical means. All the matter in the universe is made up of chemical elements. There are 92 naturally occurring elements and a further 20 or so have been produced in the laboratory. Many elements – carbon, calcium, sulphur and lead, for example – have names that are in everyday use, but others are rare and their names much less familiar.

Each element is represented by a symbol, either a single capital letter, for example C for carbon, N for nitrogen, K for potassium, or a capital followed by a single lower case letter, for example Ca for calcium and Pb for lead. Conveniently, the symbol is quite often the first letter of the element's English name, or it begins with the first letter of the English name, but this is by no means a universal rule, as just shown for potassium and lead. In fact when this scheme was originally proposed, by the Swedish chemist Jöns Berzelius in 1813, it was the initial letter of the Latin name that was used, followed where necessary by a letter from within the word. Previously it had been necessary to spell out the element name, a very cumbersome process. A full list of chemical elements is shown in Tables 3.1 and 3.2. You may like to glance at these lists to see which element names are familiar, and ask yourself what you know about them. You may be surprised that there are so many element names you have never encountered before. In Table 3.1 the elements are listed in order of increasing atomic number. The meaning of this term is explained in **4.1**. Atomic masses are also shown, a concept which is introduced in **4.2**. Table 3.2 lists elements in alphabetical order: this will be useful when you want to check an element's atomic number or mass but are not sure of its position in the order.

Table 3.1 List of chemical elements in ascending order of atomic number.

Element	***Symbol***	***Atomic number***	***Atomic mass***	***Element***	***Symbol***	***Atomic number***	***Atomic mass***
Hydrogen	H	1	1.008	Nickel	Ni	28	58.69
Helium	He	2	4.003	Copper	Cu	29	63.55
Lithium	Li	3	6.941	Zinc	Zn	30	65.39
Beryllium	Be	4	9.012	Gallium	Ga	31	69.72
Boron	Bo	5	10.81	Germanium	Ge	32	72.59
Carbon	C	6	12.01	Arsenic	As	33	74.92
Nitrogen	N	7	14.01	Selenium	Se	34	78.96
Oxygen	O	8	16.00	Bromine	Br	35	79.90
Fluorine	F	9	19.00	Krypton	Kr	36	83.80
Neon	Ne	10	20.18	Rubidium	Rb	37	85.47
Sodium	Na	11	22.99	Strontium	Sr	38	87.62
Magnesium	Mg	12	24.31	Yttrium	Y	39	88.91
Aluminium	Al	13	26.98	Zirconium	Zr	40	91.22
Silicon	Si	14	28.09	Niobium	Nb	41	92.91
Phosphorus	P	15	30.97	Molybdenum	Mo	42	95.94
Sulphur	S	16	32.07	Technetium	Tc	43	(98)
Chlorine	Cl	17	35.45	Ruthenium	Ru	44	101.1
Argon	Ar	18	39.95	Rhodium	Rh	45	102.9
Potassium	K	19	39.10	Palladium	Pd	46	106.4
Calcium	Ca	20	40.08	Silver	Ag	47	107.9
Scandium	Sc	21	44.96	Cadmium	Cd	48	112.4
Titanium	Ti	22	47.88	Indium	In	49	114.8
Vanadium	V	23	50.94	Tin	Sn	50	118.7
Chromium	Cr	24	52.00	Antimony	Sb	51	121.8
Manganese	Mn	25	54.94	Tellurium	Te	52	127.6
Iron	Fe	26	55.85	Iodine	I	53	126.9
Cobalt	Co	27	58.93	Xenon	Xe	54	131.3

Table 3.1 (continued)

Element	*Symbol*	*Atomic number*	*Atomic mass*	*Element*	*Symbol*	*Atomic number*	*Atomic mass*
Caesium	Cs	55	132.9	Mercury	Hg	80	200.6
Barium	Ba	56	137.3	Thallium	Tl	81	204.4
Lanthanum	La	57	138.9	Lead	Pb	82	207.2
Cerium	Ce	58	140.1	Bismuth	Bi	83	209.0
Praseodymium	Pr	59	140.9	Polonium	Po	84	(209)
Neodymium	Nd	60	144.2	Astatine	At	85	(210)
Promethium	Pm	61	(145)	Radon	Rn	86	(222)
Samarium	Sm	62	150.4	Francium	Fr	87	(223)
Europium	Eu	63	152.0	Radium	Ra	88	(226)
Gadolinium	Gd	64	157.3	Actinium	Ac	89	(227)
Terbium	Tb	65	158.9	Thorium	Th	90	232.0
Dysprosium	Dy	66	162.5	Protactinium	Pa	91	(231)
Holmium	Ho	67	164.9	Uranium	U	92	238.0
Erbium	Er	68	167.3	Neptunium	Np	93	(237)
Thulium	Tm	69	168.9	Plutonium	Pu	94	(244)
Ytterbium	Yb	70	173.0	Americum	Am	95	(243)
Lutetium	Lu	71	175.0	Curium	Cm	96	(247)
Hafnium	Hf	72	178.5	Berkelium	Bk	97	(247)
Tantalum	Ta	73	180.9	Californium	Cf	98	(251)
Tungsten	W	74	183.9	Einsteinium	Es	99	(252)
Rhenium	Re	75	186.2	Fermium	Fm	100	(257)
Osmium	Os	76	190.2	Mendelevium	Md	101	(258)
Iridium	Ir	77	192.2	Nobelium	No	102	(259)
Platinum	Pt	78	195.1	Lawrencium	Lr	103	(260)
Gold	Au	79	197.0				

Table 3.2 List of chemical elements in alphabetical order. Atomic number refers to the number of protons in the nucleus of an atom of the element. Relative atomic masses are with reference to carbon-12. The mass values in brackets are for the most stable isotope. Other mass values are for the isotopic ratios in which the elements are normally found.

Element	*Symbol*	*Atomic number*	*Atomic mass*	*Element*	*Symbol*	*Atomic number*	*Atomic mass*
Actinium	Ac	89	(227)	Einsteinium	Es	99	(252)
Aluminium	Al	13	26.98	Erbium	Er	68	167.3
Americum	Am	95	(243)	Europium	Eu	63	152.0
Antimony	Sb	51	121.8	Fermium	Fm	100	(257)
Argon	Ar	18	39.95	Fluorine	F	9	19.00
Arsenic	As	33	74.92	Francium	Fr	87	(223)
Astatine	At	85	(210)	Gadolinium	Gd	64	157.3
Barium	Ba	56	137.3	Gallium	Ga	31	69.72
Berkelium	Bk	97	(247)	Germanium	Ge	32	72.59
Beryllium	Be	4	9.012	Gold	Au	79	197.0
Bismuth	Bi	83	209.0	Hafnium	Hf	72	178.5
Boron	Bo	5	10.81	Helium	He	2	4.003
Bromine	Br	35	79.9	Holmium	Ho	67	164.9
Cadmium	Cd	48	112.4	Hydrogen	H	1	1.008
Caesium	Cs	55	132.9	Indium	In	49	114.9
Calcium	Ca	20	40.08	Iodine	I	53	126.9
Californicum	Cf	98	(251)	Iridium	Ir	77	192.2
Carbon	C	6	12.01	Iron	Fe	26	55.85
Cerium	Ce	58	140.1	Krypton	Kr	36	83.80
Chlorine	Cl	17	35.45	Lanthanium	La	57	138.9
Chromium	Cr	24	52.00	Lawrencium	Lr	103	(260)
Cobalt	Co	27	58.93	Lead	Pb	82	207.2
Copper	Cu	29	63.55	Lithium	Li	3	6.941
Curium	Cm	96	(247)	Lutetium	Lu	71	175.0
Dysprosium	Dy	66	162.5	Magnesium	Mg	12	24.31

Table 3.2 (continued)

Element	*Symbol*	*Atomic number*	*Atomic mass*	*Element*	*Symbol*	*Atomic number*	*Atomic mass*
Manganese	Mn	25	54.94	Rubidium	Rb	37	85.47
Mendelevium	Md	101	(258)	Ruthenium	Ru	44	101.1
Mercury	Hg	80	200.6	Samarium	Sm	62	150.4
Molybdenum	Mo	42	95.94	Scandium	Sc	21	44.96
Neodymium	Nd	60	144.2	Selenium	Se	34	78.96
Neon	Ne	10	20.18	Silicon	Si	14	28.09
Neptunium	Np	93	(237)	Silver	Ag	47	107.9
Nickel	Ni	28	58.69	Sodium	Na	11	22.99
Niobium	Nb	41	92.91	Strontium	Sr	38	87.62
Nitrogen	N	7	14.01	Sulphur	S	16	32.07
Nobelium	No	102	(259)	Tantalum	Ta	73	180.9
Osmium	Os	76	190.2	Technetium	Tc	43	(98)
Oxygen	O	8	16.00	Tellurium	Te	52	127.6
Palladium	Pd	46	106.4	Terbium	Tb	65	158.9
Phosphorus	P	15	30.97	Thallium	Tl	81	204.4
Platinum	Pt	78	195.1	Thorium	Th	90	232.0
Plutonium	Pu	94	(244)	Thulium	Tm	69	168.9
Polonium	Po	84	(209)	Tin	Sn	50	118.7
Potassium	K	19	39.10	Titanium	Ti	22	47.88
Praseodymium	Pr	59	150.9	Tungsten	W	74	183.9
Promethium	Pm	61	(145)	Uranium	U	92	238.0
Protactinium	Pa	91	(231)	Vanadium	V	23	50.94
Radium	Ra	88	(226)	Xenon	Xe	54	131.3
Radon	Rn	86	(222)	Ytterbium	Yb	70	173.0
Rhenium	Re	75	186.2	Yttrium	Y	39	88.91
Rhodium	Rh	45	102.9	Zinc	Zn	30	65.39
				Zirconium	Zr	40	91.22

Table 3.3 The most abundant elements of the lithosphere and atmosphere.

Crust element	*Percentage (by mass)*	*Atmosphere element*	*Percentage (by volume)*
Oxygen (O)	46.6	Nitrogen (N)	78.1
Silicon (Si)	27.7	Oxygen (O)	21.0
Aluminium (Al)	8.1	Argon (Ar)	0.9
Iron (Fe)	5.0		
Calcium (Ca)	3.6		
Sodium (Na)	2.8		
Potassium (K)	2.6		
Magnesium (Mg)	2.1		

3.2 Abundance in the environment

During the course of this book it will become clear which are the dominant elements in our environment in terms of amounts, and which elements play key roles in environmental processes. We also learn how elements combine with each other, and which form molecules on their own. As an introduction, we show in Table 3.3 the eight most abundant elements in the Earth's crust (**43.2**) and the three most abundant elements in the Earth's lower atmosphere (**43.4**). What we notice is that a handful of elements dominate our environment. However, it is important not to equate element abundance with importance. For example, carbon does not appear in these lists. It is in fact the seventeenth most abundant element in the crust overall and accounts for just a small fraction of 1 per cent of the atmosphere. Yet carbon contributes nearly 50 per cent of the dry weight of the living part of the planet (i.e. the Earth's **biota**).

4 Atomic structure

4.1 Nature of the atom and atomic number

On one level, all substances are composed of the hundred or so chemical elements (**3.1**), but at another level all matter, whether solid, liquid or gas, is made up of extremely tiny entities known as **atoms**. 'Extremely tiny' is in fact a barely appropriate description because the size of atoms is of the order of one ten-billionth of a metre, i.e. 10^{-10} m, and between 1×10^{-29} and 1×10^{-31} kilograms. The important point to grasp now is that the atoms of the various elements, although conforming to a basic structure in terms of the particles they contain, differ from one another in the number, and to some extent the arrangement, of particles. Knowing something about the structure of atoms is therefore essential for appreciating what it is that makes one element different from another, and also for understanding the chemical processes dealt with in this book. The details of atomic structure were worked out in the first three decades of the twentieth century, although some far-sighted individuals had proposed the existence of atom-like particles long before. The term 'atom' means 'something tiny' and indivisible, and it was adopted for use by scientists before developments in atomic structure revealed the term to be not entirely appropriate.

First, we present a simple scheme for the structure of an atom. Each atom is made up of a central **nucleus**, which is surrounded by **electrons** which are in constant motion. The nucleus accounts for practically all the mass of an atom, but it makes up just a tiny fraction of an atom's volume. The nucleus is made up of two key particles, called **protons** and **neutrons**. The number of protons in the nucleus distinguishes the atoms of one element from the atoms of another element. This number is known as the **atomic number**. So atoms with the same atomic number are atoms of one particular element. The atomic number of each of the elements is shown in Tables 3.1 and 3.2. For now, just note that atomic numbers range from 1 (for hydrogen) to a little over 100. When it is necessary to show the atomic number of an individual element it is conventionally written as a preceding subscript. In the case of iron, for example, it is ${}_{26}Fe$. Another crucial piece of information is that each proton carries a single positive electrical charge.

4.2 Neutrons, mass numbers and isotopes

Atomic nuclei contain particles called **neutrons** as well as protons. (The single exception is the common form of hydrogen.) Neutrons have no charge, i.e. they are electrically neutral. All but a few of the elements have atomic nuclei with different numbers of neutrons: in other words, the atoms of most of the elements occur in different forms. These forms are known as **isotopes**. Most elements therefore consist of a mixture of isotopes. There is a certain number of isotopes of each element, but one is usually overwhelmingly predominant. Furthermore, the isotopes usually occur in characteristic ratios. To refer to a particular isotope of an element we use the **mass number**, which is the sum of the number of protons (i.e. the atomic number) and the number of neutrons. This number appears either after the element name (for example carbon-12, carbon-13 and carbon-14) or as a superscript preceding the element symbol (${}^{12}C$, ${}^{13}C$, ${}^{14}C$). As a carbon atom has six protons in the nucleus, the number of neutrons in these three isotopes is six, seven and eight respectively. Sometimes it is necessary to show both the atomic number and mass number when referring to a particular isotope. Using carbon-14 as an example, the conventional way of presenting this is as ${}^{14}_{6}C$. As in the case of carbon, mass numbers for the isotopes of a particular element are similar, usually in sequence in fact. This indicates that the number of neutrons actually varies little between the isotopes of a single element. A few elements occur naturally in only one isotopic form – sodium, aluminium and gold are examples – but two to seven isotopes is more typical. As a general rule, all the isotopes of a particular element behave chemically in the same way. However, because of differences in mass (on account of differences in neutron number) physical properties vary a little.

It is worth introducing the term **nuclide** here. Nuclide is used to refer to a particular type of atom or nucleus as determined by the number of protons and neutrons. It might appear that isotope and nuclide can be used almost interchangeably, which in certain circumstances is true. The difference is that the term 'isotope' is most frequently used when referring to a member of a series of atoms with the same atomic number but different numbers of neutrons, as in the case of ${}^{12}C$, ${}^{13}C$ and ${}^{14}C$.

Isotopes (and nuclides) are categorized as either **stable** or **radioactive** (**radioisotopes**, **radionuclides**). The essential features of radioactivity, and some of its applications for environmental study, are discussed elsewhere (see **24** and **25**). We may just note here that some radioisotopes are natural but the majority are produced artificially.

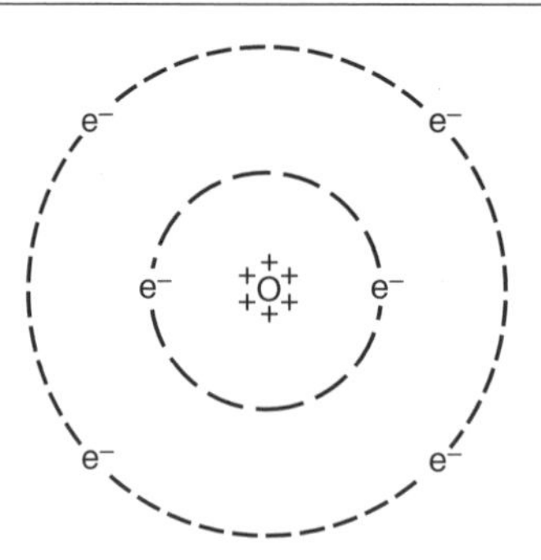

Figure 4.1 The simplest way of representing atomic structure. The example is an atom of carbon. The nucleus, with six protons, is surrounded by six electrons arranged in two shells (represented by the dashed circles).

4.3 Electrons

Surrounding the nucleus are the electrons, each of which carries a single negative charge. Each one has a mass approximately 1/1840 that of a proton. At this stage we can regard the number of electrons as being equal to the number of protons in the nucleus, which means that an atom is an electrically neutral entity. It follows also that the number of electrons in an atom is also the atomic number because it is equal to the number of protons. The configuration of electrons is depicted in various ways. It is simplest to regard the electrons as moving around the nucleus in one or more circular orbits or **shells**, as in Figure 4.1. This is appropriately known as the planetary model: it was the earliest way of representing atomic structure but has been superseded by more realistic models. The atoms of each element have a characteristic number of shells, and a characteristic number of electrons within each. In fact there are certain rules governing the number of electrons which are found in each shell (see **7**).

5 Ions and ionization

In **4.3** atoms are described as electrically neutral entities. Under some circumstances, however, they can gain or lose electrons, and thus become electrically charged. This process is called **ionization** and it gives rise to entities known as **ions**. Some of the very common elements in the environment exist principally in ionic form. When one or more electrons are lost from an electrically neutral atom, the number of protons in the atomic nucleus exceeds the number of electrons. The result is a positively charged entity, called a **cation**. Conversely, when one or more electrons is acquired by an electrically neutral atom, the number of protons in the nucleus is less than the number of electrons. The result is a negatively charged entity, called an **anion**.

The number of excess positive or negative charges is low, usually between one and three, and is shown after the element symbol. Thus potassium, which commonly occurs as a cation, with one unit of positive charge, is represented as K^+. Calcium, which commonly occurs as a cation with two units of positive charge, is shown as Ca^{2+} (or Ca^{++}). The element chlorine commonly occurs in anionic form, known as chloride, with one unit of negative charge, and is therefore denoted by Cl^-. As with the elements, the symbols can be used to indicate not only the type of ion but also the number of ions. So Cl^- could represent a single chloride ion while $(Cl^-)_2$ would represent two chloride ions. A few elements occur naturally in more than one ionic form. Iron (Fe) is a good example, and one of great environmental importance. If an iron atom loses two electrons, it carries two positive charges (represented as Fe^{2+}), but under certain conditions such ions can lose another electron so that the resulting cation has three excess positive charges (represented as Fe^{3+}). Each of these species of iron has its own name: ferrous ions and ferric ions, respectively. Ions are described as monovalent, divalent, trivalent or tetravalent according to the number of excess charges. Thus K^+ and Br^- are monovalent ions while Mg^{2+} is a divalent ion.

Groups, or aggregates, of atoms of different sorts may also be positively or negatively charged as the result of losing or acquiring electrons. They are known as **polyatomic ions** to distinguish them from the **monoatomic** ions which contain atoms of only one type. One example is the anion known as nitrate (NO_3^-); another is the cation called ammonium (NH_4^+). In your reading about the environment you will continually encounter ions. The way in which ions are represented in symbolic form may suggest that ions can exist as independent entities, and could be stored as such. This is not the case, however. While a cation is not linked to a specific anion, the cations and anions will balance electrically.

6 Molecules and compounds

6.1 Molecules

The atoms of nearly all the elements combine with atoms of certain other elements, and in some cases just with atoms of the same element, to form **molecules**. Formally defined, a molecule is the smallest part of a substance that can exist independently and retain the properties of that substance. To appreciate what this means we provide some examples. First we consider molecules formed from atoms of one type. Nitrogen is one of the key elements in the environment; for example, nitrogen gas accounts for nearly 80 per cent of the Earth's atmosphere. Individual nitrogen atoms are very unstable, but they readily combine with each other to form 'nitrogen pairs'. These entities, which are very stable, are called molecules, and are represented as N_2. The term dinitrogen is also used. Similarly, the most common form of 'free' oxygen is the molecule dioxygen, represented by O_2 to denote that it consists of a pair of oxygen atoms. Incidentally, oxygen forms another molecular gas of critical environmental significance. This is ozone, which is represented by O_3 to denote that it is composed of three oxygen atoms.

6.2 Compounds

Only a handful of elements can exist on their own in molecular form. Most elements, and this includes nitrogen and oxygen, occur naturally in combination with other elements. When more than one element is involved in a fixed ratio in the formation of a stable substance, that substance is referred to as a **compound**. Some examples are methane (CH_4), sulphuric acid (H_2SO_4) and glucose ($C_6H_{12}O_6$). In some compounds the constituent elements occur in ionic form (see **5**), in which case the substance is known as an **ionic compound**. Common salt (NaCl) is a good example of an ionic compound.

When this substance is dissolved in water the sodium and chlorine ions exist more-or-less as independent entities, i.e.

$$NaCl \rightarrow Na^+ + Cl^-$$

6.3 Molecular and empirical formulae

The **molecular formula** of a compound shows the numbers of atoms of the various elements of which one molecule is composed: for example, in methane, which has a molecular formula of CH_4, there are four hydrogen atoms per carbon atom. But consider the molecular formula for glucose ($C_6H_{12}O_6$). We could divide the number of atoms of each element by six to derive the **empirical formula** (CH_2O), and thus show only the ratios of the atoms in a molecule of the compound, not the absolute number of atoms. For methane (CH_4), the molecular and empirical formulae are obviously the same. Incidentally, glucose is interesting in another respect. It is one of several sugars (fructose is another) which share the same molecular formula. The difference between these compounds lies not in the numbers of the three types of atom of which they are composed but in the geometric arrangement, or configuration, of the atoms (see **16.6**).

By itself, the molecular formula tells us little about the configuration of atoms in a compound: for this we need to see the arrangement of atoms in diagrammatic form, i.e. the structural formula. However, it should be borne in mind that to represent this on the page we have to collapse what in reality is a three-dimensional form to just two dimensions. So, for a more realistic impression, we need to construct a three-dimensional model.

While there is general agreement concerning the naming of the chemical elements, there is less uniformity regarding the naming of compounds. Also, many names in common use give no clues as to which elements the compound contains. One example is methane. In other cases, and this is particularly true of binary compounds (containing two elements), the element names are indicated in the compound name. Examples are carbon dioxide and hydrogen chloride. In the former case it is not only the element names that are provided, but the ratio of oxygen to carbon atoms is shown by the prefix 'di-'.

A final, and key, point is that the chemical characteristics of a compound can rarely be predicted by the properties of the constituent elements. For example, we could not predict from their appearance that the metal sodium and the green gas chlorine would combine to form a white crystalline solid.

7 The configuration of electrons

7.1 Numbers of electrons and energy levels

Electrons can be portrayed as negatively charged entities moving around the atomic nucleus in orbits, or shells (see **4.3**). Important additional points here are, first, that the atoms of each element have a characteristic number of shells, and second, that within each of these shells there is a characteristic number of electrons. Knowing these fundamentals is vital, but in order to understand better some important chemical phenomena we need to look more closely at electrons and their shells.

Depending on the element, the number of shells varies between one and seven. The individual shells are denoted by letters or by numbers. The letters used run from K (the innermost or first shell), through L, M, N, O, P to Q (the seventh shell). The more recent scheme uses shell number from 1 to 7 in numerical order. Shell number 1 is the same as shell K, shell 2 is the same as shell L and so on. Now the electrons in a given shell possess approximately the same amount of energy, and the greater the distance of an electron from the nucleus then the greater is its energy level.

First we look at the principles of 'filling' shells with electrons. The shell 'closest' to the nucleus is said to 'fill' first. This shell holds a maximum of two electrons. The elements hydrogen and helium, which respectively have one and two electrons per atom, thus have just one electron shell. The element lithium, however, because it has three electrons per atom, has a second electron shell, and this contains a single electron. However, the second electron shell of an atom can hold up to eight electrons, and it is filled next. A look at the list of elements (refer back to Table 3.1) will quickly reveal the element which contains two electrons in its first shell and eight electrons in its second shell (i.e. atomic number 10): it is neon, an unreactive gas which is found in tiny concentrations in the atmosphere. Clearly, the atoms of elements with atomic numbers exceeding 10 must have three or more shells to accommodate all the electrons because the first two shells are full. (The atomic number is the number of protons or electrons in an atom; atoms exist with all possible atomic numbers from 1 to over 100.)

Another key point is that in all shells other than the first the electrons are distributed among **subshells**, each of which also holds a maximum number of electrons. The second shell has two subshells, the inner holding up to two electrons and the outer holding up to six electrons. Obviously, if there are only two electrons in the second shell only one subshell will be occupied. The next shell, the third, has three subshells and these hold up to two, six and ten electrons respectively.

We have seen that shells are designated by letter (K, L, M, N, O, P, Q) or by number (1, 2, 3, 4, 5, 6, 7). There is also a system of labelling the subshells. These are denoted by the letters *s*, *p*, *d* and *f*. So we can specify precisely a particular subshell, as for example 2*p* and 5*s*. Such designations are shown in Table 7.1. This table presents the configuration of electrons for elements with an atomic number of 36 and less. It is also possible to summarize the electron configuration of an element, using a superscript to denote the number of electrons associated with each subshell. You may care to verify that an atom of phosphorus has an electron structure of $1s^2 2s^2 2p^6 3s^2 3p^3$. When we are dealing with ions, the number of electrons in the outer shell will be slightly greater or fewer than for atoms of the same element.

The filling of subshells with electrons according to the number they can hold is relatively straightforward for elements with atomic numbers between 1 and 18. But now look at the element potassium (K) with an atomic number of 19. Following the pattern of elements with a lower atomic number, we would expect the 3*d* subshell of a potassium atom to hold just one electron. However, this is not the case. Rather, it is the 4*s* subshell that fills first, and thus holds one electron. And the next element in the series, calcium (Ca), has two 4*s* electrons.

In order to understand the filling of subshells we again need to think of electrons in terms of their energy levels and not simply in terms of their numbers. First, for each type of subshell (*s*, *p*, *d*, *f*) the energy level increases sequentially with an increase in shell number, i.e. from the innermost shell outwards. So the energy level of a 4*s* electron is greater than that of a 1*s* electron. Secondly, there are differences in energy levels between subshells *within* each shell: the order is $f > d > p > s$ (although in fact it is only shells 4 and 5 of certain elements that have four subshells).

Crucially, the order in which subshells fill with electrons is determined by their relative energy levels. Subshells of relatively low energy levels fill before subshells of higher energy level. Now, in shell 4 the energy of the *s* (inner) subshell is slightly less than the energy level of the most highly energetic subshell in the underlying shell, which is 3*d*. Now look again at potassium (K) in Table 7.1. The reason the atoms of this element have a lone electron in the 4*s* subshell, and none in the 3*d* subshell, is that the energy level of the 4*s* subshell is less than that of the 3*d* subshell. The 4*s* subshell thus fills first. The next element after potassium, which is calcium (Ca), has two electrons in the 4*s* subshell, but none in the 3*d* subshell. (This does not explain everything because you will see that for elements with less than ten

Table 7.1 The configuration of electrons for elements with an atomic number of 36 and less.

	Shell	***1***	***2***		***3***			***4***			
	Subshell	***1s***	***2s***	***2p***	***3s***	***3p***	***3d***	***4s***	***4p***	***4d***	***4f***
Element	***Atomic no.***										
Hydrogen	1	1									
Helium	2	2									
Lithium	3	2	1								
Beryllium	4	2	2								
Boron	5	2	2	1							
Carbon	6	2	2	2							
Nitrogen	7	2	2	3							
Oxygen	8	2	2	4							
Fluorine	9	2	2	5							
Neon	10	2	2	6							
Sodium	11	2	2	6	1						
Magnesium	12	2	2	6	2						
Aluminium	13	2	2	6	2	1					
Silicon	14	2	2	6	2	2					
Phosphorus	15	2	2	6	2	3					
Sulphur	16	2	2	6	2	4					
Chlorine	17	2	2	6	2	5					
Argon	18	2	2	6	2	6					
Potassium	19	2	2	6	2	6		1			
Calcium	20	2	2	6	2	6		2			
Scandium	21	2	2	6	2	6	1	2			
Titanium	22	2	2	6	2	6	2	2			
Vanadium	23	2	2	6	2	6	3	2			
Chromium	24	2	2	6	2	6	5	1			
Manganese	25	2	2	6	2	6	5	2			
Iron	26	2	2	6	2	6	6	2			
Cobalt	27	2	2	6	2	6	7	2			
Nickel	28	2	2	6	2	6	8	2			
Copper	29	2	2	6	2	6	10	1			
Zinc	30	2	2	6	2	6	10	2			
Gallium	31	2	2	6	2	6	10	2	1		
Germanium	32	2	2	6	2	6	10	2	2		
Astatine	33	2	2	6	2	6	10	2	3		
Selenium	34	2	2	6	2	6	10	2	4		
Bromine	35	2	2	6	2	6	10	2	5		
Krypton	36	2	2	6	2	6	10	2	6		

3*d* electrons – the maximum number – there may be either one or two electrons in the 4*s* subshell.)

7.2 Limitations of the planetary model

The planetary model of atomic structure (see **4.3**) was proposed early in the twentieth century and remains useful because it is easy both to retain it in the brain and to represent it diagrammatically. However, it is not a very faithful representation of reality. The model in more general use today was proposed in the 1920s by the distinguished Austrian scientist Erwin Schrödinger and is intimately related to developments in quantum mechanics. The current model uses the concept of the **electron orbital**, which may be interpreted in terms of the energy level of an electron, but also geometrically and statistically in terms of the *probability* of finding an electron in a particular area around the nucleus.

The orbitals can be thought of as the pathways which are circumscribed by the electrons. The four types of subshell which we have met are the same as the types of orbitals. So we may speak of the 2*s* subshell or the 2*s* orbital. The orbital of lowest energy level is designated 1*s*: it is the only orbital associated with the first shell, and in atoms of all elements except hydrogen it holds an electron pair. (Recall that a hydrogen atom has only one electron.) The geometric form of the 1*s* orbital is spherical (*s* does not stand for sphere incidentally), which is because the electrons follow a circular path of constantly changing plane around the nucleus. The second shell holds up to four orbitals. (Recall that the second shell contains up to eight electrons and note that electrons move in pairs.) One of these, the 2*s* orbital, is similar to the 1*s* orbital. For all elements apart from lithium, which has one 2*s* electron, the 2*s* orbital holds two electrons. The other orbitals of the second shell are called 2*p* orbitals. Like the *p* orbitals of other shells they are often likened to dumbbells, or figures-of-eight. Their geometric arrangement is such that the electron pairs are as far away from each other as possible.

The distinction between subshell and orbital can be confusing. Remember that electrons usually ‘travel’ in pairs. In effect, a particular orbital is associated with an electron pair, unless there is a single ‘spare’ electron. Now the *s* subshell, regardless of shell number, has no more than two electrons, and so there can only be one *s* orbital associated with any *s* subshell. The other subshells, however, potentially hold more than one electron pair: three pairs in the case of the *p* subshell, five pairs in the case of the *d* subshell, and as many as seven pairs in the case of the *f* subshell. So, potentially, there are three, five and seven orbitals, respectively, associated with each of these subshells.

8 Linking atoms: chemical bonding

8.1 The valence shell

In this section we introduce the **chemical bonds** which link atoms together to form molecules and other chemical aggregates. A slightly more advanced treatment of this topic is to be found later (see **37**). In attempting to understand the various forces by which atoms are linked to each other, it is vital to keep the outline framework of atomic structure in mind (see **4**), particularly the configuration of electrons (see **7**). This is because chemical bonding phenomena are determined largely by the behaviour of electrons, particularly those in the outer shell of the atom. The outer shell of an atom of most of the elements is known as the **valence shell**. (We say 'most' here because for some elements the valence shell includes the underlying subshells.) Because of the key role of the valence shell, individual atoms and ions are frequently depicted by the element symbol and a number of surrounding dots which is equal to the number of electrons in the outer shell, as shown here:

H·   :N·   :O·   Na·   :Cl·

Such depictions are referred to as Lewis structures, after the American chemist Gilbert Lewis, or simply electron-dot structures. Knowing the number of electrons in the valence shell permits reasonable predictions about how an element will behave chemically. Furthermore, elements with the same valence shell configuration behave chemically in broadly similar ways.

To appreciate the nature of chemical bonding we need to introduce a group of elements known as the **noble gases**. These are helium, neon, argon, krypton, xenon and radon. All are found in the atmosphere in trace concentrations, and all are very stable. In fact if these elements combine with others they do so only under very special circumstances. Because of this feature these elements were formerly known as the inert gases. The main point about the noble gases in the present context is that, with the exception of helium, their atoms have eight electrons in the outer shell. For argon this can be represented as : $\ddot{\underset{\cdot\cdot}{\text{Ar}}}$:. This turns out to be a very stable configuration for the electrons of an atom (and explains the extreme reluctance of the noble gases to form compounds). Now it is a characteristic of many other elements that they too have a tendency to attain the same configuration. This principle is sometimes known as the octet rule; it is not inviolate but it is a fairly reliable guide. In the case of helium, which is also a noble gas, the stable configuration is associated with just two electrons in its valence (and indeed only) shell. The tendency of elements to attain a noble gas configuration of electrons is the key to understanding the chemical bonds which are now described.

8.2 The ionic bond

First we consider the **ionic**, or **electrovalent bond**. If we look at sodium (atomic number 11) in Table 7.1, we see that the loss of the sole electron in the outer shell (shell 3) provides a new outer shell, with eight electrons. This has two effects. First, the new outer shell has the noble gas number of electrons. Secondly, the sodium atom becomes a sodium ion because the number of protons exceeds (by one) the number of electrons. This can be represented as:

$$Na \rightarrow Na^+ + e^-$$

Now consider a chlorine atom, which has an outer shell of seven electrons. By acquiring an electron, the outer shell attains the stable configuration of eight electrons, and an anion (known as chloride) is formed, thus:

$$Cl + e^- \rightarrow Cl^-$$

The simultaneous loss and gain of electrons between atoms is the basis of ionic (electrovalent) bonding.

Sodium and chlorine thus combine ionically to form sodium chloride (NaCl), which is the common salt for culinary use. However, in crystalline form, and in aqueous solution, this substance consists of sodium and chlorine *ions*, not as molecules. Thus the term **ionic compound** is applied here. In the solid, or crystalline state, sodium chloride forms an **ionic lattice**. In essence, a lattice is a three-dimensional structure made up of 'cells'. The lattice of a particular substance has a physical structure which is defined by the dimensions of its constituent cells and by the angles of the bonds between the ions.

Calcium can also combine with chlorine, to form calcium chloride ($CaCl_2$), but in this case two chlorines are held individually to one calcium atom. This is because a calcium atom holds two electrons in its outer shell (not one electron as with sodium), and also eight electrons in the shell beneath. Atoms of calcium therefore have two electrons to donate. These may be donated to an atom which has six electrons in its valence shell, or alternatively, one electron can be donated to each of two atoms which have seven electrons in the valence shell.

It is worth noting here that elements with atomic numbers which fall just short of those of one of the noble gases form negative ions (anions), while elements with atomic numbers that just exceed those of a noble gas form positive ions (cations).

8.3 The covalent bond

In addition to forming bonds by the simultaneous donation and acquisition of electrons, atoms may *share* electrons in their outer shell. In so doing, each atom attains a stable configuration. This sharing of electrons is the basis of an exceedingly important type of bond, referred to as a **covalent bond**. If there are just two atoms involved in a covalent bond, the shared electrons can be thought of as circulating around both atomic nuclei; in effect they are 'members' of the valence shell of both atoms. For example, an atom of chlorine has seven electrons in its valence shell. Thus if two chlorine atoms come close together, six electrons belong to each chlorine atom while a pair of electrons can be shared between the two atoms. If this electron pair contributes to each chlorine atom then each has in effect an outer shell of eight electrons, as shown here:

$$:\ddot{\underset{..}{Cl}}\cdot \quad :\ddot{\underset{..}{Cl}}\cdot \longrightarrow :\ddot{\underset{..}{Cl}}:\ddot{\underset{..}{Cl}}:$$

It is the force of attraction between the two nuclei and the shared electrons which forms the bond that holds the entity together. Covalent bonds form between atoms of different elements to produce compounds, and the atoms of a few elements, such as oxygen, nitrogen and chlorine, form stable molecules as a result of covalent bonds between them. For simplicity we have shown bonds between just two atoms, but a single atom may well be linked covalently with more than one other atom.

Covalent bonds are conventionally represented by dashes, as in Cl–Cl. A single dash means that one electron pair is shared between the two atoms. Two dashes represent the sharing of two electron pairs: this is known as a double bond, as in molecular oxygen, O=O. Three dashes represent the sharing of three electron pairs, known as a triple bond, as in dinitrogen (N≡N). These examples have been used to demonstrate the principle of covalent bonding, but in fact it is only in a few cases that molecules consist of covalently bonded atoms of the same type. The vast majority of molecules consist of atoms of different elements. Some examples of covalently bonded chemical compounds are shown in Figure 8.1. Recall that the dots between any two element symbols indicate the number of electron pairs that are 'serving' the two atoms.

8.4 The coordinate covalent bond

The covalent bond outlined above is based on each atom contributing electrons which are thus shared between atoms. In addition, bonds can be

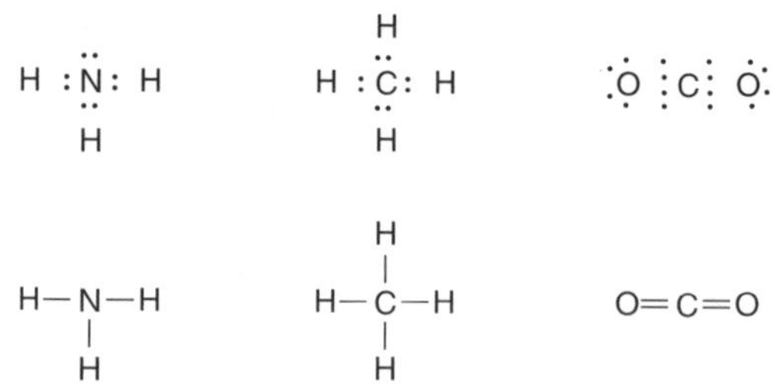

Ammonia (NH_3) Methane (CH_4) Carbon dioxide (CO_2)

Figure 8.1 Examples of chemical compounds in which the covalent bonds are represented by electron dots and dashes.

formed when only one atom provides the electrons which serve two atoms: this is called a coordinate covalent bond.

8.5 The metallic bond

As the name suggests, **metallic bonds** are associated with metals (see **15.1**). In the metallic elements the atoms occur as positively charged ions (cations) which are fixed in place to form a lattice structure. The outer electrons from each atom are loosely bound and consequently move around quite freely to form what is referred to as an electron 'gas'. (The capacity of metals to conduct electricity and heat effectively is owing to these 'free' electrons.) The metallic bond is formed by the electrostatic attraction between the metallic ions and the electrons which, because of their movement, are shared between the ions.

8.6 The hydrogen bond

The **hydrogen bond** is due to weak electrostatic forces between atoms. Such forces form between strongly electronegative atoms (for example oxygen, nitrogen and fluorine) in one molecule and a hydrogen atom which is itself bound to an electronegative atom in another molecule. In a water molecule (H_2O), for example, the oxygen attracts the electrons from the two hydrogen atoms to which it is covalently bound. The hydrogen is thus electropositive, and therefore exposed to the electronegative forces of oxygens in neighbouring molecules. Hydrogen bonds are comparatively weak, being only around 10 per cent of the strength of covalent bonds; hence they are usually depicted by dashes, as in H----O or N----H. Despite their comparative weakness, hydrogen bonds are extremely important. Water provides a good example because hydrogen bonding explains the many (often anomalous) properties of water (see **41.4.3**). In addition, hydrogen bonds are important in determining the physical structure of large biological molecules such as proteins (**32.3.4**) and nucleic acids (**32.3.5**).

8.7 Van der Waals' force

Although not usually referred to as a type of bond, it is convenient to mention **van der Waals' force** here because this too is a force of attraction. It arises because the pattern of movement of electrons from neighbouring atoms or molecules is coupled in such a way as to generate weak electrostatic forces between them. These forces are very weak compared with the bonds mentioned above, and operate only over short distances, but in gases particularly they provide a force of attraction between molecules which are otherwise moving freely.

9 Chemical substances: values and amounts

9.1 Relative atomic and molecular mass

The defining feature of each element is the number of protons in the atomic nucleus (see **4.1**). In addition, each element has a unique **relative atomic mass** or **atomic weight**. Since 1961 it has been agreed that this should be the ratio of the average mass per atom of an element to one-twelfth of the mass of an atom of carbon-12 (the predominant isotope of carbon). For this purpose, carbon-12 has been assigned a value of 12. Note that because relative atomic masses are based on ratios, they have no units. The atomic masses of the elements are shown in Tables 3.1 and 3.2, and you will see that for most elements they are not whole numbers.

If we are given the formula of a molecule, we can use the atomic mass values to calculate its **relative molecular mass** or **molecular weight**. This is simply the sum of the atomic masses of the atoms in one molecule. Inspection of Table 3.2 will permit you to verify the following calculation which is used to determine the relative molecular mass of one molecule of H_2SO_4 (sulphuric acid):

$$(2 \times 1.008) + 32.064 + (15.999 \times 4) = 98.076$$

The percentage elemental composition of a compound can also be readily determined. We have just seen that the relative molecular mass of one molecule of sulphuric acid is 98.076. We also showed the relative atomic masses of the three elements involved, together with the numbers of the atoms in one molecule. So the percentage contribution of hydrogen (relative atomic mass 1.008, two atoms) is simply:

$$\frac{2 \times 1.008}{98.076} \times 100 = 2.056$$

Using the same approach you should be able to confirm that the percentage contributions of sulphur and oxygen are 32.693 and 65.251, respectively.

9.2 Gram atomic mass, Avogadro constant, moles

The values introduced in the previous paragraphs have no units, but chemistry has a system of expressing actual amounts of substances. For any element, that weight in grams which is numerically equal to its atomic mass (or atomic weight) is called the **gram atomic mass** (or **weight**), or the **gram-atom**. So, for example, a gram-atom of calcium (atomic mass 40.08) is 40.08 grams and a gram-atom of copper is 63.55 grams. Now, a gram-atom of each element contains exactly the same number of atoms. So 40.08 grams of calcium contains exactly the same number of atoms as 63.55 grams of copper.

The number of atoms in a gram-atom, which is known as the **Avogadro constant**, or **Avogadro's number**, is 6.023×10^{23}. This huge quantity is now known as one **mole** (abbreviated as mol.). The mole is very important in chemistry, and has largely replaced the gram atomic mass. It is the SI unit of amount of a substance. Formally defined, a mole is the amount of substance that contains as many elementary particles as there are atoms in 0.012 kilograms of carbon. (Note 0.012 kg rather than 12 g because the kilogram is the basic SI unit for mass.) This definition covers atoms, molecules, ions, and even electrons.

Examples will show that the use of moles is straightforward in practice. One mole of atoms has a mass that is equal to the relative atomic mass expressed in grams. So one mole of carbon atoms has a mass of 12.01 grams and one mole of magnesium atoms has a mass of 24.31 grams. Similarly, one mole of a molecule has a mass that is equal to the relative molecular mass (molecular weight) expressed in grams. Thus the masses of one mole of the following molecules are as shown here:

Oxygen (O_2) 32 g (16×2)
Carbon dioxide (CO_2) 44.01 g ($12.01 + 16 \times 2$)
Ammonia (NH_3) 17.04 g ($14.01 + 1.01 \times 3$)

If we want more or less than one mole (say 5 moles or 0.6 mole) we simply multiply the mass value by the appropriate number. Thus:

$$\text{Number of moles} \times \text{Mass of 1 mole} = \text{Mass (g)}$$

A simple rearrangement allows us to convert grams to moles. Thus:

$$\frac{\text{Mass (g)}}{\text{Mass of 1 mole}} = \text{Number of moles}$$

9.3 Empirical formulae

The **empirical formula** of a compound shows the simplest ratio of the numbers of atoms of each element. Often, as in carbon dioxide (CO_2), this is the same as the molecular formula. In other cases, though, and this applies particularly to organic compounds (see **16**), the numbers of atoms of the elements in the molecular formulae are divisible by the same whole number to give simple ratios. For example, the gas ethane, which has the molecular formula C_2H_6, has the empirical formula CH_3.

If we know the percentage elemental composition of a substance we can readily work out its empirical formula. Suppose we are told that a certain substance is made up of 79.89 per cent by mass of carbon and 20.11 per cent by mass of hydrogen. In other words, in a 100 g sample of this substance there are 79.89 g of carbon and 20.11 g of hydrogen. We divide each of these values by the relevant relative atomic mass, which is 12.01 for carbon and 1.008 for hydrogen. This gives us the number of moles, which in the case of carbon is 6.65 and in the case of hydrogen is 19.95. By dividing the larger of these values by the smaller we get the ratio of hydrogen to carbon atoms, which is 3 to 1. Thus the empirical formula is CH_3.

10 Chemical reactions

10.1 Introduction

A chemical reaction can be said to occur whenever there is any change in a chemical substance. Reactions are at the heart of chemistry. If you have studied any chemistry at all you will have been asked to learn a number of chemical reactions. Certainly we cannot proceed very far in environmental study without dealing with chemical reactions, because they are occurring around us continually. First here, we introduce the ways in which chemical reactions are conventionally expressed. Consider the following expression:

$$CaSO_4 + C \rightarrow CaS + CO$$

This simply tells us that when calcium sulphate is combined with carbon, calcium sulphide and carbon monoxide are produced, the arrow indicating the direction of the reaction. In other words, symbols for elements and compounds are used rather than the written name, which would be cumbersome. But the two sides do not balance: there are four oxygen atoms on the left-hand side of the arrow and only one on the right. (The subscript 4 indicates the number of atoms of the preceding element that is present in the molecule or compound. If no subscripted number is shown, this implies that one atom is present. Thus, SO_4 contains one sulphur atom and four oxygen atoms.) To provide an *equation*, in which the total numbers of atoms of each element are equal on both sides, numbers are placed *before* the appropriate compound symbols, thus:

$$CaSO_4 + 4C \rightarrow CaS + 4CO$$

Quite often, when chemical reactions are described – in diagrams for example – the full details are not shown. Instead, just the key chemical entities, usually only the starting materials and certain products, are presented. This is done simply to focus attention on the chemical groups of interest.

10.2 Rates of reactions

10.2.1 Basic principles

A key feature of a chemical reaction is how quickly it proceeds. Some reactions occur almost instantaneously, in some cases explosively. Other reactions take time to evolve, and the rate of the reaction can be more easily measured. Measurement of the reaction can be in terms of the rate at which a **product** accumulates or the rate at which a **reactant** (i.e. one of the starting materials) disappears. A further fundamental point is that some reactions are reversible, depending on the conditions such as temperature and pressure. An example is the reaction involving two oxides of nitrogen, N_2O_4 and NO_2. This reversible reaction can be represented by:

$$N_2O_4 \rightleftharpoons NO_2$$

In this reaction a change in pressure shifts the equilibrium. In this case the shift in equilibrium is reversible; in other cases a change in conditions may bring about an irreversible directional change.

The rate of many reactions can be increased by the addition of a specific substance which is not actually consumed during the reaction. The general term for such a substance is **catalyst**. Many reactions in effect do not proceed in the absence of a catalyst. Catalysts are employed very widely in industry. **Enzymes**, which are principally made of protein, comprise a special category of catalysts. They are produced within living cells and are largely responsible for the organization and function of cells and organisms. Because of their importance, enzymes are treated separately elsewhere (see **33**).

10.2.2 Factors governing reaction rates

The rate of a chemical reaction is governed by three general factors:

- The rate at which potential reactants are brought into contact.
- The probability that once in contact they will react with each other.
- The rate at which the reaction products are dispersed.

Probabilities of making contact are influenced by concentration. In a mixture of hydrogen (H_2) and oxygen (O_2) molecules in gaseous form, doubling the amount of H_2 will double (approximately) the number of collisions between O_2 and H_2 molecules. However, by raising the temperature, and thereby increasing the velocity of the molecules, the rate of collision can also be increased. Collision does not, however, guarantee that a reaction will occur. The H_2 and O_2 mixture is highly unstable (that is, the mixed gases have far more chemical energy than the equivalent amount of their reaction product, H_2O). Despite this, they do not readily react because forces of repulsion between the molecules do not allow them to get close enough. They need to have some energy (termed the **activation energy**) to overcome these forces and initiate the reaction. The activation energy is small compared with the energy that will be released by the reaction, but it must nevertheless be found before reaction can take place. A physical analogy to this situation is water in a mountain-top lake. It clearly has vast gravitation potential energy, but before that can be utilized, energy has to be given to the water to lift it out of its basin. In the case of H_2 and O_2, the activation energy can be provided by heat. At a high enough temperature the kinetic energy of the molecules is sufficient to overcome any repulsion, and the gases react explosively. Seen from this perspective it is easy to understand why concentration and temperature are key factors in reaction rate.

10.2.3 Orders of reaction

Expressions for reaction rate take on a number of different forms depending upon the influence of reactant concentrations on the reaction rate. A **zero-order reaction** is one in which the rate is not influenced by the concentration of the reactant(s). A **first-order reaction** is one in which the rate of reaction is proportional to the concentration of the reactant(s). This is typical for unimolecular reactions, i.e. where there is a single reactant. In second- and higher-order reactions the reaction rate is a more complex function of the various reactants.

10.2.4 Rate-limiting steps

Many reactions take place in a series of steps. Some of these occur rapidly, and others rather slowly. The overall reaction rate will be governed mainly by the slowest step in the series.

10.2.5 Environmental factors, including catalysts

Chemical reaction rates are not independent of their environment. Earlier we discussed the role of temperature in activating reactions. However, there are other environmental factors that are significant. Of great importance in natural systems is *moisture content*. Liquid water permits reactants to be transported via aqueous diffusion, and allows dispersal of the reaction products. Also, by taking ions into solution, water provides a means of bringing the reactants into close contact. Also, many chemical reactions proceed more rapidly in the presence of a catalyst, as mentioned above. A catalyst is a substance that increases the rate of a chemical reaction, commonly by reducing the activation energy. Catalysts can be particle surfaces, which remain unchanged by the chemical reaction, or they can be chemical species that become involved in the reaction. There are many catalytic reactions that are of great importance in the atmosphere (see **11**).

10.3 Energetics of reactions

A key issue in chemical reactions is their energy relations. Energy is the capacity to do work, and the greater the amount of energy put into a system then the greater the amount of work that can be carried out. As discussed earlier, chemical reactions are responsive to temperature, i.e. they proceed at different rates at different temperatures, tending to be faster at higher than at lower temperatures. Temperature is simply a measure of the amount of heat in a system, and heat is a form of energy. So when substances are heated, reactions proceed more quickly. The explanation for this is the greater movement of molecules at higher temperatures, which increases the probability that molecules will encounter other molecules, and increases the probability that interactions will take place once they have met. For any reaction there is a minimum amount of energy needed before it may proceed: this is the activation energy referred to earlier.

11 Gas phase reactions: heterogeneous reactions and free radicals

11.1 Introduction

The most abundant species in the atmosphere are nitrogen (N_2), oxygen (O_2), water (H_2O), argon (Ar) and carbon dioxide (CO_2). However, in terms of variations in atmospheric chemistry, there is a series of less abundant gases (trace gases) which have a greater impact than their concentrations might suggest. The most important of these are ozone (O_3), methane (CH_4), nitric oxide (NO), nitrogen dioxide (NO_2), and hydrocarbons. The chemical reactions of these trace gases are influenced by many factors, including the concentration of particle surfaces, the amount of water available, and intensity of solar radiation. All these factors vary widely, not only within the lower atmosphere, but also between the major layers of the atmosphere (see **43.4**). The stratosphere is dry and exposed to very high energy photons from the Sun. It is also relatively free of hydrocarbons and dust. In contrast, the troposphere has more water and dust, and it contains a cocktail of hydrocarbons released by biological processes. The troposphere is also protected from high-energy photons because these are absorbed by matter in the stratosphere above.

The variety and complexity of chemical reactions in the atmosphere preclude a full account here; however, there are some crucial reactions which it is useful to understand.

11.2 Photolysis

Photolysis means the breaking up of a molecule owing to interaction with a photon. An example is the photolytic destruction of dioxygen:

$$O_2 + h\nu \rightarrow O + O$$

where $h\nu$ is a photon of specified energy. This reaction leads to the generation of atomic oxygen. We do not encounter atomic oxygen in solution because it is too reactive. However, in the atmosphere it can briefly survive before reacting with other chemical species. The reaction shown can only take place in the stratosphere because it needs a photon with a wavelength of less than 242 nm (remember, the shorter the wave length, the higher the energy – see **22.1**). By the time solar radiation reaches the troposphere, all of the sufficiently energetic photons have been absorbed by substances such as ozone (O_3).

In the troposphere there are other photolytic reactions of importance. Nitrogen dioxide is photolysed to atomic oxygen and nitric oxide by photons longer than 424 nm:

$$NO_2 + h\nu \rightarrow NO + O$$

The O then reacts with dioxygen to form ozone.

11.3 Free radicals and other high-energy species

We do not encounter free, unattached, oxygen atoms in solutions or solids. In solution, and in solid compounds, any oxygen atoms are either O^{2-} ions (in ionic lattices) or they are covalently bonded to other atoms. Clearly, then, there is something different about the atmosphere. Furthermore, a number of other chemical species not normally found in solutions and solids also occur as free entities in the atmosphere (for example OH and HO_2). These species, which are the building-blocks of well-known molecules or compounds but which are not normally found separately, were termed radicals (meaning component parts) by early chemists. In the early twentieth century experimental chemists started to show that some of these radicals could be separated, if only very briefly. These separate forms were named **free radicals**.

Free radicals are so reactive that they exist for only a very short time. This is why their presence in solutions and solids is so hard to detect. However, in the atmosphere they can be sufficiently abundant and long-lived to have a major impact on the atmosphere's chemistry.

The reactivity of free radicals like atomic oxygen (O) is due to their high chemical energy. Some of the free radicals have more than one energy state, and this property affects the way in which they react. The atomic oxygen formed in the photolysis of O_2 can be in the ground state (that is, all electrons are in their lowest possible orbitals). This is not, however, the only possibility. Consider again the photolysis of dioxygen (O_2), this time with a more energetic photon (wavelength less than 180 nm):

$$O_2 + h\nu \rightarrow O + O(^1D)$$

This time one of the O atoms is not in the ground state (the ^1D qualifier indicates the precise electron structure of the atom). Some of the electrons are in higher orbitals, and thus hold more energy. This excess of energy can be lost only by giving it to another molecule in the form of heat:

$$O(^1D) + M \rightarrow O + M$$

where M is N_2 or O_2 and is simply a bit warmer at the end of the reaction. Or the energy can be used to make chemical changes:

$$O(^1D) + H_2O \rightarrow 2OH\bullet$$

where OH• is the hydroxyl free radical. Free radicals play a crucial role in governing reactions in the atmosphere. Chains of reactions can be set in motion this way. Consider the following series of reactions, starting with carbon monoxide and the hydroxyl radical:

1. $CO + OH\bullet \rightarrow CO_2 + H\bullet$
2. $H\bullet + O_2 + M \rightarrow HO_2\bullet + M$ (M is O_2 or N_2 and simply mops up some excess energy)
3. $HO_2\bullet + NO \rightarrow NO_2 + OH\bullet$
4. $NO_2 + h\nu \rightarrow NO + O$
5. $O + O_2 + M \rightarrow O_3 + M$

Note that we begin by destroying OH•, and at some stage generate the hydrogen radical and the peroxyl radical ($HO_2\bullet$). However, in the end we have the same amount of OH• as at the start. If you look carefully through this series of reactions you will see that the net effect is photolytic oxidation of carbon monoxide to produce ozone and carbon dioxide. This reaction could not, however, have taken place without the OH• free radical, which has acted as a catalyst in the reaction. The cycle of reactions catalysed by free radicals is termed a **catalytic cycle**. Understanding atmospheric chemistry is, to a large extent, a question of understanding the interactions of various catalytic cycles.

12 Acids and bases

12.1 Concepts and definitions

The term **acid** is in everyday use. We associate the word particularly with food. For example, we say that lemon juice and vinegar are acidic. And if we recall school chemistry, we probably know that the terms 'basic' and **alkali** are somehow associated with the concept of acidity. But what do we mean by acidity? It would be convenient at this point to offer short, simple definitions of the terms. However, the fact is that the gradual evolution of the concepts of acidity has led to definitions that are hard to appreciate. A better approach is to introduce you to the development of thought about acidity, before seeing what useful definitions can be applied.

From as early as the seventeenth century it was noticed that a number of chemicals, when in solution, had shared properties. They tasted sour, they were capable of dissolving many mineral materials, and they commonly reacted with metals to release hydrogen gas. This behaviour was termed **acidic**, and the substances which behaved this way were termed **acids**. Another class of chemicals, again in solution, shared a different set of properties. They were soapy to the touch, and were caustic (attacked organic tissues). This behaviour was termed **basic**, and the substances which behaved in this way were termed **bases**. It was also recognized that there were some mutual properties of acids and bases. They reacted with each other to produce salt solutions (that is, solutions that lack the acidic/basic properties of the parent solutions, and on evaporation give rise to crystalline salts). And they could change the colour of plant dyes such as litmus.

During the nineteenth century, chemists demonstrated that a shared characteristic of acid solutions was their enrichment with hydrogen ions (H^+), and it was hypothesized that the properties of the acidic solutions stemmed from their presence. Likewise, bases were shown to derive their properties from an abundance of hydroxide ions (OH^-). By the end of the nineteenth century the Swedish chemist Svante Arrhenius had concluded that acids were hydrogen-bearing substances that could break up on reaction with water, releasing H^+ ions into the solution (in contrast with the many organic substances that are rich in hydrogen yet do not behave as acids). In a similar way, bases were seen as substances that released OH^- ions into solution.

In many ways this interpretation remains useful. However, there are two ways in which it is inadequate. The first problem is relatively trivial from our point of view. Early in the twentieth century it became clear that H^+ ions, which are no more than bare protons, are too reactive to exist alone in solution. Instead, they bind firmly to water molecules to form the **hydroxonium** ion (H_3O^+). This truth does not, however, mean that hydrogen ions in water cannot be *treated* as if they were in the form H^+. In most chemical equilibrium calculations, H^+ can be assumed to exist, without loss of accuracy, although it is clearer to treat it as H_3O^+ when considering acid/base properties.

The second problem is that all of the properties we have looked at relate to reaction with water. Chemists soon discovered that acidic and basic reactions occur in other media, and that the acidity and 'basicity' of substances vary with the medium. In other words, acidity and 'basicity' are not absolute and can only be defined in relative terms. In 1923, J.N. Brønsted and T.M. Lowry formalized this approach. They showed that the materials that were defined as acids by Arrhenius are better described as being substances that are more acidic than water. Let us see what is meant by this idea, and then arrive at a usable definition for acids and bases.

Brønsted and Lowry showed that water can be both acidic and basic. Consider the reaction:

$$H_2O + H_2O \rightarrow H_3O^+ + OH^-$$

In this reaction, which is known to take place to a limited extent, water is supplying H^+ ions (and is therefore definitively an acid according to Arrhenius), and is also supplying OH^- ions (and is therefore definitively a base according to Arrhenius). Thus, part of the water acts as an acid, while part acts as a base. Brønsted and Lowry showed that this pairing of acids and bases always occurs, and that in aqueous solutions one half of the pair is always water. Consider the reaction:

$$HCl + H_2O \rightarrow H_3O^+ + Cl^-$$

The hydrochloric acid (HCl) is more acidic than water. Thus the HCl behaves as an acid and donates its H^+ ion, while the water behaves as a base and accepts the H^+ ion. Such compounds that can donate H^+ ions to water are termed **protonic acids** (the H^+ ion being simply a proton).

Now consider this reaction:

$$H_2O + NH_3 \rightarrow NH_4^+ + OH^-$$

Ammonia (NH_3) is less acidic than water, and thus acts as a base by accepting an H^+ ion, while the water acts as an acid by donating an H^+ ion.

This describes the Brønsted–Lowry definition of acids and bases:

An acid is a substance with a tendency to donate protons (hydrogen ions). A base is a substance with a tendency to accept protons.

In aqueous solutions this can be slightly refined:

An acid is a substance with a stronger tendency to donate protons than water. A base is a substance with a stronger tendency to accept protons than water.

Let us see how well this works with a number of common acids and bases. With all of the mineral acids – hydrochloric acid (HCl), nitric acid (HNO_3), sulphuric acid (H_2SO_4), etc. – it works very well, as illustrated for hydrochloric acid above. But what about the two following reactions?

$$CO_2 + 2H_2O \rightarrow HCO_3^- + H_3O^+$$

$$Al^{3+} + 2H_2O \rightarrow Al(OH)^{2+} + H_3O^+$$

(You should note that the reaction of Al^{3+} with water is typical for metal ions, but is significant only where the charge is high (2+ or greater) or the ion is small (such as Be^{2+}).) In these two reactions, CO_2 and Al^{3+} are both behaving as acids, yet neither is donating H^+. It was for cases like these that G.N. Lewis formulated more generalized definitions of acids and bases in terms of electron transfers. According to Lewis, an acid is an electron acceptor, and a base is an electron donor. This ties in with the Brønsted–Lowry definition because the H^+ ion is a strong electron acceptor. Thus HCl is an acid because the H^+ ion it releases is a powerful electron acceptor. CO_2 is an acid because it is itself an electron acceptor: it binds to the electron-bearing OH^-, freeing the H_3O^+ ions from water. Al^{3+} is an acid for the same reason: it is an electron acceptor, binding up OH^- ions.

It has been the practice in most basic geochemistry texts to disregard the Lewis interpretation of acids as unnecessary when considering aqueous solutions, which is what environmental scientists are concerned with most of the time (the Lewis interpretation is often essential in non-aqueous cases). However, there are good reasons not to do this. The acidic properties of carbon dioxide, for example, can be simply explained by the Lewis interpretation. The Brønsted–Lowry interpretation can do this only by invoking the process of **hydrolysis**. Hydrolysis is the reaction of a substance with water leading to its break-up, i.e. the binding of either H^+ or OH^- and consequent freeing of the other part of the water. The essential part of hydrolysis is the formation of a bond between the hydrolysing ion and a water molecule. (This is in contrast to hydration which involves a loose attraction of an ion to water, but without real bond formation – see **42.3**). CO_2 undergoes hydrolysis in water, thus:

$$CO_2 + H_2O \rightarrow H_2CO_3 \text{ (carbonic acid)}$$

The carbonic acid is then free to donate hydrogen ions, and thus fits the Brønsted–Lowry definition of an acid. Al^{3+} also undergoes hydrolysis, first binding with water, which can then subsequently donate a hydrogen ion:

$$Al^{3+} + H_2O \rightarrow Al(H_2O)^{3+} \rightarrow Al(OH)^{2+} + H^+$$

Negatively charged ions can also behave this way, with strength increasing with ionic charge. The most important example is the O^{2-} ion. This is present in many minerals, but is too reactive to remain in solution. Instead, it rapidly hydrolyses, freeing OH^- ions:

$$O^{2-} + H_2O \rightarrow 2OH^-$$

The O^{2-} ion is an exceptionally strong base (proton acceptor and electron donor). Most other common anions are less reactive, and the comparable reaction proceeds only to a very limited extent.

The Lewis definition of acidity does not need to invoke hydrolysis. Instead, the reactions termed hydrolysis are simply acid/base reactions: they can be viewed in terms of electron donation or acceptance. Thus, O^{2-} is an exceptional electron donor (base), while small or highly charged cations are good electron acceptors (acids). This also explains why H^+ is strongly acidic.

So, although we cannot offer a short, crisp definition of an acid, we can make a number of statements which sum up the concept and, importantly, allow us to understand the associated phenomena:

- Acids are substances that have a positive tendency, either by donation of a positive species (H^+ in the case of water) or by accepting a negative species (OH^- ion, oxide ion, or other electron-rich molecule/atom).
- Bases are substances that have a negative tendency, either by donation of a negative species or by accepting a positive species.
- H^+ and OH^- ions play central roles in acidic and basic properties of aqueous solutions.
- The Brønsted–Lowry definitions of an acid and a base (shown above) can be usefully applied to water if the role of hydrolysis is taken into account.

Now that we have dealt with definitions and the basic phenomena we can address the link between acidity and other chemical processes.

12.2 Acidity and oxidation/reduction

The emphasis on the role of charge transfer in acid/base reactions shows their similarity to the **redox** reactions described below. An acid has a strong tendency to accept electrons, and a base has a strong tendency to donate them. As we will see below, the property of being electron-rich can also be described as being **reducing**. Thus an electron donor is not just a base but also a **reducing agent**, though not necessarily a very good one. Likewise, an electron acceptor is not just an acid but also an **oxidizing agent**, though again not necessarily a strong one. Consider this reaction:

$$2HCl + 2Na \rightarrow 2Na^{2+} + 2Cl^- + H_2 \uparrow$$

where the arrow ↑ indicates that H_2 escapes as a gas. The hydrochloric acid reacts with the base sodium to form the salt NaCl. At the same time the hydrogen is reduced to H_2 gas, and sodium is oxidized to Na^+.

While the definitions of oxidation, reduction and acidity share similarities, they are measured in different ways. The pH scale focuses entirely on the single acidic component H^+, whereas Eh and pE (the common measure of redox status) focus on the effective concentration of electrons.

12.3 Alkali or base?

There is considerable confusion between the terms **alkali** and 'base'. An alkali is a member of a class of metals which includes sodium (Na) and potassium (K) (see **40.3**). In elemental form these metals react vigorously with water to form very basic solutions. These solutions can be termed **alkaline**, because they are solutions of alkali elements. However, the term 'alkali' is also commonly, though not universally, applied to basic solutions even when alkali metals are absent. More widely used is the term **alkalinity**. This is the opposite to acidity and is simply a measure of what might more appropriately be, but is not, termed the 'basicity' of a solution.

12.4 Oxides and acidity

Metallic elements (sodium etc.) form ionic oxides comprising regular lattices of cations and O^{2-} ions. The O^{2-} ions are basic in character, while the metal ions are acidic. The overall effect is usually strongly basic because O^{2-} is a strong base while the metal ions are only weakly acidic. However, the acidity of metal ions varies with their size and charge, which has a variable modifying effect on the net 'basicity' of the oxide. This effect can be predicted from the electronegativity. Thus, rubidium oxide (Rb_2O) is highly basic because Rb^+ is very electropositive (see **15.3**) and is thus only weakly acidic. On the other hand, beryllium oxide (BeO) is much less basic because the very small Be^{2+} ion is much less electropositive, and is a stronger acid.

Oxides of the more electronegative elements such as sulphur and nitrogen are not ionic. The bonding in sulphur dioxide and nitrogen dioxide is covalent. The resulting stable polyatomic molecules are good electron donors and are therefore acidic.

12.5 Strengths of acids

Acids (and bases) can be described as *weak* or *strong*. This leads to confusion: the strength of an acid has nothing to do with its concentration. Both strong and weak acids can be either concentrated or dilute. A strong acid is simply one which is highly dissociated at the acidity of interest. A weak acid, conversely, is little dissociated. Let us see what this means.

Hydrochloric acid (HCl) is a strong acid; in water it dissociates (splits) almost entirely into its component parts:

$$HCl + H_2O \rightarrow H_3O^+ + Cl^-$$

Thus the aqueous solution contains almost only H_3O^+ and Cl^- (HCl is present, but in trivial amounts). Carbonic acid, in contrast, is a weak acid, and in water forms an equilibrium between its undissociated molecular form and its component parts:

$$H_2CO_3 + H_2O \rightleftharpoons H_3O^+ + HCO_3^-$$

Thus carbonic acid in solution at near neutral pH values has all three possible species present in significant amounts (H_2CO_3, H_3O^+ and HCO_3^-).

When quantifying the strength of an acid, it is the strength *relative to the acidity of water* that is considered. Thus, HCl is a strong acid relative to water because there is a complete transfer of the H^+ ion to the water (to create the H_3O^+ ion). In the case of the carbonic acid, only a part of the H^+ is transferred to water; some is retained as molecular carbonic acid.

Acidity strength is quantified using the acidity constant (K_A). This is a special form of the equilibrium constant (see **35**): it tells us the acidity at which the concentration of undissociated acid equals the concentration of dissociated acid. Thus, it is the acidity corresponding to 50 per cent dissociation. Consider carbonic acid as an example. The equilibrium expression for the carbonic acid dissociation, treating the free H^+ ion as if it existed (thus, not bound to water as H_3O^+) is:

$$H_2CO_3 \rightleftharpoons H^+ + HCO_3^-$$

The corresponding equilibrium expression is:

$$\frac{[H^+][HCO_3^-]}{[H_2CO_3]} = K_A$$

This is commonly expressed in $\log_{10}$ form, as this can be applied more simply. If we convert the expression above to log form (and convert $\log_{10}[H^+]$ to $-pH$), we get:

$$-pH + \log_{10}\frac{[HCO_3^-]}{[H_2CO_3]} = -pK_A$$

or, rearranged,

$$pH = \log_{10}\frac{[HCO_3^-]}{[H_2CO_3]} + pK_A$$

where pK_A is the negative $\log_{10}$ of K_A.

When the acid is 50 per cent dissociated, such that $[HCO_3^-] = [H_2CO_3]$, the expression can be simplified as:

$$\log_{10}\frac{[HCO_3^-]}{[H_2CO_3]} = \log_{10}[1] = 0 \text{ and } pH = pK_A$$

pK_A for carbonic acid is 6.35 at 25 °C. Thus, carbonic acid is half dissociated at pH 6.35.

12.6 Acidity buffering

If we look at different lakes or streams we find that they differ in their capacity to resist changes in pH. This is particularly important in the context of 'acid rain' impacts. What we are seeing here is the effect of differences in **acidity buffering**.

Recall that neutral, gas-free, water has a pH of 7. What happens to the pH if we add 10 micromoles of HCl to one litre of this water? The dissociation of the strong acid is complete, thus the concentration of H^+ ions will also be 10 micromoles. What will the pH be? If we have 10 micromoles, then we have 1×10^{-5} moles. Thus $pH = -\log_{10}(1 \times 10^{-5}) = 5$. Therefore the addition of 10 micromoles of acid causes a 2 unit change in the pH, from 7 to 5.

But what if we allow the water to come into equilibrium with atmospheric carbon dioxide? This acidic gas will reduce the pH of the pure water down to pH 5.6 (see **42.4**). If we want to return the pH to 7 we must add base (such as $Ca(OH)_2$). We must in fact add 47 micromoles of base to each litre of water (in equilibrium with present-day atmospheric CO_2) to bring the pH value back up to 7. What happens if we add 10 micromoles of HCl to this water? We find that the pH falls from 7 to only 6.89. More dramatically still, we have added 10 micromoles of acid, but the hydrogen ion activity has increased by only 0.03 micromoles. Something in the solution is consuming added H^+ ions. The solution is said to have some buffering capacity. So what is this buffer?

In the first, 'unbuffered', case we have:

$$HCl + H_2O \rightarrow H_3O^+ + Cl^-$$

where all of the added strong acid expresses itself as an increased H^+ concentration. In the second case we have:

$$HCl + Ca^{2+} + HCO_3^- \rightarrow H_2CO_3 + Ca^{2+} + Cl^-$$

Nearly all (99.7 per cent) the added H^+ goes into converting HCO_3^- into H_2CO_3, leaving only a tiny amount to add to the free H^+ concentration.

Of course, such buffering effects work only as long as their capacity to work is not exceeded. If, for example, more HCl is added than there is HCO_3^- in the water, then all of the bicarbonate will be consumed, and any excess HCl will then contribute directly to free H^+ ions, just as in the first case above.

Natural waters that are rich in HCO_3^- are said to be **bicarbonate buffered**. If bicarbonate is absent the water is said to be acidic (sometimes referred to as 'acidified', but this term should be avoided as it implies that the water has changed its acidity, which need not be the case).

12.7 pH, measurement of acidity, and strengths of acids and bases

It is clear from the definitions given above that there are a number of different ways of looking at acidity; this gives rise to more than one approach to measuring it. The link between acidic properties and the hydrogen ion leads to the best-known method: pH. However, it is often more convenient to treat H^+ concentration as a derivative property of a solution and to determine acidity by the balance of acids and bases present.

12.7.1 Hydrogen ion concentration

The hydrogen ion concentration can be measured directly in solution by means of appropriate electrodes (though this is not to imply that such measurements are simple). Conventionally, such measurements are expressed not in terms of ion concentrations but as the negative $\log_{10}$ of the hydrogen ion concentration (strictly this should be in terms of its activity – see **42.2.1**). The main advantage of using the log scale is that H^+ concentrations are highly skewed in nature towards rare high values whereas pH is more normally distributed.

The pH and the H^+ concentration can be readily converted:

$$\mathrm{pH} = \log_{10}^{-[\mathrm{H}^+]}$$

and, therefore,

$$[\mathrm{H}^+] = 10^{-\mathrm{pH}}$$

The pH scale is conventionally regarded as ranging between 0 and 14, though it can reach beyond these limits. Natural waters, however, have a narrower range. Here are some guidelines:

- Unpolluted rain and snow typically range between pH 5 and 7, with the lower values in marine areas, and higher values in continental interiors.
- Lakes and rivers tend to range between pH 6 and 8.4 (which is the highest pH of a solution in equilibrium with both calcium carbonate and atmospheric carbon dioxide). Higher pH values in lakes are found only where:
  - closed basins allow the build-up of sodium, in which case the pH can rise to 11;
  - nutrient pollution and high rates of photosynthesis lead to carbon dioxide depletion, and temporarily higher pH values (up to 11).
- The sea has a pH value of about 8.1.
- Naturally acidic water is found on rain-fed peat bogs (down to pH 3).
- Most soils are within the range pH 4.5–8.5.

Always remember that the pH scale is logarithmic (see **26.6**). This means that small differences in pH correspond to large differences in hydrogen ion concentration, and thus may have great significance for chemical and biological properties.

12.7.2 Balance of acids and bases: titration

An alternative way of looking at acidity and alkalinity is to consider the balance between acids and bases in solution. Indeed, this is how measurement of acids and bases started, using vegetable dyes such as litmus as indicators of acidity. If a solution has more acids than bases, then it is acidic, and if we add a drop of litmus solution the colour turns red. Conversely, if there is more base than acid, the solution is basic and adding litmus turns the solution blue. Done this way, the method is qualitative: it distinguishes acidic and basic solutions, but does not measure the *degree* of acidity. To quantify the acidity using this approach we must use **titration**.

The principle of titration is very simple. If we have an acidic solution of unknown acidity, we can measure its acidity by following these steps:

1. Add an indicator solution like litmus.
2. Gradually add known amounts of a base until the litmus turns blue.

The point at which the solution changes from red to blue is known as the **end-point** of the titration. It is the point when the amount of added base exactly equals the amount of acid that was present in the unknown solution. Because we know exactly how much base was needed to reach the end-point, we can use this to calculate exactly how much acid was in our unknown acidic solution. The concentration of an unknown base can be found in the same way, using a known acid.

It must be remembered that end-points are slightly influenced by the choice of indicator. Thus, if you measure alkalinity, or acidity, by indicator titration, it is necessary to say which indicator you used.

13 Ions, particles and solutions: ion exchange and adsorption

13.1 Introduction

Anywhere that natural water comes into contact with solid surfaces, some of the dissolved ions adhere, or **adsorb**, to the solid. In soil or sediment, in which the small particles provide a particularly large surface area, the amount of adsorbed ions per unit of volume greatly exceeds the amount in solution. This is illustrated in Figure 13.1, which shows a cube of soil water 10 nm across containing four small idealized clay crystals. The 33 000 water molecules are not shown, but cations (mainly Ca^{2+}) are shown both in the water and adsorbed to the surface of the crystals. The drawing is to scale, and is accurate for the adsorbed ions. However, it exaggerates the dissolved concentration of cations: four are shown, while a typical soil water of this volume would only have a 0.25 probability of containing a single ion at any given time.

Ions in the solution can displace those adsorbed to the surfaces. If the adsorption sites are occupied predominantly by one cation type, as illustrated in Figure 13.1, then the ion displaced is likely to be of that type, regardless of the nature of the newly adsorbed ion. This effect is termed **ion exchange**. The reservoir of exchangeable ions bound to the mineral surfaces is typically two to three orders of magnitude greater than the dissolved reservoir. This means that the concentration and type of cations in soil water are governed by the exchangeable reservoir rather than by the nature of any inflowing water. The soil and sediments act as **ion exchangers**, buffering the water composition against short-term change. In the longer run, 'new' ions, supplied by mineral weathering, rainfall, or as fertilizers, may gradually replace the ions bound to the soil. This is particularly the case for well-drained soils in humid areas where hydrogen ions in the rainwater gradually displace the metal ions (calcium, magnesium and potassium, etc.) in the soil, flushing them down the soil profile. This effect reduces the concentration of key nutrients in the soil and leads to acidification of the upper soil. To counter these effects, farmers add lime (as $CaCO_3$ or $Ca(OH)_2$) to soils, to replace the Ca leached

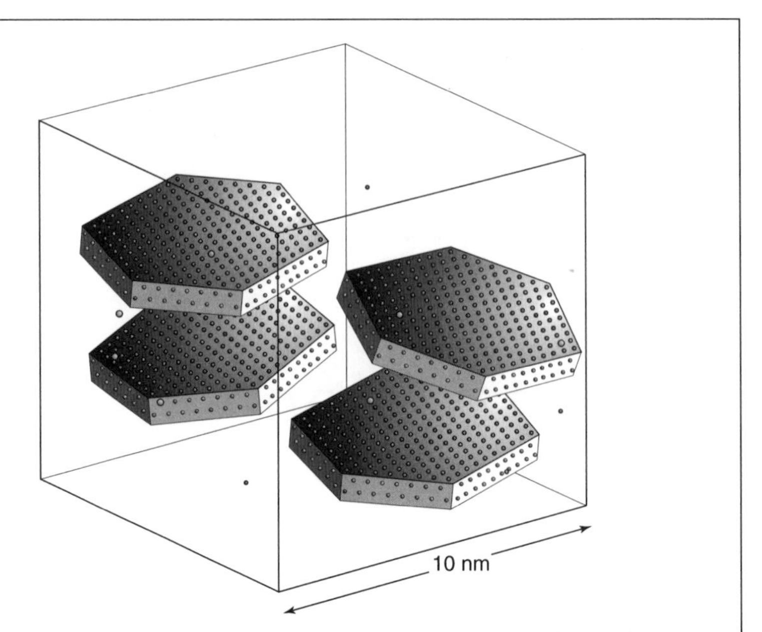

Figure 13.1 A cube of soil water 10 nm across is shown containing four idealized clay crystals. The small spheres are Ca^{2+} ions (shown to scale), which are mostly adsorbed to the clay crystal surfaces. Six ions of another (unspecified) type are shown by larger spheres. Adsorption of one of these to a clay surface will displace an ion, which is most likely to be of the dominant ion type (Ca^{2+}). This effect is termed ion exchange.

from the system. In this case, it is the Ca which exchanges for the H^+ ions adsorbed to the soil particles.

The term **cation exchange capacity** (CEC) is used to indicate the maximum concentration of cations that can be bound by a particular soil. It is conventionally expressed in milliequivalents (meq) per 100 g of dry soil, and is typically in the range 10–100 meq/100 g. Because the crucial factor in adsorption is the area of particle surface, fine-grained soils or sediment have the highest cation exchange capacities. Soil organic matter also has a very high surface area for a given mass. Thus, clay-rich or highly organic soils or sediments have high CEC values, while sandy soils have very low CEC values. You should note, though, that CEC values are critically dependent upon the method used to measure them. So if different methods are used, the results are difficult to compare.

Ion exchange processes, for both positive and negative ions, have very wide application. Special resins with exceptionally high CEC values are used to adjust the ion concentrations in water for many industrial purposes.

13.2 The chemistry of particle surfaces

In discussing ion exchange processes, the nature of the adsorbing surfaces, and mechanisms for that adsorption, have not been considered. The blanket

Consider a fragment of clay (looking down on the sheet)

At the margin are broken bonds

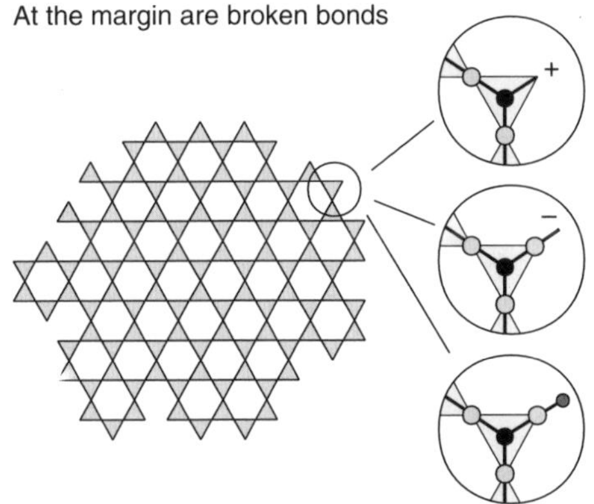

Case 1: Outer O atom is missing. The broken silica tetrahedron has a net positive charge

Case 2: Outer O atom is in place. This leaves the silica tetrahedron with a net negative charge

Case 3: A positively charged ion, usually a hydrogen ion, 'bonds' to the silica tetrahedron. There is no net charge

Figure 13.2 Broken bonds at the surface of a crystal make the surface electrically charged. The actual charge varies with the nature of the chemical species attached to the outer bonds.

term 'adsorption' actually arises from a number of different processes that take place at particle surfaces. These processes must be considered in more detail before the chemistry can be properly understood.

All phase boundaries (liquid–solid, liquid–gas or solid–gas) are complex. This is because some of the intermolecular bonds that are intact within the phase are inevitably broken at the surface. For solid particles these broken bonds lead to a high amount of chemical energy around the surface, and also make the surface electrically charged (Figure 13.2). In this sense, particles can be treated as any other charged object, i.e. they display some ionic characteristics. Thus, a small fragment of mineral will attract a mantle of positively charged ions in exactly the same way as a Cl^- ion does. This leads to what is termed the 'electrical double layer' (Figure 13.3).

The electrical double layer has two parts. First there is an inner charged layer at the surface of the mineral owing to broken bonds. This may be positively or negatively charged, or electrically neutral, depending on the nature of the mineral. This part of the charge is fixed, predetermined by the nature of the particle. Second, there is an outer layer of opposite charge, formed by the ions attracted to the mineral surface. As stated earlier, such ions are said to be adsorbed. The total amount of charged ions in this layer is fixed; it must exactly balance the permanent charge of the particle surface. However, this layer can be subdivided into a compact charged layer (ions bonded with specific atoms or molecules in the particle surface, and said to be specifically adsorbed), and a diffuse layer which is loosely bound to the surface by relatively weak electrostatic forces. Ions in the diffuse layer are termed **counter ions**. H^+ ions and transition metal ions (Cu^{2+}, Zn^{2+} etc.) tend to bind specifically, while the larger cations (Ca^{2+}, Mg^{2+} and K^+) tend to be in the diffuse layer. This double layer has a profound

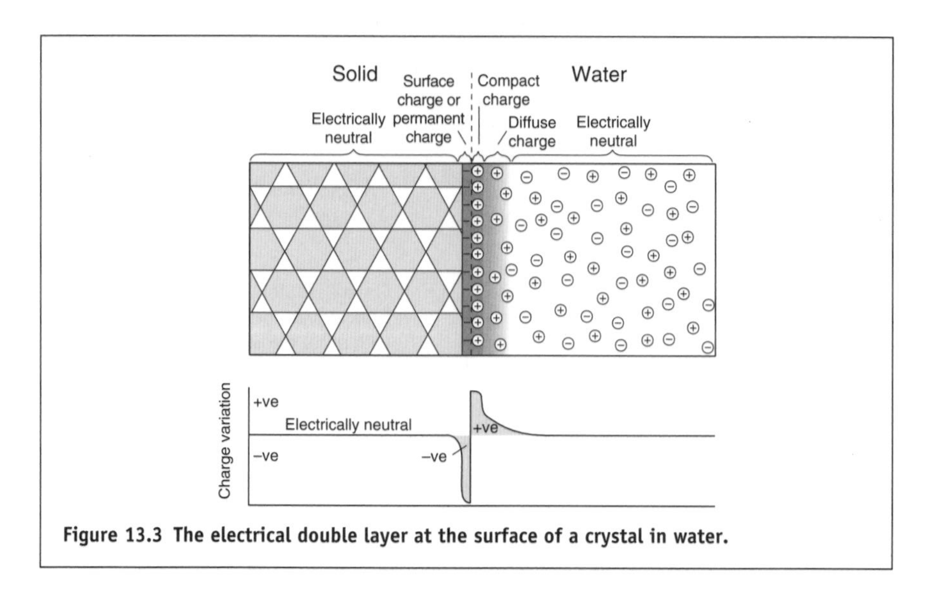

Figure 13.3 The electrical double layer at the surface of a crystal in water.

influence on the chemistry of both aqueous systems and soils. Let us consider a few key aspects.

13.2.1 Acid/base properties of particle surfaces

In acidic conditions, H^+ ions are concentrated in water. This favours the binding of H^+ to the surface of particles. In the case of a particle of silica (SiO_2) this can be seen as an oxygen atom at the surface of the oxide accepting additional H^+ ions (e.g. Figure 13.2, case 3). A shorthand used by chemists to illustrate this kind of reaction represents the solid SiO_2 as $\equiv$Si–O. The surface-bonded oxygen ion can be shown by linking it with a dash: $\equiv$Si–O^-. The whole reaction can thus be represented in two successive additions of H^+ ions:

$$\equiv \text{Si–O}^- + \text{H}^+ \rightleftharpoons \equiv \text{Si–OH}$$

$$\equiv \text{Si–OH} + \text{H}^+ \rightleftharpoons \equiv \text{Si–OH}_2^+$$

Invoking le Chatelier's principle (see **35**), it is clear that the addition of H^+ ions to this system will move the expressions to the right (thus favouring $\equiv$Si–OH_2^+), while addition of base (removal of H^+) will move the reactions to the left (thus favouring $\equiv$Si–O^-). From these reactions it is clear that the particle is behaving as a weak acid (see **12.5**). In basic solutions it is a proton donor and is thus an acid, whereas in acidic solutions it is a proton acceptor, and thus a base.

This property of particle surfaces, which is common to most naturally occurring particles, including organic matter, has an important effect in aqueous systems. The effective surface charge (the permanent charge + the compact charge) of the particle varies with pH. In acidic solutions, particles tend to have a positive charge, while in basic solutions they are negatively charged. This variation in the total surface charge impacts upon the diffuse layer. The charge of the diffuse layer, and hence the ability of the particle to bind ions like Ca^{2+}, Mg^{2+} and K^+, varies with solution pH.

13.2.2 Binding of metals by particle surfaces

Considering the account above, it should be no surprise to learn that there are two mechanisms involved in the binding of metals to particle surfaces. These are, first, specific binding to atoms in the surface of the mineral, and second, non-specific electrostatic binding to the diffuse layer.

Specific binding can be treated like any other equilibrium reaction. Thus the binding of copper with an organic particle in water can be represented as:

$$\equiv\text{HUMIC–OH} + \text{Cu}^{2+} \rightleftharpoons \equiv\text{HUMIC–COOCu}^{+} + \text{H}^{+}$$

The equilibrium can be expressed as:

$$\frac{[\equiv\text{HUMIC–OCu}][\text{H}^{+}]}{[\equiv\text{HUMIC–OH}][\text{Cu}^{2+}]} = \text{K}$$

In recent years there has been considerable progress in determining the constants for reactions between metals and humic substances. This is very useful because most particles in freshwaters are coated by humic substances and thus exhibit the same chemical behaviour (42.3.4).

13.3 Ion exchange equilibria

For the major dissolved nutrients calcium, magnesium and potassium, adsorption is dominated by the diffuse electrostatic effects. This means that the more complex chemical interactions can be disregarded, and a simple equilibrium approach to adsorption can be adopted. Central to this is the concept of the cation exchange capacity, which we introduced earlier. Typical values for the CEC lie in the range 10–100 meq/100 g of dry materials, the precise value depending upon particle type, grain size and pH. In most soils that are not acidic, Ca^{2+} is usually the dominant cation. Under acidic conditions, H^+ or Al^{3+} dominate. Indeed, the fraction of all possible sites that is occupied by Ca^{2+}, termed the **base saturation**, is a useful measure of the acidity of the soil. Thus:

$$100 \times \frac{\text{Ca}_{\text{exch}}}{\text{CEC}} \approx \%\ \text{base saturation}$$

The importance of cation exchange lies in the way that water and soil come into equilibrium with each other. Thus:

$$\frac{[\text{Ca}^{2+}_{\text{aq}}]}{[\text{Other aqueous ions}]} = K_{\text{S}} \frac{[\text{Ca}^{2+}_{\text{soil}}]}{[\text{Other soil ions}]}$$

where K_S is termed the selectivity coefficient. This shows that soil composition and water composition are not independent, but rather that water and soil can influence each other. However, the capacity of soils to carry ions is orders of magnitude greater than that of water. For example, typical river water contains around 300 $\mu eq\,l^{-1}$ (microequivalents per litre) of calcium. Yet a soil at only 10 per cent base saturation is likely to contain more than 20 000 $\mu eq\,dm^{-3}$. Thus in practice, at least in the short term, it is soil which controls the calcium concentration of water which comes into contact with it.

14 Oxidation, reduction and redox reactions

14.1 Introduction

The terms **oxidation** and **reduction** are very important in chemistry, and you will frequently encounter them in studies of the environment. 'Oxidation' suggests that oxygen is involved, and in fact the term was once used only to refer to chemical reactions involving the addition of oxygen to a substance. In contrast, reduction was used to refer to the loss of oxygen from a substance. This usage is still commonly employed: after all, oxygen is a very abundant element and it reacts readily with many of the elements. Nearly all the familiar minerals, and many gases contain oxygen. Environments may be described as oxidizing (or **aerobic**) if free oxygen (O_2) is present, and reducing (or **anaerobic**) if oxygen is absent. Similarly, we use the expressions aerobic and anaerobic metabolism depending on whether or not oxygen is involved, and aerobic and anaerobic organisms depending on whether or not they use oxygen (see **34.3.3**). The use of the terms 'oxidation' and 'reduction' was later extended to cover hydrogen. A substance is said to be oxidized if it loses hydrogen, and reduced if it acquires hydrogen.

Nowadays, however, the terms 'oxidation' and 'reduction' are used in a more general way to cover *all* reactions in which there are changes in the number of electrons. An atom, ion or molecule is said to be oxidized when it loses electrons and it is said to be reduced when it gains electrons. Such reactions are in contrast to acid/base reactions (see **12**) which simply involve the rearrangement of atoms, rather than changes to the atoms themselves. In the next few sections we describe the processes involved.

14.2 Oxidation/reduction reactions

Consider these examples. If we burn metallic calcium in air, we get an intense reaction, which results in the production of calcium oxide:

$$2Ca + O_2 \rightarrow 2CaO$$

where CaO is an ionic substance, and is therefore a lattice composed of Ca^{2+} and O^{2-} ions. We have started with electrically neutral calcium and have generated positively charged Ca^{2+} ions. Clearly, each calcium atom has lost two electrons leaving a net positive charge for the whole atom.

This kind of reaction led to the term **oxidation**, literally meaning reaction with oxygen. However, later research showed that the crucial part of the reaction is the transfer of the electrons, and oxygen provides only one way of doing this. For example:

$$Ca + Cl_2 \rightarrow CaCl_2$$

is also an oxidation reaction, because the calcium still loses its two electrons.

The essence of the oxidation reaction can thus be expressed:

$$Ca \rightarrow Ca^{2+} + 2e$$

where e is an electron. What we see is an increase in the electrical charge on the calcium. Metallic calcium has no electrical charge and thus has a valence value of zero. Ionic calcium, however, has a charge of +2, and thus a valence of +2. This provides us with a simple definition of oxidation: *oxidation is an increase in valence.*

Reduction is the opposite:

$$2e + Ca^{2+} \rightarrow Ca$$

In this reaction calcium is reduced, the valence value having fallen from +2 to 0.

To consider further the nature of oxidation and reduction reactions, let us look closer at the reaction of calcium with chlorine. We notice that two reactions take place simultaneously:

$$Ca \rightarrow Ca^{2+} + 2e$$

$$2e + Cl_2 \rightarrow 2Cl^-$$

Thus, while the valence of calcium has increased from 0 to +2, the valence of Cl has decreased from 0 to −1. This reaction is, therefore, not the simple oxidation reaction it appeared to be, but a simultaneous oxidation of calcium and reduction of Cl. Oxidation cannot occur without an accompanying reduction. The term **redox** is used for a reaction in which one part is reduced and the other oxidized.

14.3 Oxidation states

As a result of the configuration of electrons in their outer shells, the elements have characteristic **oxidation states**, or **oxidation numbers**. This value is simply the difference between the number of electrons in an atom when present in a compound and the number present in its elemental form. Elements with atomic numbers which fall just short of one of the noble gases form negative ions, while elements with atomic numbers that just exceed those of a noble gas form positive ions. So the oxidation state of sodium is +1 (because the lone electron in the outer shell is donated during formation of sodium chloride),

Table 14.1 Oxidation states of some abundant elements.

Element	*Ionic charge*	*Element*	*Ionic charge*
H	+1	P	+5
C	−4 to +4	S	−2 to +6
N	−3 to +5	I	−1
O	−2, 0	K	+1
Na	+1	Ca	+2
Mg	+2	Mn	+2, +4
Al	+3	Fe	+2, +3
Si	+4		

while the oxidation state of chlorine is −1 (because it has one more electron in compounds than when it exists as an atom).

Ions of the same element may exist in various oxidation states. A good example, and one of great environmental importance is iron (Fe). If an iron atom loses two electrons then a ferrous ion (Fe^{2+}) is produced, and if a ferrous ion is oxidized by the loss of another electron then ferric ions (Fe^{3+}) are produced. Oxidation numbers, or states, of some very important elements in the environment are shown in Table 14.1. Ordering is by increasing atomic number from left to right, and by row. Note that it is not only the elements and their ions which are assigned oxidation numbers or oxidation states, but also complex ions and compounds.

As we stated earlier, environments are often described as oxidizing if free oxygen (O_2) is available and as reducing if oxygen is absent. However, we should be aware that although it is the presence or absence of oxygen which is critical, oxidation and reduction do not necessarily involve oxygen.

14.4 Oxidizing and reducing agents

Now we know what oxidation and reduction mean, we must address the question of why a substance would either oxidize or reduce. It comes down to the electronic structure of elements. Generally, non-metals have a tendency to gain electrons (become negative, and thus reduced), while metals have a tendency to lose electrons (become positive, and thus oxidized). A substance which has a strong tendency to be reduced (say O_2) will seek out a substance which can be oxidized. This is because the redox constraint prevents it from becoming reduced unless something else becomes oxidized. Such a substance is referred to as an **oxidizing agent**. Conversely, a substance which has a strong tendency to be oxidized (say metallic zinc) will seek out a substance to reduce, and is termed a **reducing agent**.

The most important oxidizing agent in the natural world is dioxygen (O_2). Other common oxidizing agents include Cl_2, Mn^{4+} and Fe^{3+}. The metals are reducing agents, of which zinc and cadmium are familiar. However, by far the most important reducing agent overall in the natural world is organic matter.

14.5 Redox status

The redox status of a system is analogous to its acidity/alkalinity. The acidity of a system is governed by the balance between acids and bases; a highly acidic solution is one in which acids dominate. In a similar way, the redox status of a solution is governed by the balance between oxidizing and reducing agents. Thus, a highly oxidizing solution is one which is dominated by oxidizing agents.

Redox status is important because of its impact on the chemical properties of some elements. As we have just learnt, some elements, for example iron, can exist in more than one oxidation state under the conditions normally encountered in the natural environment. Iron may be present as either Fe^{2+} or Fe^{3+}. The former is favoured by reducing conditions, while the latter is favoured by oxidizing conditions. The difference is that the Fe^{3+} is virtually insoluble. Thus iron has very low solubility under oxidizing conditions. If the system becomes reducing, this will encourage reduction of Fe^{3+} to Fe^{2+}, which brings iron into solution.

Another important example is sulphur. In reducing conditions, sulphur is commonly present as sulphide. This causes precipitation of many metal ions, particularly toxic metals such as lead and cadmium, and is therefore very important in influencing the environmental fate of these metals. Further, the dominant sulphide is the gas H_2S, which is highly toxic to aqueous animals. Clearly, then, redox conditions are of great importance in the chemistry of the environment.

14.6 Eh and the measurement of redox status

If we are to predict the behaviour of metals in the environment, we must be able to quantify the redox conditions. This can be done by considering the role of electrons. Recall that substances are reduced if they accept electrons

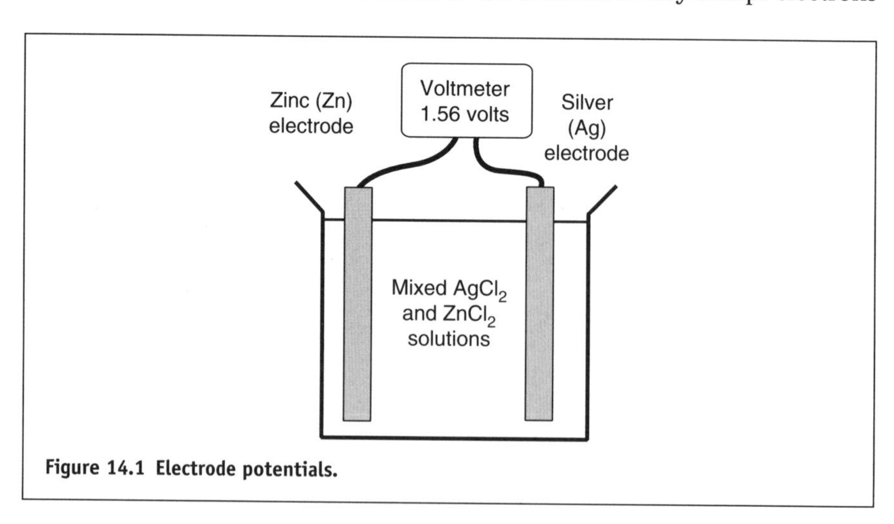

Figure 14.1 Electrode potentials.

from their surroundings, as in:

$$O_2 + 4e \rightarrow 2O^{2-}$$

The greater the concentration of reducing agents the more reducing is the environment, and the greater the possibility of a substance acquiring electrons. In just the same way that acidity can be viewed as the free hydrogen ion concentration, the redox status can be viewed in terms of free electron concentration. Conditions are thus reducing if there are lots of 'free' electrons around, and

Table 14.2 Redox status in oxygen-rich and oxygen-depleted environments.

Eh (volts)	***Environment***	***Redox status***
c. 0.4	Rainwater, rivers, ocean surface	Oxidizing
c. −0.4	Waterlogged soil, organic sediment	Reducing

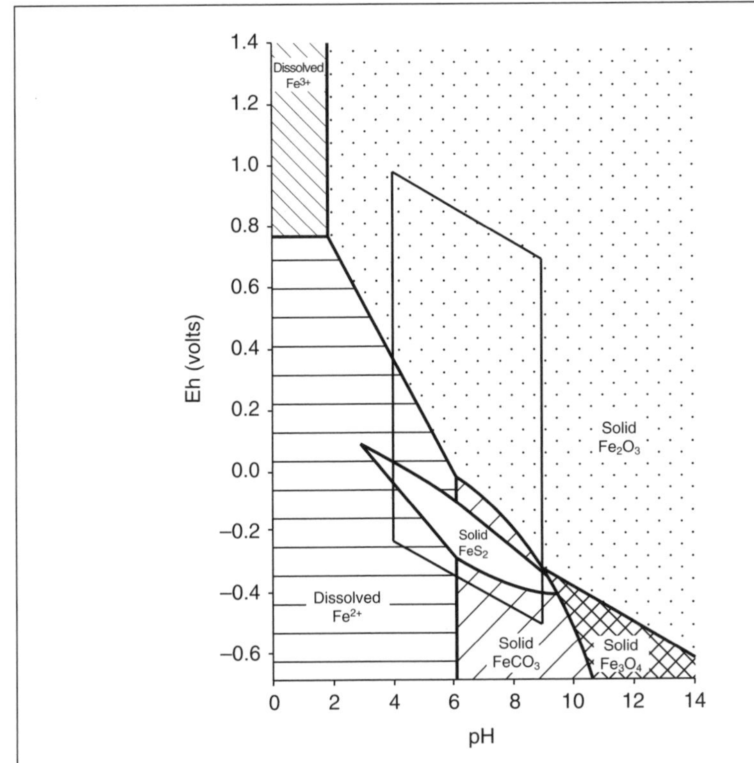

Figure 14.2 Eh/pH diagram for iron species in the presence of sulphur and carbonate. This Eh/pH diagram is constructed assuming: dissolved carbonate = 1 M, dissolved sulphur = 10^{-6} M, dissolved iron = 10^{-6} M. The quadrilateral in the centre of the figure shows the normal range of naturally occurring pH and Eh. If the Eh and pH of a soil or water body are known, and the concentrations of iron, sulphur and carbonate conform to the assumed values, this diagram can be used to predict the predominant stable form of iron.

oxidizing if there is a scarcity of free electrons. Actually, the electrons are not truly free (just as 'free' protons are not actually present in acidic solutions), but their potential availability can be quantified as if they were free.

This electron concentration can be measured as an electric potential (Figure 14.1). The more positive the potential, the fewer the electrons that are present, and the more oxidizing the environment. The more negative the potential, the greater the concentration of electrons, and the more reducing the environment. This potential (referred to as the **redox potential**) is written Eh. Natural water shows a range of Eh values, as shown in Table 14.2.

Eh and the pH are convenient measures of the status of the water and soil. You will commonly meet Eh/pH diagrams which illustrate the stability of chemical species in the environment (Figure 14.2).

15 Families of elements: the periodic table

15.1 Overview

Each chemical element has a unique number of protons and electrons (see **3** and **4**), and the electrons are arranged in a particular way within shells (see **7**). However, while each element is unique, groups of elements do share aspects of their general behaviour. This applies particularly to the way that they react with other elements. The shared behaviour is owing largely to similarities in electron configuration, and not surprisingly it is the outer, or valence, shell which is the key feature. From the mid-nineteenth century, some chemists attempted to design schemes which presented the elements in a way which reflected their similar properties. The most notable of these schemes was proposed by the Russian chemist Dmitri Mendeleev (sometimes spelt Mendeleyev) in 1869. It was given the name **periodic table**. In his scheme the elements were arranged in columns in terms of increasing atomic weight (atomic particles had yet to be discovered so atomic numbers were unknown). When the table was proposed, only about sixty elements were known, but Mendeleev correctly predicted the existence of other elements, their place in the scheme, and their properties. Although Mendeleev's scheme has been refined with the discovery of other elements, and some elements have been moved to a new position, the basic structure has endured. A modern periodic table of the elements is shown in Figure 15.1.

The columns, which are numbered from 1 to 18, are known as **groups**. (Earlier, but still widely used versions use a slightly more complicated labelling system with Roman numerals followed by either A or B to denote the groups.) The elements within a particular group have similar chemical properties because they have the same number of electrons in their outer shells. However, aspects of their behaviour, for example their degree of reactivity with certain other elements, change in a systematic way. The behaviour of the elements in the various groups in the periodic table is discussed in more detail elsewhere (see **40**).

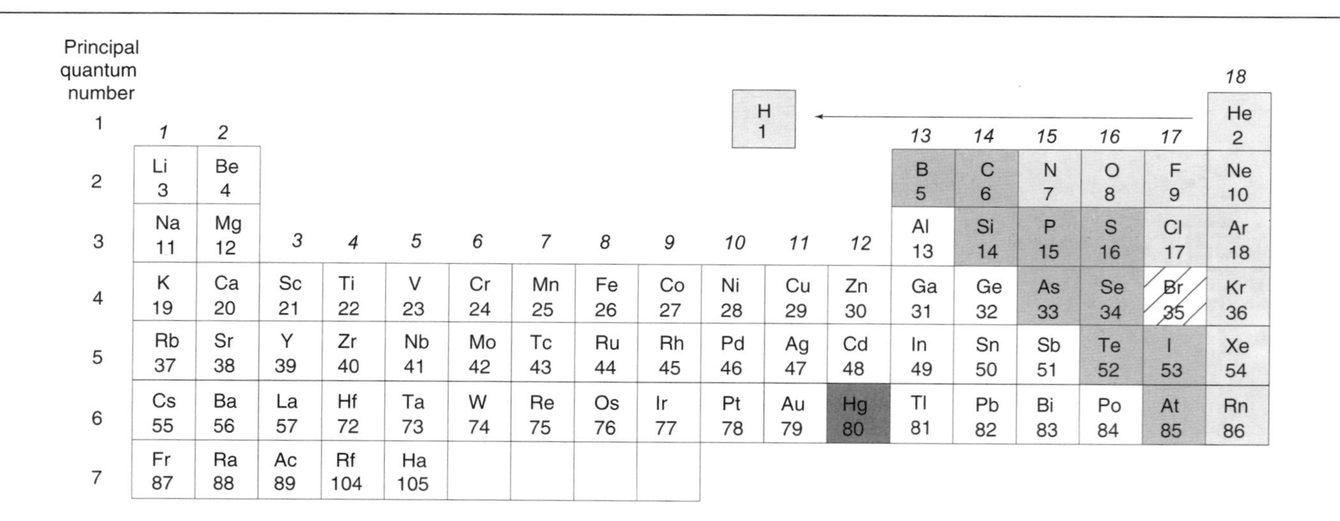

Figure 15.1 The periodic table. The atomic number of each element is shown below the element symbol.

Names are applied to particular groups of elements in the periodic table, which is convenient when referring to elements with similar behaviour. One such group – the noble gases (group 18) – was referred to while introducing chemical bonding (see **8**). Another example is the halogen group (chlorine, fluorine, bromine, iodine, astatine); these comprise group 17. Group 1, with the exception of hydrogen which is a special case, is made up of the alkali metals: all of these react quite violently with water, and the higher the atomic number the greater the rate of reaction. Elements in groups 1 and 2, and in groups 13 to 18 are called **main-group elements**, while those in groups 3 to 12 inclusive are known as the **transition elements**.

There are seven main rows in this scheme, and these are referred to as **periods**. Notice that the periods tend to get longer as we move down the table. In fact, periods 6 and 7 require two outlier rows which we mention again a little later. The atomic number increases from left to right across each period. Notice also that each complete period ends with a noble gas, each of which has eight electrons in its outer shell and reacts with other elements only under extreme conditions. Incidentally, the ordering of elements by atomic number is not exactly the same as the ordering of elements by atomic mass (refer back to Table 3.1), which is what Mendeleev used, and this explains some of the positional changes in the periodic table which have occurred since it was originally proposed.

The uppermost of the outlier rows beneath the main body of the table contains fifteen elements which are collectively called the lanthanoids, or the rare earth elements. These belong to period 6 and form a subset of group 3. We can see their atomic numbers run from 57 (lanthanum) to 71 (lutetium). The elements in the lower row, which are termed the actinides or actinoids, belong to period 7, and are another subset of group 3 elements. With the exception of uranium and thorium, their names will probably be unfamiliar to non-chemists. All the isotopes of the elements in this series are radioactive.

Returning to the table as a whole, a useful initial distinction is drawn between metals and non-metals, with the latter being located towards the right-hand side of the table. In practice, though, the distinction between metals and non-metals is not absolute. The elements become increasingly metallic as we move from right to left across a period and from top to bottom within a group. An important feature of metals is their tendency to lose electrons, and thus form positive ions (cations), in chemical reactions. (They also tend to conduct electricity and heat effectively.) In contrast, the non-metals tend to be electronegative, forming negative ions (anions) by gaining electrons during chemical reactions.

15.2 Trends in ionization potential and electron affinity

The energy needed to remove an electron from an atom or ion is measured by the **ionization potential** (see **5**). When an atom loses an electron it becomes a cation, i.e. a positively charged ion. The general rule is that the further an outer electron is from the nucleus, the less tightly it is held and thus the more easily it is removed. As we move from top to bottom within each group in the periodic table there is a large increase in the number of electrons associated with the atomic nucleus. In effect this means that there are more electron shells, so the outer electrons are further and further away from the nucleus. They are thus held less and less tightly, so ionization potential decreases from top to bottom in each group.

However, as we move from left to right within a period the atomic radius decreases: therefore the forces of attraction between electron and nucleus increase, and it follows that ionization potential also increases. The decrease in atomic radius from left to right in a period may seem counterintuitive as electron number is increasing in this direction. However, the radius decreases because of the increasingly large forces exerted by the atomic nucleus as proton number increases, and these cause a reduction in atomic radii.

Electron affinity refers to the ability of an element to capture electrons. By acquiring an electron an electrically neutral atom is converted to an anion, i.e. a negatively charged ion. As we move from top to bottom in a group in the periodic table, electron affinity tends to decrease. The reason is outlined above: with an increase in the number of shells the outer electrons are located further and further from the nucleus. Thus the forces of attraction between nucleus and electrons decline, and so too does the ability to capture electrons. As we move from left to right in the table, electron affinity increases. Recall, atomic radius declines as we move in this direction, so the nucleus has increasing pulling power for negatively charged electrons.

15.3 Trends in electronegativity

Electronegativity refers to the relative attractiveness of an atom for electrons which it shares with another atom. (Electropositivity is the same thing viewed back to front.) Fluorine (F), as the most electronegative element, is accordingly assigned the maximum value, which, in the most commonly used scheme, is arbitrarily set at 4. All other values are therefore less than 4, and all are positive. These values are shown in Figure 15.2. (The noble gases, with their complete valence shells, and accordingly low attractiveness for electrons, are not assigned an electronegativity value.) Notice the position of the most electronegative element, fluorine (F), in the top right-hand corner of the table, and the fact that as we move towards the table's bottom left-hand corner the values gradually decrease, to 1 or less. You may recall the distribution of metals and non-metals in the periodic table (Figure 15.1): as we move from right to left across series, and from top to bottom within groups, electropositivity or 'metallicity' increases. And remember also that a key expression of increased 'metallicity' is the tendency to lose electrons and thus form positively charged ions.

The table of electronegativities can be used as a rule-of-thumb to ascertain whether the bond between any two elements is likely to be ionic or covalent in nature (see **8**). This involves determining the difference in value between the

Principal quantum number	1	2	3	4	5	6	7	8	9	10	11	12	13	14	15	16	17	18
1										H								He
2	Li 1.0	Be 1.5											B 2.0	C 2.5	N 3.0	O 3.5	F 4.0	Ne
3	Na 0.9	Mg 1.2											Al 1.5	Si 1.8	P 2.1	S 2.5	Cl 3.0	Ar
4	K 0.8	Ca 1.0	Sc 1.3	Ti 1.5	V 1.6	Cr 1.6	Mn 1.5	Fe 1.8	Co 1.8	Ni 1.8	Cu 1.9	Zn 1.7	Ga 1.6	Ge 1.8	As 2.0	Se 2.4	Br 2.8	Kr
5	Rb 0.8	Sr 1.0	Y 1.2	Zr 1.4	Nb 1.6	Mo 1.8	Tc 1.9	Ru 2.2	Rh 2.2	Pd 2.2	Ag 1.9	Cd 1.7	In 1.7	Sn 1.8	Sb 1.9	Te 2.1	I 2.5	Xe
6	Cs 0.7	Ba 0.9	La 1.1	Hf 1.3	Ta 1.5	W 1.7	Re 1.9	Os 2.2	Ir 2.2	Pt 2.2	Au 2.4	Hg 1.9	Tl 1.8	Pb 1.8	Bi 1.9	Po 2.0	At 2.2	Rn
7	Fr 0.7	Ra 0.9	Ac 1.1	Rf	Ha													

Ce 1.1	Pr 1.1	Nd 1.1	Pm 1.1	Sm 1.1	Eu 1.1	Gd 1.1	Tb 1.1	Dy 1.1	Ho 1.1	Er 1.1	Tm 1.1	Yb 1.1	Lu 1.2
Th 1.3	Pa 1.5	U 1.7	Np 1.3	Pu 1.3	Am 1.3	Cm 1.3	Bk 1.3	Cf 1.3	Es 1.3	Fm 1.3	Md 1.3	No 1.3	Lr

Figure 15.2 Periodic table showing electronegativity values for the elements.

electronegativity scores for the two elements concerned. Values greater than 2 indicate an ionic bond while values less than 2 indicate a covalent bond. Thus, sodium and chlorine ($3.0 - 0.9 = 2.1$) form an ionic bond, while oxygen and carbon ($3.5 - 2.5 = 1$) form a covalent bond. However, the distinction between covalent and ionic bonds is true only up to a point. Chemists talk in terms of *degrees* of the two types within a single bond. In the case of diatomic molecules (atoms of one type), as in hydrogen (H_2), oxygen (O_2), nitrogen (N_2), and the halogens (Cl_2 etc.), the bond is 100 per cent covalent, 0 per cent ionic. This is because the electronegativities are equal. In other cases, though, the electronegativities are not equal. The difference in electronegativities between two elements can thus be used to determine the percentage of the two bond types. What we are really saying here is that in all but the sorts of example we have just presented, the behaviour of electrons is such that one atom in the bond is positively charged and the other is negatively charged, even though that difference in charge may be extremely slight. The substance hydrogen chloride (HCl) provides an example. In this case the bonding between the hydrogen and chlorine atoms is predominantly covalent (a pair of electrons is shared), but the chlorine atom is more electronegative than the hydrogen atom. The hydrogen atom thus carries a slight positive charge while the chlorine carries a slight negative charge. Such bonds are appropriately referred to as **polar covalent bonds**.

16 Organic chemistry

16.1 Introduction

Organic chemistry is loosely defined as that branch of chemistry dealing with compounds containing carbon, while inorganic chemistry is concerned with all other compounds. We say 'loosely defined' because the boundary is diffuse: for example, an inorganic chemist may well be interested in the chemistry of small carbon-containing compounds such as carbon dioxide or calcium carbonate. However, for most of the time, organic and inorganic chemists focus on very different types of chemical substances. The term 'organic' was originally applied because the substances of which living organisms are made contain carbon. Indeed, until the 1820s it was assumed that such substances could not be synthesized artificially, but required some unknown 'vital' force. Of course, we still use the term 'organic' when referring to biotic substances and processes. (To add some confusion, however, the expression 'inorganic nutrition' is often used when referring to mineral elements such as calcium, phosphorus and iron which are required for living organisms.)

Organic chemistry now has a very much broader remit than the chemistry of life, which is usually termed *bio*chemistry. This is because a huge number of compounds containing carbon have been synthesized artificially. Many of these feature prominently in our everyday lives, and some pose serious problems for the environment because of their resistance to biological decay processes. The subject of organic chemistry is an important one for environmental scientists because it underpins an understanding of life processes and also because organic chemicals are of great significance in the physical environment.

16.2 The uniqueness of carbon

Carbon is found in carbonate minerals in sedimentary rocks, in the bicarbonate ion (HCO_3^-) in water, in some very important gases (for example carbon

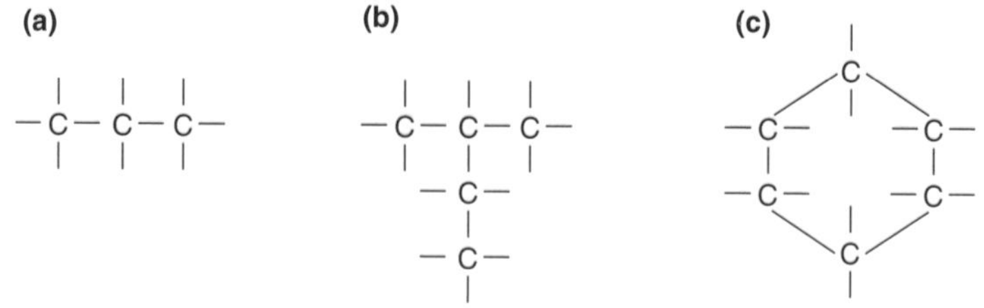

Figure 16.1 Typical arrangements of carbon atoms in organic substances. (a) Straight chain; (b) branched chains; (c) ring structure.

dioxide, carbon monoxide and methane), and in literally millions of organic substances. Carbon also occurs as a pure substance, but in different forms, a phenomenon known as **allotropy**. Two elemental forms (**allotropes**) of carbon occur as crystals. One of these is diamond, which is among the hardest substances known. The other is graphite, a relatively soft substance which provides the 'lead' for pencils. Here, however, our interest is focused on the capacity of carbon to form so many different compounds. Carbon atoms have an almost limitless capacity to link covalently (see **8.3**) with each other to form straight chains, branched chains and closed rings (Figure 16.1). This behaviour is due to the configuration of electrons. Carbon has an atomic number of 6; there are two electrons in the inner shell and four electrons in the outer, or valence shell (see **7**). Now, carbon atoms cannot readily give up electrons, because with only six protons in the nucleus this would lead to a large imbalance of charges; and neither can carbon atoms readily acquire electrons because they are only weakly electronegative (see **15.3**). Hence, to attain stability, carbon atoms must *share* electrons, i.e. form covalent bonds. This they do readily with each other, and to just a handful of other elements, notably hydrogen, oxygen, nitrogen sulphur and the halogens (for example chlorine). In fact, the vast number of organic substances is based on comparatively few elements. Many contain just carbon and hydrogen (hydrocarbons), and many more contain only carbon, hydrogen and oxygen. In contrast, inorganic compounds, which collectively make use of a hundred or so elements, are very much fewer in number.

Carbon displays a tendency to share four electrons to form four covalent bonds. Often, each of these bonds is orientated from a central carbon atom to the four corners of a tetrahedron (a geometric form consisting of four triangles). Such an arrangement is shown in Figure 16.2. In the case of a methane molecule the central carbon atom attaches to four hydrogen atoms. In diamond, the central carbon attaches to four other carbon atoms. This arrangement gives great strength and rigidity to the mineral. (Graphite, however, is composed of sheets of carbon atoms, but each atom is covalently bonded to just one other carbon atom and the sheets are held together by weak bonds which readily break when energy is applied.) The tetrahedral arrangement of atoms just described is owing to the tendency of the negatively charged electrons to repel one another. Thus the four outer electrons are positioned as far as possible

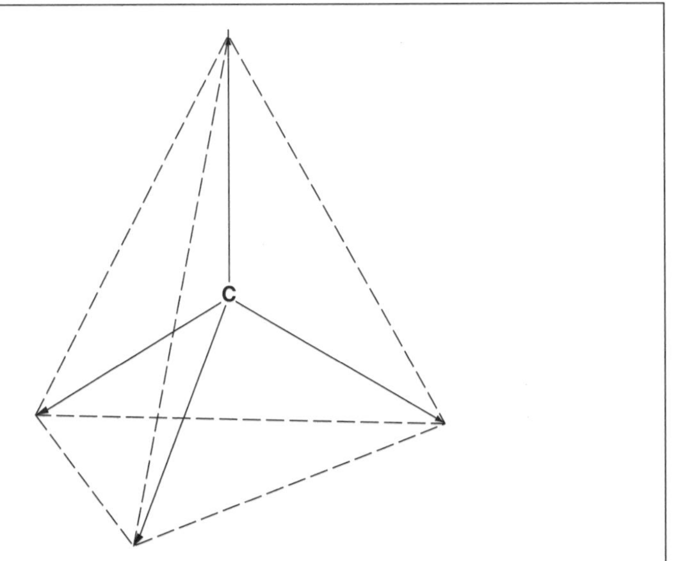

Figure 16.2 Representation of the tetrahedral structure of a methane molecule. The carbon atom is in the centre, the solid lines indicate the bonds between the central carbon and the four hydrogen atoms.

from each other. On the page we usually represent this arrangement by bonds subtending angles of 90°, but of course the structures are really three-dimensional. The actual bond angle for such an arrangement is 109.5°.

In the tetrahedral arrangement just described, carbon forms four single covalent bonds, indicated on the page by single dashes, or a pair of dots. In addition, carbon atoms commonly form double covalent bonds with other carbon atoms and also with oxygen, nitrogen and sulphur atoms. (Such double bonds are shown as a pair of dashes or two pairs of dots.) Carbon can also be linked by triple covalent bonds (denoted by three dashes or three pairs of dots) to other carbon atoms and to nitrogen. A carbon atom linked to four other atoms (by four single covalent bonds) is said to be **saturated**, and it is a very stable state. If the carbon atom is linked to less than four other atoms, which means that it is linked by double or triple bonds, it is said to be **unsaturated**.

Finally here, it is important to be aware that particular chemical groups are very common in organic compounds. Because they largely determine the properties of compounds they are often termed **functional groups**. Important examples of functional groups include CH_3 (methyl), OH (hydroxyl), CO (carbonyl), COOH (carboxyl), NH_2 (amino), SH (sulphydryl), and phosphate (H_2PO_4).

16.3 Representing organic substances

It is customary to represent *inorganic* compounds by molecular formulae (see **6**), as in H_2SO_4 for sulphuric acid. This is because the number of atoms in a molecule is quite small and the configuration of atoms is fixed and can usually be worked out with a little understanding of chemistry. For the vast majority of organic substances, however, molecular formulae are not particularly informative because the atoms can usually be arranged in a variety of ways. In fact there may be several, perhaps dozens of possible arrangements of the atoms for a particular molecular formula. In other words, it is common for the same molecular formula to be shared by several different compounds, but, in addition, some substances can exist in different forms according to the orientation of groups relative to the carbon atoms. The phenomenon whereby compounds with the same molecular formula have different **structural formulae** is known as **isomerism**. Compounds with the same molecular formula but different geometric formulae are called **isomers**. The topic is discussed further below.

To cope with these difficulties, it is standard practice to use structural formulae for organic compounds. These can be presented in various ways, but the objectives are to show the approximate positions of the atoms relative to each other, and the nature of the bonds between them. It is, of course, important to remember that diagrams of organic compounds are distortions because the atoms are actually arranged in three dimensions. So when we talk about 'straight chain' carbon compounds we really mean they are unbranched, not that the carbon bonds are aligned in one dimension. In diagrammatic representations of compounds you will frequently see that groups attached to carbon atoms are written as they would be in text, for example COOH for a carboxyl group, and to show its linkage to a carbon atom as COOH–C. However, this is merely a convention and it does not mean the carbon and hydrogen atoms are linked. A more accurate representation of the structure would in fact be HOOC–C, and sometimes it is written like this. Expressing the formula of a compound in two dimensions is never completely satisfactory. A more accurate geometric representation of the configuration of atoms in a compound can be obtained by constructing models, using balls and rods to represent the atoms and bonds respectively. You may have encountered such models if you have taken any organic chemistry classes; they are particularly useful in aiding understanding of functional aspects of organic molecules.

One way, then, to represent the structural formula of a compound is to show *all* the atoms and *all* the bonds between them. When there are more than just a few atoms in a compound, however, such diagrams become rather complicated. Even a simple sugar with molecular formula $C_6H_{12}O_6$ requires twenty-four atoms. Fortunately there are conventions for condensing and simplifying structural formulae (Figure 16.3). We say that carbon forms the 'backbone' of organic substances. Often, as in Figure 16.1, only the carbon atoms are shown in a diagram, or else only the bonds are shown to give just an indication of a compound's structure. This is particularly common in ring structures. If an element other than carbon occurs within the ring then that element might be shown while the carbons are omitted. Another convention is to show only particular functional groups, or only the atoms involved in the bonds which link two formerly separated molecules. It is also standard practice to use the

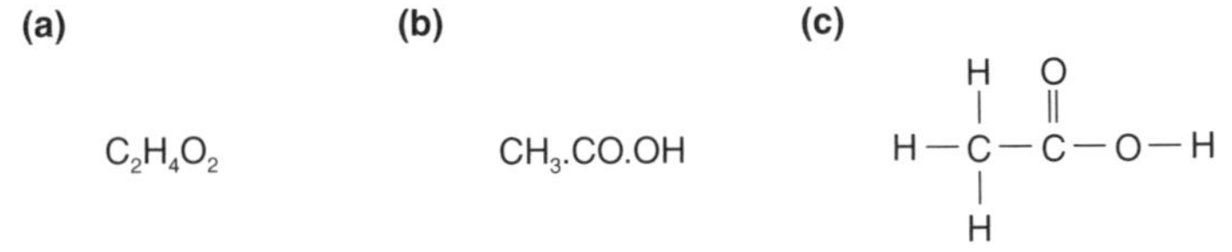

(a)	(b)	(c)
$C_2H_4O_2$	$CH_3.CO.OH$	H–C(H)(H)–C(=O)–O–H

Figure 16.3 Ways of depicting acetic acid. (a) Molecular formula; (b) condensed structural (or rational) formula; (c) structural formula.

letter R to denote a certain structural group (which may be very large). We adopt this approach when considering the major classes of biochemicals (see **32**). None of these approaches to simplification presents a problem for practitioners because they are aware of the identity of the atoms or groups of atoms in the various positions in the more common compounds.

In the condensed, or rational form, a formula is shown in a straight line. This is useful because it removes the need for a diagram and allows incorporation of formula in text. In this scheme, atoms are grouped according to their observed chemical properties. The molecular formula for lactic acid is $C_3H_6O_3$. The condensed formula for this substance is $CH_3CH(OH)COOH$. Drawing a structural formula from this condensed formula would not be a problem because the various chemical groups in the molecule are so well known. The brackets here indicate that the OH group is linked to the preceding carbon, along with a hydrogen atom. When a compound contains more than one of a particular group, the number of such groups can be represented by n, as in $(CH_2)_n$.

16.4 Classifying and naming organic compounds

Although there are millions of organic compounds, they are grouped into major classes, and there are conventions concerning the naming of compounds according to their structure. Without going too deeply into this subject, a few key terms and concepts are introduced here to facilitate further exploration of the important world of organic chemicals. You need not be put off by the formidably long names of many organic chemicals. It is unlikely that you would need to commit such names to memory. They are named according to a logical system, and many have common names by which they are much better known. The formal naming of organic compounds is governed by internationally agreed rules which ensure consistency and also permit the structure of a compound to be deduced.

In many classes of organic substances the individual members form a **homologous series**. An example will make it clear what this term means, and also introduce some important terminology. The **alkanes** are structurally simple hydrocarbons of general formula C_nH_{2n+2}. Individual alkane members differ incrementally in the number of CH_2 groups. The first ten members of the alkanes are shown in Table 16.1: some of these names are probably quite familiar

Table 16.1 Prefixes for the first ten alkanes, illustrating the naming convention.

Prefix	*Number of carbons*
Methane	1
Ethane	2
Propane	3
Butane	4
Pentane	5
Hexane	6
Heptane	7
Octane	8
Nonane	9
Decane	10

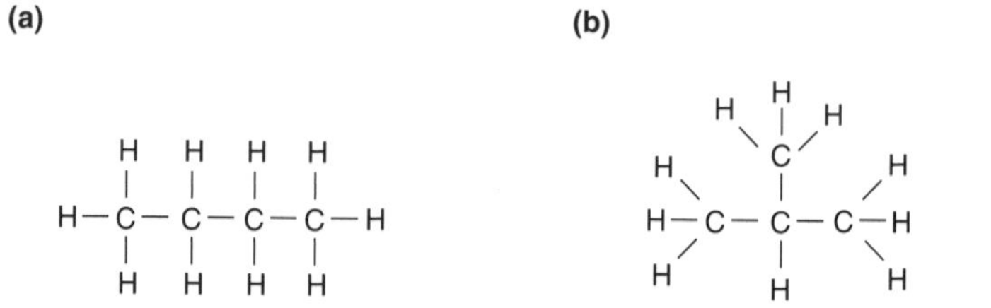

(a) (b)

Figure 16.4 Structure of (a) straight-chain (*n*), and (b) branched-chain (*iso*) forms of butane.

to you. Note that the prefixes ('meth-', for one carbon, 'eth-' for two carbon atoms, etc.) in the names in this list are in general use to denote the number of carbon atoms in a compound, or in the main carbon chain. The suffix is also important because it denotes the class of organic compounds. For example, while '-ane' denotes an alkane, '-ene' denotes an alkene, and '-ol' denotes an alcohol. These are examples of the rules set by the International Union of Pure and Applied Chemistry (IUPAC) for naming organic compounds.

Butane, C_4H_{10}, is the first member of the alkane series to form **isomers** (defined in **16.3**). Butane occurs in both straight-chain and branched forms. Straight-chain forms are always denoted by an italicized *n*- (for 'normal') as in *n*-butane, whereas the branched form is commonly called *iso*butane. The structural formula for each is shown in Figure 16.4. The number of isomers increases markedly with an increase in carbon atoms in a homologous series: decane, which has ten carbon atoms, has 75 isomers.

16.5 Reactions in organic chemistry

Chemical reactions are introduced elsewhere (see **10**). Here we mention some of the key reactions in organic chemistry. First, **addition** simply involves the combination of two compounds with no release of chemical by-products. Second, **substitution** involves one atom or radical taking the place of another of equal valency so that two new compounds are formed. Third, **condensation** involves two molecules becoming chemically bound to each other with the release of a small molecule, which is usually water. Condensation reactions are extremely important in **polymerization**. A **polymer** is a molecular chain of the same small molecule (generally called a monomer). For example, starch (**32.3.1**) is a polymer based upon glucose molecules, and most plastics are polymers. While condensation reactions give rise to larger molecules from smaller molecules, **hydrolysis** results in the splitting up of molecules into smaller components. During hydrolysis a water molecule is added to the two products of the reaction. One product acquires a hydrogen, the other product acquires a hydroxyl group. So when starch is completely hydrolysed, individual molecules of glucose are formed.

16.6 Isomerism

The molecular formula of a chemical compound informs us of the number of atoms of the various elements contained in a molecule of the substance, but it tells us nothing of the arrangement of atoms. Two quite different compounds could share the same molecular formula, or else two forms of the same compound could occur, differing only in the spatial arrangement of atoms around one carbon atom. In such cases the different forms which share the same molecular formula are called **isomers**, while the phenomenon itself is referred to as **isomerism**. It is important to remember, though, that the structural formula of a particular isomer can frequently be drawn in alternative ways. So initially it can be a bit confusing to differentiate between different diagrammatic representations of the same isomer and true isomers.

There are different sorts of isomerism. **Functional isomerism** applies to situations in which either the functional groups on the molecule are actually different, or a particular functional group occupies different positions within the molecule (Figure 16.5). These differences generally result in such isomers having different physical and chemical properties.

In **stereoisomerism** (Figure 16.6) the isomers share the same formula and the same functional groups. They differ, however, in the spatial arrangement of atoms. One form of stereoisomerism is **optical isomerism**, so called because the different isomers of a compound (known as optical isomers) differ in their effects on a beam of polarized light. Specifically, they differ in the direction to which they rotate a beam of polarized light. One of a pair of optical isomers rotates polarized light to the right with respect to the direction of the light. It is said to be dextrorotatory, and is denoted by the prefix + (formerly by *d*-). The other isomer rotates polarized light to the left. It is said to be laevorotatory and is denoted by the prefix – (formerly by *l*-). Molecules that exhibit such optical activity have no plane of symmetry. This is most commonly found in organic molecules which have a central carbon atom to which are attached four different atoms or groups of atoms, as in Figure

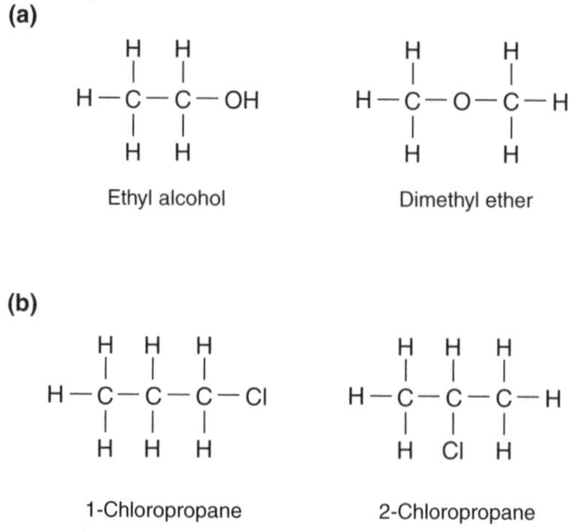

(a)

Ethyl alcohol

Dimethyl ether

(b)

1-Chloropropane

2-Chloropropane

Figure 16.5 Functional isomerism. (a) These two compounds share the same molecular formula but the functional groups are different. (b) In this compound the functional group occupies different positions.

16.6a. The carbon atom is described as asymmetric, although it is actually the molecule that is asymmetric. It means that if we constructed a model of one isomer and then a model of its mirror image we would find it impossible to superimpose one upon the other, just as it is impossible to superimpose a left-hand glove on a right-hand glove. The optical isomers of a compound have the same physical and chemical properties: it is only with respect to their effect on polarized light that they differ. Many naturally occurring compounds exhibit optical isomerism, although usually only one form is found in nature. The other form may be produced artificially but is not

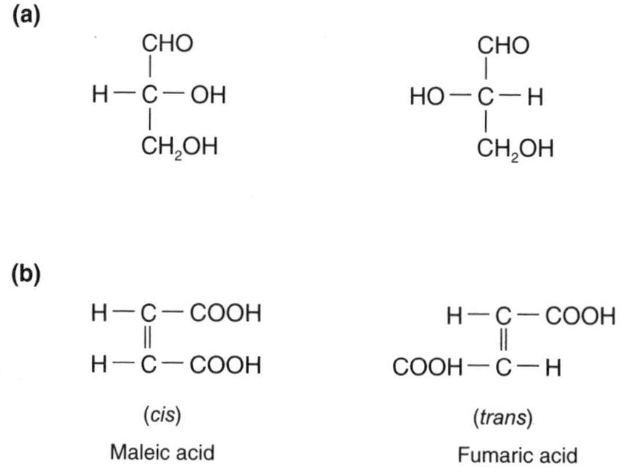

(a)

(b)

(*cis*)
Maleic acid

(*trans*)
Fumaric acid

Figure 16.6 Stereoisomerism. (a) Optical isomerism; (b) *cis–trans* isomerism.

normally biologically 'active', in other words it cannot substitute for the natural optical isomer in metabolic reactions.

Another form of stereoisomerism is termed **geometric**, or **cis–trans** isomerism (Figure 16.6b). It occurs when two dissimilar groups (for example H and OH) occupy different positions with respect to two carbon atoms which are linked by a double bond, or to a central carbon atom within a ring structure. In the former case the double bond does not permit the rotation of the two carbon atoms, thus enabling different forms to exist. In the latter case the free rotation of atoms is also restricted (although double bonds between carbon atoms are not required), and the groups may be situated above or below the plane of the ring.

Finally, note that there are certain types of enzymes (see **33**) which can bring about the conversion of a particular isomer to another by rearrangement of the atoms. Any such enzyme is known as an **isomerase**.

17 Physics in words

17.1 Key skills

We live in a world where physical laws accurately predict events and phenomena. Even if we know little about physics, we tend to have an intuitive grasp of the consequences of the laws of physics. For example, not only do we know that a bicycle travelling downhill will gradually increase its speed, but if it deviated significantly from the expected pattern, it would 'look odd'. Indeed, it is a combination of the fact that such laws are generally obeyed, and that we have this instinctive grasp of them, that makes modern computer games appear so realistic. All that the programmers need to do is follow the laws of motion, and the events portrayed in the program will look right. However, this intuitive understanding is deeply buried, and it is a difficult journey to get from this to a full understanding of the physical laws. A further difficulty for teachers is the progressive nature of an individual's development in this understanding, which makes it hard to appreciate the barriers that face students of physics. It is difficult for teachers to identify how and when they themselves first acquired an insight to the problems that they are attempting to convey to their students. Whereas most of this book presents aspects of the physical laws that describe our world, here we look at the skills and approaches that are useful when you attempt to improve your understanding of these laws. While it is very difficult, if not impossible, to put into words something that must be learnt through experience and time, we can at least suggest a few key skills which will help. We have decided to break these down into five headings:

- *Critical observation*. A combination of careful observation, constructive scepticism, and repetition.
- *Seeing the big picture*. Getting to know how to identify the most important factors.
- *Curiosity*. Curiosity is habit forming. Developing your curiosity about the world is probably the single most important thing you can do.

- *Mathematics as a tool.* This is less important than most of the other skills. However, a good practical grasp of how to apply simple mathematical procedures is a great help.
- And finally, in addition to these skills there is the boring practical issue of overcoming any difficulty you have with technical terminology. Sadly, this is simply a matter of hard work.

The remainder of this section explores the first four skills described above. However, as an example to show how they may be applied, let us first consider a very simple experiment. Imagine that you drop a ball from shoulder height on to the ground. Picture in your mind exactly what would happen, right from when the experiment starts up to when the ball ceases to move. Make a detailed written description of what you think will happen. Finally, do the experiment: drop the ball and observe what happens. Now you have two lists, and they will not be in complete agreement. How many of the things you predicted actually came true? How many things happened that you did not predict? Where you find a mismatch between expectation and reality, try to decide why. Repeat the experiment and check for changes in the outcome. You could try different heights and different surfaces. If you do this experiment diligently you will get a great deal more out of the remainder of this section.

17.2 Critical observation

Simple observation is probably our single most important method when assessing an experiment, at least at the qualitative level. We can elaborate using a story told by Richard Feynman, who won a Nobel prize for physics in 1965. As a young boy he had a toy cart inside which he placed a ball. As he pulled the cart forward the ball hit the cart's back wall. Fascinated, the young Feynman repeated the experiment several times, always obtaining the same result. He then asked his father, who was looking on from a distance, why the ball had moved backwards. His father pointed out that it was not the ball that had moved, but rather that the cart itself had moved forwards hitting the ball. In fact, until the impact, the ball remained in its original position, or even moved forward slightly.

The fact that the boy got the same result when the experiment was repeated under similar conditions suggested that what was being observed was not happening by chance but was being governed by physical laws. However, the boy's frame of reference, from near the cart, was localized: thus he concluded that the ball was moving backwards. From further away, however, the father had a different perspective (and, of course, more experience): he could see that the cart was moving under the ball. Developing the ability to observe a situation from one perspective and then considering how it would look from another perspective is very important when attempting to understand the physical world.

Returning to our ball experiment we can see parallels. The ball always falls to the ground; this aspect of the experiment is extremely repeatable, and indeed is what we predicted. The ball nearly always bounces; only the softest surface prevents this. Predicting the height of the bounce is difficult, and it varies greatly between surfaces. However, a number of attempts would show that bounce height on specified surfaces is quite repeatable. On the other hand, final resting place is generally unpredictable. With a small number of repetitions, one might get the illusion that a pattern is being followed, but with more repeats this pattern generally breaks down. There are exceptions, however. A sufficiently soft surface may 'catch' the ball in a very repeatable way, and a sufficiently biased surface (at a steep angle for example) may lead to very consistent patterns.

There are two key points we can learn from our ball experiment. First, things do not always do what we expect. Second, repeated observation is an extremely powerful investigative tool: it helps us to see both where patterns are real, and where they are not.

17.3 Seeing the big picture

This skill might as well be described as learning to see the wood for the trees in physical systems. For example, after repeating the ball experiment enough times, and *thinking* about the patterns, we reduced the problem to a few key issues (highly consistent fall, variable bounce depending on surface, unpredictable endpoint). These were not so obvious at the start of the experiment, and we could easily have focused on less significant factors.

This skill may also be described as the ability to define the problem you are facing. Solving problems in the natural environment is hard work because the real world is so much more complex than the laboratory. Learning to see the big picture is therefore very difficult. The pathway to it is through critical observation. Probably the most important tip is to learn not to rush into problem solving.

17.4 Curiosity

At first glance, curiosity seems an odd attribute to include in this list. Perhaps this is because we associate ideas such as enthusiasm and level of interest with curiosity. These last two are certainly helpful, although they are not easily taught. However, curiosity is something more. Its essence is expectation. We find curious those things that do not match our expectation. In this way, curiosity is very much a learnt skill, because it increases with our understanding of the world around us. Thus, the experienced scientist will be curious about aspects of the ball experiment that the inexperienced observer failed even to notice. It is here that repetition is so important. When you repeat an experiment you have an expectation against which to judge the repeat result; no experience is required. Thus, the question which might derive from curiosity is, 'why did it behave differently this time?'

17.5 Mathematics as a tool

With questions like, 'how long does it take for the ball to reach the ground?' and 'how high does the ball bounce?', we enter the realm of mathematics, albeit at a fairly basic level. Prediction, above all, is what science using mathematics allows us to do. While not underestimating the importance of qualitative observation, science is ultimately quantitative in character. We want to describe precisely the details of the changing velocities of the ball before and on striking the floor, and to do this we must be comfortable in applying the basic algebraic expressions of the equations of motion. At the other extreme, if we want to describe mathematically the distribution of possible resting places for the ball at the end of the experiment, then we engage with mathematical theory that could lie at the cutting edge of the discipline. This much said, it is important to stress that no amount of mathematical skill will make up for failure to understand the system at a qualitative level.

Perhaps the overriding concept in this section is curiosity. Always be curious: ask yourself why the ball behaves differently from expected when it hits the floor. Ask yourself why things work, and what stops them from working. Above all, be curious about your environment: you can learn much from your own critical observations.

18 An introduction to matter

18.1 Meaning

In a scientific sense, **matter** is that which occupies space. The purpose of this section is to introduce some of the terminology used in describing matter, because this is very useful in describing the world around us. The properties of matter, and the explanations for them, are dealt with in more detail elsewhere (see **29**). Very familiar is the concept of the **state** in which matter exists. We conventionally recognize three states of matter, namely **solid**, **liquid** and **gas**. Here we attempt to characterize each of these states in as simple a way as possible, although in other sections we show that this task is not quite as simple as it seems.

18.2 Solids

The usual elementary definition of a solid is that it has a fixed shape and fixed volume. A solid tends to retain its shape when external forces change. The atoms, ions or molecules of which a solid is composed may vibrate, but they do not shift the positions they occupy: they are said not to exhibit translatory motion. **Crystalline** solids (see **38**) have a well-defined structure in which there is a regular arrangement of molecules, atoms or ions. They also have a fixed melting point. (You may note here that many crystalline substances – most metals for example – do not occur as crystals.) Crystalline solids are sometimes called 'true solids'. Solids which do not have a regular internal arrangement of constituent molecules or ions are termed **amorphous**. For example, when glass is heated it simply becomes more pliable, or plastic. Its behaviour appears to be more like that of a liquid than a solid. In fact, glass has no fixed melting point.

18.3 Liquids

The usual working definition of a liquid is a substance with a definite volume but a shape which is determined by the vessel in which it is contained. This simple definition is not entirely accurate, however: for example, a drop of water on a leaf does not conform to the shape of any vessel. In general a certain amount of liquid occupies a fixed volume. The properties of liquids are explained by the degree of movement between the constituent molecules and the cohesive forces between them.

18.4 Gases

In a gas, the molecules are much further apart than they are in either a liquid or a solid: accordingly the density of gases is much lower than that of solids and liquids. The molecules move about more-or-less freely in a gas because cohesive forces between them are either absent or very weak. The only cohesive forces present in gases are known as **van der Waals' forces**. (Note that such forces are not confined to gases.) The more-or-less independent movement of molecules means that a gas occupies all the space available to it; accordingly a gas has no shape of its own. At the same time, the low density of a gas means that it is amenable to compression.

18.5 Other useful terms and concepts

Matter may also be described as a **pure substance** or a **mixture**. A pure substance, oxygen for example, has a fixed composition, whereas a mixture contains two or more substances in variable proportions. The various components of a mixture can be separated by physical means, for example by heat, but this is not the case for a pure substance.

Matter may also be categorized as either **homogeneous** or **heterogeneous**. Pure substances are homogeneous, but mixtures may be homogeneous or heterogeneous. A familiar example of a homogeneous mixture is that of sugar or salt in water. In a heterogeneous mixture, in contrast, the constituent components retain their own properties. A combination of 'dry' salt and sugar, for example, forms a heterogeneous mixture.

The term **solution** is used for homogeneous mixtures. Probably the most familiar solutions are those formed when solids (for example sugar or salt) are dissolved in liquids such as water. But solutions also form when gases dissolve in liquids and when liquids dissolve in other liquids. Other solutions arise from liquids or gases in gases, and even gases, liquids and solids within solids. The components of a solution are termed either the **solute** or the **solvent**. The former is the substance that is dissolved while the latter is the substance in which the solute is dissolved. Water is an excellent solvent, producing **aqueous solutions**, but many other substances, the alcohols for example, are also common liquid solvents.

When extremely tiny particles (arbitrary range 10^{-9} to 10^{-6} m, i.e. one-billionth to one-millionth of a metre) are dispersed through another medium (the continuous phase), the mixture is referred to as a **colloid**. There are many different types of colloid. Some colloidal solutions adopt jelly-like characteristics known as **gels**. Gels vary in the extent to which their shape can be modified (i.e. degree of plasticity). The behaviour of a gel approximates that of a solid, and in fact is sometimes described as a pseudo-solid. Other important types of colloid are: aerosols (liquid or solid dispersed in a gas), foams (gas dispersed in a liquid or a solid), emulsions (liquid dispersed in a liquid), and sols (solid dispersed in a liquid). A mixture in which small solid or liquid particles are held within a gas or a liquid is called a **suspension**.

19 Changes of state

19.1 Basic processes

The general features of solids, liquids and gases are introduced elsewhere (see **18**). Individual substances can exist in different states, however, depending on external conditions. Here we are concerned with changes in state. To describe these processes we use water and its response to temperature as an example. Not only is this a theme to which we can readily relate, but state changes of water are fundamental to understanding many key environmental processes.

We consider the effects of changing the temperature of a container of pure water over a range of 120 °C or so (Figure 19.1). First we will imagine our water in solid form, as ice, at a temperature somewhat less than 0 °C. Now we raise the temperature, i.e. increase the thermal energy in the system. Raising the temperature increases the kinetic energy (see **28.2.6**) of the constituent particles. The greater vibration eventually causes the particles to break apart. What we are now witnessing is **melting**. All pure substances have characteristic melting points: for water it is 0 °C. Observation suggests that melting does not occur instantaneously: its rate depends on the rate at which external energy is applied as heat. For a while, the added energy is used in melting the ice, not in increasing its kinetic energy. So for a period there is an ice–water mixture in our container.

At another critical temperature, pure liquids are freely converted to gases. As before, adding heat to a substance increases its kinetic energy. If you carry out such an experiment with water you will see bubbles appear in the liquid. These are actually dissolved gases (nitrogen and oxygen for example). Further heating will produce bubbles which contain water vapour. (The term **vapour** is in general use for the gaseous form of a substance which is normally present as a liquid.) These bubbles appear first near the bottom of the container, where the temperature is highest. If the rate of heat increase is moderate, the first bubbles do not escape from the liquid because they condense as they encounter the cooler water above. Eventually

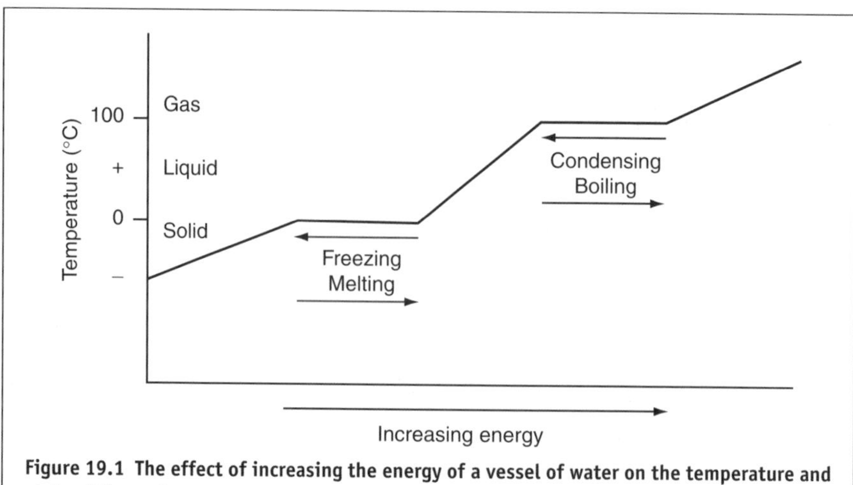

Figure 19.1 The effect of increasing the energy of a vessel of water on the temperature and state of the water.

though, water does reach the surface – we say the water is boiling. Formally defined, the **boiling point** of a liquid is the temperature at which the **equilibrium vapour pressure** of the liquid is equal to the atmospheric pressure. The boiling point of a liquid is thus dependent upon the atmospheric pressure. So when we say that the boiling point of water is 100 °C, a 'normal', or sea-level atmospheric pressure is assumed. This value is 760 mm of mercury (760 mmHg) or 101 325 pascals (Pa). If atmospheric pressure is increased, the boiling point is raised. Conversely, if atmospheric pressure is reduced the boiling point is lowered. The reason for this, simply put, is that with a lowering of pressure the water molecules break apart more easily.

You should note that as long as pressure is constant, the temperature of the water in your container will not rise above its boiling point, even though you may continue to apply heat. All the additional heat is used for weakening the structure of the substance so that molecules are converted to the gaseous (vapour) state.

Condensation refers to the transition from the gaseous to the liquid state; for pure water it is brought about by a reduction in temperature below 100 °C. The temperatures at which certain pure substances change state are shown in Table 19.1. It is possible for substances to undergo a change of state, for example from gas to solid or solid to gas, without passing through the liquid state. This process is called **sublimation**. One example from the natural world is the formation of ice from water vapour on the surface of frozen tree branches.

A **vapour** is formally defined as a gas with a temperature below its **critical temperature**, which is the temperature above which it is not possible to liquefy the gas. More informally, vapour is used for the gaseous form of a substance which is normally present in liquid form (as in water vapour) or as a solid, while vaporization is used to mean the release of molecules in gaseous form from a liquid.

Table 19.1 Melting point (Mp) and boiling point (Bp) of some common substances in degrees Celsius. Atmospheric pressure is assumed.

Substance	*Mp*	*Bp*
Water	0	100
Nitrogen	−210	−129
Ethene (ethylene)	−169	−105
Potassium	64	774
Sodium hydroxide	318	1390

19.2 Evaporation and vaporization

Although water, like all liquids, boils at a critical temperature, observation tells us that water can also be converted from the liquid to vapour phases (i.e. it evaporates) at much lower temperatures. In fact all liquids (and solids) vaporize. The rate at which they do so depends on the nature of the substance, its temperature, its surface area and the conditions in the surrounding atmosphere. Consider a closed container of liquid. The liquid molecules are in constant vibrational motion, i.e. they possess kinetic energy through this vibration. Energy is transferred from molecule to molecule when they collide. Molecules near to the surface may thus attain sufficient energy to raise them clear of the liquid's surface. In other words, such molecules have changed their state, from liquid to vapour. It follows that the greater the energy in the liquid, the more rapid will be the rate of change of molecules from liquid to vapour state: by applying more heat we can raise the rate of vaporization.

The movement of molecules from liquid to vapour state is not one-way: molecules will also be returned to the water surface. Evaporation means there is a *net* movement of molecules from the liquid to vapour state. In our experiment this change of state increases the **vapour pressure** of the atmosphere, which is most readily appreciated in terms of the pressure exerted on the container's surfaces and the liquid surface. In our enclosed container, evaporation will continue until the air can hold no more water. At this point the air is said to be saturated, and the pressure exerted by the (vaporized) water molecules is called the **saturation vapour pressure**. Although the reciprocal transfer of water molecules will continue at the saturation vapour pressure, there is no *net* transfer from liquid to the atmosphere: the system is therefore in a state of **dynamic equilibrium**.

Now if we raise the temperature of the container still more, molecular movement increases, and once again there is a net transfer of molecules from liquid to vapour state, until a new dynamic equilibrium is established. This confirms what we know intuitively: water tends to evaporate more quickly in hot weather than in cold weather.

Substances which are readily converted to the vapour phase, i.e. have a high vapour pressure, are described as **volatile**. The reason why DDT and other pesticides can be distributed so widely is their high vapour pressure, which ensures their long-distance dispersal in the atmosphere. The term 'volatilization' is sometimes used to refer to the process by which substances are converted to the vapour phase, as various chemical elements are when vegetation is burnt.

20 Heat and heat transfer

20.1 Introduction

Heat is very important in all environmental systems. Heat, which is the most common form of energy, is formally defined as the kinetic energy possessed by a substance as a consequence of the movement of its constituent atoms and molecules. Heat should not be confused with temperature. Temperature is a manifestation of the amount of heat contained in a system. Different objects of the same temperature will contain, mass for mass, different amounts of heat.

Heat as a form of energy is discussed more fully elsewhere (**28.2.6**), as are the units in which heat is measured (**2.4.5**), and its role in bringing about changes in the state of matter, for example from solid to liquid, or liquid to gas (see **19**). In this section we focus on one aspect of heat: its transfer. The reason for this is that many environmental processes can be better appreciated with a little understanding of the ways in which heat is transferred from place to place. The rules to remember are, first, that a flow of heat occurs wherever there is a difference in temperature, and secondly, that heat flows from hotter to colder regions. The difference in temperature per unit distance is called the **temperature gradient**. The steeper the gradient the more rapid is the rate of heat transfer. There are three ways by which heat is transferred, namely **conduction**, **convection** and **radiation**.

20.2 Processes of heat transfer

20.2.1 Conduction

Most people have felt the discomfort of holding the metal handle of a hot saucepan. The temperature rise of the handle is caused by the flow of heat by conduction, in this case from the heated container towards the end of the handle. Different substances conduct heat at different rates: in general, metals are better heat conductors than wood and plastic, while metals differ between themselves in this respect. Rock types differ in heat conductivity because of differences in their chemical and physical structures. Liquids and gases also conduct heat but at much lower rates than do solids. Materials with a high air content are therefore used for insulation, in clothing for example, while cavity walls are used to insulate buildings. The poor heat conductivity of air means that just a few centimetres of the air above the ground are heated by conduction, at least under calm conditions. The **thermal conductivity** of a substance is a measure of how efficiently the substance conducts heat.

The conduction process is based on an increase in kinetic energy of atoms and/or electrons. For all substances, an increase in their heat content raises the rate of vibration of atoms. In other words, their kinetic energy increases and this is passed on to neighbouring atoms. In metals, however, an additional process operates because of the large numbers of freely moving electrons they contain (see **8.5**). The rate of movement of these electrons is increased by heating, which results in a much faster rate of heat conduction than is the case with non-metals. It follows also that good 'solid' insulators are those substances with a relatively fixed and stable molecular structure.

20.2.2 Convection

The transfer of heat by convection is confined to liquids and gases, and it involves the movement of the fluid itself. (The term **fluid** applies to both liquids and gases here.) Convective heat transfer is associated with pressure differences in a fluid, brought about most commonly by differences in density. It is caused by the expansion of the hottest region of a fluid and its consequent movement as its density is lowered relative to other regions of the fluid. As 'parcels' of fluid are replaced by cooler gases or liquids, a **convection current** is created. Probably the most well-known example of convection occurs when water is heated in a pan. When heated from below, the temperature of the water near the bottom becomes greater than that of the overlying water. This results in a lowered density of water near the bottom. This relatively warm, low-density water rises to the top and is replaced by cooler, higher-density water from above. The continuous flow of water constitutes a **convection cell**. Convection phenomena are extremely important for understanding the behaviour of the atmosphere and bodies of water. As a parcel of air is heated, it expands, and its lowered density causes it to rise to a level at which its density is equal to that of the surrounding air. Conversely, a parcel of air which is cooler than the surrounding air sinks because of its relatively great density. Similar phenomena also occur in water bodies (**41.4.3**).

20.2.3 Radiation

Radiation phenomena appear in a number of places in this book. We introduce the uses of the term elsewhere (see **22**) and confine our remarks here to radiation as a process by which heat is transferred. Unlike conduction and convection, matter is not required for the transfer of heat by radiation: in other words, heat can be transferred in a vacuum. For example, our planet receives its heat (and light) from the Sun by radiation. The transfer of heat by radiation occurs by electromagnetic waves, which are discussed in more detail elsewhere (see **28.4**).

Although we can readily appreciate that hot objects release heat by radiation, the fact is that *all* bodies at a temperature above absolute zero emit heat by radiation. Now, some surfaces absorb radiation better than other surfaces: matt black surfaces are very good absorbers (but poor reflectors) of radiation, while white surfaces are poor absorbers (but good reflectors). Surfaces also behave differently with respect to the emission of heat by radiation. Again, matt black surfaces are efficient emitters while shiny surfaces are poor emitters.

20.3 Temperature and heat

We think of temperature as a very familiar concept. We regularly use temperature scales (for example Celsius) and we describe how hot or cold an object

In a gas at a given temperature the molecules have a characteristic range of velocities. It is convenient to estimate a 'typical' velocity as this is constant for any particular temperature. An arithmetic mean cannot be used because the distribution of velocities is highly skewed. An alternative way of expressing this typical speed is the root mean square velocity (v_{rms})

The equation is:

$$v_{rms} = \sqrt{\frac{3RT}{M}}$$

where R is the universal molar gas constant (8.31 J K^{-1} mol^{-1}), T is the temperature (kelvin), and M is the molecular mass of the gas (see **9.2**) expressed in kilograms (0.032 for dioxygen and 0.028 for dinitrogen).

What is the typical speed of oxygen molecules in air at a temperature of 293 K?

$$v_{rms} = \sqrt{((3 \times 8.31 \times 293)/32 \times 10^{-3})}$$

The brackets make it clear in what order the calculations should be done:

Step 1: complete the multiplication	$v_{rms} = \sqrt{((7304.49)/32 \times 10^{-3})}$
Step 2: complete the division	$v_{rms} = \sqrt{(228\,265.31)}$
Step 3: calculate the square root	$v_{rms} = 477.77$ m s^{-1}

Try the same calculation for dinitrogen molecules at the same temperature. You should find the answer is 510.76 m s^{-1}

Figure 20.1 How to calculate the speed of typical molecules within a gas.

feels to the touch. However, few of us have tried to work out what temperature represents in a physical sense (see **28.3**). Temperature is based on the movement of molecules: temperature should not be confused with the amount of heat that an object contains. For example, a litre of water at 100 °C contains far more heat than a litre of air at the same temperature.

At, say, 293 K (20 °C), i.e. 'room temperature', the average nitrogen or oxygen molecule in the air around you is moving in excess of $400\,m\,s^{-1}$. Given that the speed of sound in air is about $330\,m\,s^{-1}$, this is very fast indeed. Figure 20.1 shows how you can calculate the speed of typical molecules within a gas. In solids and liquids the individual atoms and molecules move at a similar speed, but with vibrational motion rather than in one direction.

20.4 Heat capacity

After discussing temperature we need to think about the **heat capacity** (or **thermal capacity**) of a substance. The substance can be a solid, a liquid or a gas. Heat capacity is the ratio of the heat supplied to a substance to the substance's rise in temperature. In practice it means the amount of heat (in joules) required to raise the temperature of the substance by one degree (K). It thus has the units of $J\,K^{-1}$.

It is more usual to see heat capacities of substances quoted as per unit quantity. This may be per unit of mass, in which case it is known as the **specific heat capacity**, i.e. the amount of heat needed to raise the temperature of one kilogram of an object by one degree. The units are thus $J\,K^{-1}\,kg^{-1}$. The specific heat capacity of a substance is different in its different states or phases: thus the specific heat capacity of ice is different from that of liquid water, which is different in turn from that of water vapour (see **41.2.2**). The unit of substance can also be expressed in moles, in which case it is known as the **specific molar capacity** and expressed as $J\,K^{-1}\,mol^{-1}$.

An object with a high heat capacity (if we assume no energy loss or transfer from the object) will warm more slowly for a given input of energy than an object with a lower specific heat capacity. The specific heat capacity of copper is $21\,J\,kg^{-1}\,K$, whereas the specific heat capacity for water is $4190\,J\,kg^{-1}\,K$. So it is easy to see why copper warms up so much more quickly than an equivalent mass of water for the same amount of energy input.

21 Mechanics

21.1 Introduction

A rudimentary understanding of mechanics is essential if we are to understand many of the processes operating within an environmental system. Its particular application is in the movement of fluids and solids. An understanding of mechanics makes you think about the processes and what is driving them within the system you are studying. Finally, and perhaps most importantly, it makes you approach what you observe in a methodical, or scientific, manner.

21.2 Force, work and power

If you move a body (i.e. any object) through a distance then you must subject it to a force. Movement of the body tells us that work has been done on the body. Work has the same units as energy, i.e. joules, which is force (newtons) multiplied by the distance it has moved, which is metres. So work in joules has the units of N m, or in dimensions $MLT^{-2}L$ or ML^2T^{-2}. (The dimensions of a physical quantity refer to its relations to mass (M), length (L) and time (T).) Power is the rate at which work is done and therefore the units are $J\,s^{-1}$, or watts.

21.3 Weight and mass

People frequently use weight and mass interchangeably, but this is not correct (see **2.4.4**). Weight is mass multiplied by the acceleration due to gravity, which on Earth is about $9.8\,m\,s^{-2}$. The unit of weight is the newton whereas the unit of mass is the kilogram. So a person who has a mass of 70 kg has a weight of 686 newtons. Now, weight is relative to the local acceleration due to gravity.

So on the moon, where the acceleration due to gravity is only one-sixth that on Earth, the same person would weigh only 114.3 newtons but would still have a mass of 70 kg. Even in weightless conditions you have mass. (It is good to see an everyday expression such as weightless is physically sound.) So what do you measure when you weigh yourselves on a bathroom scales? Well, strictly it is kilogram force rather than the kilograms shown on the scales.

21.4 Mechanics and motion

The laws that form the basis for the science of mechanics are contained within *Philosophiae Naturalis Principia Mathematica* (Mathematical Principles of Natural Philosophy) written by Sir Isaac Newton and published in 1687. The book is often referred to simply as the *Principia*. The ideas developed in this great work can explain almost all of the phenomena that we can observe today. The exceptions are phenomena which occur at very high speeds (towards the speed of light), for which we have to use Einstein's special theory of relativity, and for atomic scales, for which we have to use quantum mechanics. So, for physical interactions in our environment Newtonian mechanics are what we need to consider.

21.5 Newton's laws

Newton's first law states that every body continues in its state of rest or in uniform (i.e. unaccelerated) motion in a straight line unless acted upon by some other external force.

Newton's second law states that the rate of change of momentum of a body is directly proportional to the external force acting on the body and takes place in the direction of the force.

Newton's third law states that if a body exerts a force on another body, then that body exerts an equal and opposite force on the first body.

You will see that these laws are written in slightly different ways in different texts but in essence they convey the same principles. Newton's first law introduces the idea of **inertia**. Inertia is what you experience when you attempt to push a large, filled supermarket trolley: getting the trolley to move is hard work because first you must overcome the inertia. The larger the mass of a body then the larger is its inertia, and the greater the force required to make it move. Likewise, once your trolley is moving it is difficult to stop it, and it can be hard to make it change direction. These are examples of inertia and thus of Newton's first law.

The second law describes what happens as you exert some force on your stationary supermarket trolley: it accelerates. But the acceleration of the trolley will depend on its mass and how much force you exert. Remember that acceleration is the rate of change of velocity with time. When discussing force, textbooks normally examine the change of momentum with time, written as $\mathrm{d}(mv)/\mathrm{d}t$. Momentum is mass times velocity, which is mv.

Momentum therefore has the units $kg\,m\,s^{-1}$. When we look at the rate of change of momentum per unit time as discussed just above, we have the units of $kg\,m\,s^{-2}$. The units for acceleration are $m\,s^{-2}$, and force is mass times acceleration:

$$F = ma$$

Recall, the SI unit of force is the newton (N). The definition of a newton is the force required to produce an acceleration of one metre per second per second on a mass of one kilogram.

Back to our trolley again for the third law. When you apply a force (push) on the trolley an equal and opposite force pushes back on you. This force is applied even if the trolley does not move. If you have ever tried to push a large box across the floor and your feet have slid backwards, just consider the effect of pushing a trolley wearing roller blades. Often the third law is misinterpreted as the forces cancelling out. However, this does not happen as the two forces are acting on different bodies.

21.6 The equations of motion

The equations of motion show how time and movement are related. The equations below allow you to accurately predict the motion of an object.

1. The present velocity (v) of an object is equal to its initial velocity (v_0) plus its uniform acceleration (a) and the duration of the acceleration (t):

$$v = v_0 + at$$

2. The square of the current velocity is equal to the square of the initial velocity plus twice the acceleration times the displacement (its current position (x) minus its initial position (x_0)), thus:

$$v^2 = v_0^2 + 2a(x - x_0)$$

3. The displacement of an object is equal to the product of its initial velocity and the time it has been moving, plus half of the acceleration times the square of the time:

$$x - x_0 = v_0t + \tfrac{1}{2}at^2$$

4. The displacement of an object is equal to half of the time it has been moving times its initial velocity plus its final velocity:

$$x - x_0 = \tfrac{1}{2}(v_0 + v)t$$

Let us try out some of the equations. Imagine you have dropped a 1 kg bag of sugar from a tall building. How fast would this bag be travelling after 5 seconds? In all these examples the initial point (x_0) is zero and the initial velocity (v_0) is zero. These equations do not take into account the effects of air friction and they assume constant acceleration. Use the first equation

($v = v_0 + at$). As the acceleration due to gravity is $9.8\,\mathrm{m\,s^{-2}}$:

$$v = 0 + 9.8 \times 5$$
$$= 49\,\mathrm{m\,s^{-1}}$$

In other words, after 5 seconds, in the absence of air friction, the bag of sugar would have a velocity of $49\,\mathrm{m\,s^{-1}}$.

How long would it take for an object to fall to the ground from a 10 m wall? For this problem we use the third equation ($x - x_0 = v_0t + \frac{1}{2}at^2$). We know that the distance the object can fall through is $x - x_0$ which is 10 m. As the object is initially at rest then v_0 is also 0, so we can write out the equation as:

$$10 = 0 + \tfrac{1}{2} \times 9.8t^2$$
$$= \tfrac{1}{2} \times 9.8t^2$$

To get rid of the $\frac{1}{2}$ we can multiply both sides by 2. Thus:

$$20 = 9.8t^2$$

We can eliminate the 9.8 if we divide both sides by 9.8. Thus:

$$2.04 = t^2$$

To get from t^2 to t we need to take the square root of both sides. Thus:

$$t = \sqrt{2.04}$$
$$= 1.43 \text{ seconds}$$

So it takes less than 1.5 seconds for a dropped object to fall 10 m.

How fast will the object be moving when it hits the ground? Using the first equation we can find that it would be travelling at $14\,\mathrm{m\,s^{-1}}$.

21.7 Potential and kinetic energy

Two other concepts in terms of movement and energy that you will come across are **kinetic** and **potential energy**. (In fact we deal with these topics in other places, see **28.2.4** and **28.2.1**.) Kinetic energy is the energy of moving bodies, while potential energy is the energy of a body while it is at rest but with the potential to move downwards. A good example of the conversion between potential and kinetic energy is provided by a falling raindrop. A raindrop held (by upward movement of the air) within a cloud has potential energy because of its position above a surface, in this case the ground. While it is not falling its potential energy is at a maximum and it has no kinetic energy. As it begins to fall, its velocity increases and it therefore gains kinetic energy, but it loses potential energy as the height above the surface lessens. When it reaches the ground we can assume that it will be travelling at its greatest velocity, and therefore has its greatest kinetic energy. When the raindrop has reached the ground it has no potential energy (or no potential to fall). Other examples are water falling over a cliff or a boulder that may start to move on a slope. In fact an avalanche is also a good example of the conversion of potential to kinetic energy.

The equations for kinetic and potential energy are:

$$\text{Kinetic energy (KE)} = \tfrac{1}{2}mv^2 \text{ (joules)}$$

$$\text{Potential energy (PE)} = mgh \text{ (joules)}$$

where m is the mass, v velocity, g acceleration due to gravity and h the height above the surface or, in the case of a slope, the potential surface.

Returning to the examples above, we can work out what the potential energy of our object was before it started to fall, and what it was at the end of its fall. Our 1 kg bag of sugar at 10 m will have a potential energy of:

$$\begin{aligned} \text{PE} &= 1 \times 9.8 \times 10 \\ &= 98\,\text{J} \end{aligned}$$

When the bag reaches the ground (from above, the velocity = $14\,\text{m s}^{-1}$) it will have a kinetic energy of:

$$\begin{aligned} \text{KE} &= \tfrac{1}{2} \times 1 \times 14^2 \\ &= \tfrac{1}{2} \times 196 \\ &= 98\,\text{J} \end{aligned}$$

You can see that the kinetic energy gained at the end of the fall equals the potential energy held by the bag before its fall.

21.8 Momentum

The final concept in this section is momentum (or more strictly linear momentum). Momentum is found by multiplying together the mass and velocity of a moving object (mv). In the absence of external forces, linear momentum is conserved. Consider the example of a 5 kg object moving at $2\,\text{m s}^{-1}$ on a frictionless surface. It collides with a stationary object of 10 kg, and after the collision both the objects join together. We can calculate the velocity at which they will move once they have joined as the momentum has to be conserved. The momentum before and after the collision is:

$$\begin{aligned} \text{Momentum} &= mv \\ &= 5 \times 2 \\ &= 10\,\text{kg m s}^{-1} \end{aligned}$$

To calculate the final velocity once the objects have joined we need to find the final mass by adding together the masses of the two objects. This value is 15 kg. So:

$$\begin{aligned} 10 &= 15v \\ \therefore \quad v &= \frac{10}{15} \\ &= 0.67\,\text{m s}^{-1} \end{aligned}$$

If the initially moving object was halted by the collision and the second object began to move this object would have a velocity of:

$$10 = 10v$$

$$\therefore \quad v = 1\,\mathrm{m\,s^{-1}}$$

Actually, some energy is lost at collision through sound, heat or deformation of the objects, and in practice surfaces are not friction-free. However, the momentum is conserved even under realistic conditions.

22 Electromagnetic radiation

22.1 Some basics

The term **radiation** covers a variety of very important phenomena in the environment. An important point is that all radiation involves the exchange of energy. **Nuclear radiation** involves the transmission of particles from a radioactive source. This type of radiation is familiar because of its associations with atomic weapons, nuclear power, and the harmful effects of exposure to radioactivity. The expression **electromagnetic radiation** may be less familiar but the names of the phenomena to which it applies, for example X-rays, light, infra-red (IR) and ultraviolet (UV) radiation, and radio waves, are in common use. The energy associated with electromagnetic radiation is termed **radiant energy**. In this form energy can exist in the absence of matter. Above, we might have implied that there is a clear distinction between nuclear and electromagnetic radiation but this is not the case. For example, gamma rays are a type of electromagnetic radiation, but they are also associated with radioactivity (see **24**), being emitted by the nuclei of radioactive atoms under certain conditions; they are highly energetic and have considerable penetrating power.

All types of electromagnetic radiation, whatever their source, are transmitted, or propagated, as waves (see **30** for a general introduction to wave phenomena). The two fundamental features of electromagnetic waves (as with all waves) are length and frequency. The longer the wavelength, the lower the frequency, and vice versa.

Wavelengths are expressed in various units, but particularly metres (the SI unit of length), micrometres ($1\,\mu m = 10^{-6}\,m$) and nanometres ($1\,nm = 10^{-9}\,m$). Sometimes you may see the former metric unit, the angstrom (Å) used: $1\,Å = 10^{-10}\,m$. Choice depends to a large extent on which part of the spectrum is being discussed. The derived SI unit of frequency, meaning cycles per second, is the hertz (Hz), which is a familiar one because of its use on radios. So, 1 kilohertz ($1\,kHz$) $= 10^3$ cycles per second.

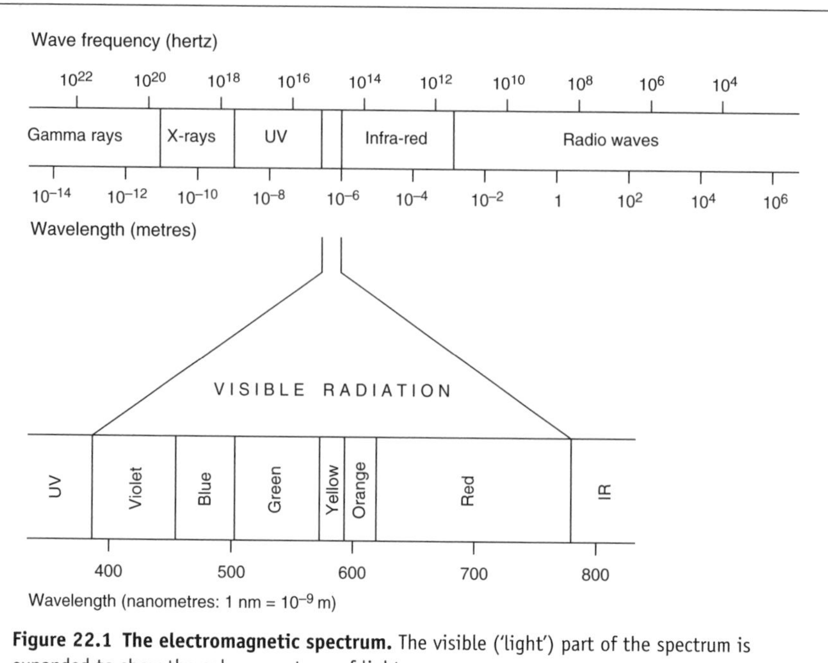

Figure 22.1 The electromagnetic spectrum. The visible ('light') part of the spectrum is expanded to show the colour spectrum of light.

The entire range of radiation wavelengths is referred to as the **electromagnetic spectrum** (Figure 22.1). Notice the enormous range of wavelengths encompassed by the electromagnetic spectrum: it is over twenty orders of magnitude. The electromagnetic spectrum extends from very short wave (high frequency) gamma rays to very long wave (low frequency) radio waves. The terms 'shortwave' radiation and 'longwave' radiation are in common use to refer to different parts of the spectrum, although we need to be aware that these expressions are sometimes used in a relative, rather than an absolute, sense. The Earth receives a very wide range of electromagnetic wavelengths from various sources in space, but our principal interest lies in the range of wavelengths consistently and continuously received from the Sun, the so-called **solar spectrum**.

As noted earlier, electromagnetic waves carry energy. A key point is that wavelength is related to energy: the shorter the wavelength, the more highly energetic the radiation. So gamma rays are enormously more energetic (and hence more biologically damaging) than radio waves. And over a smaller part of the electromagnetic range, ultraviolet radiation is more highly energetic than infra-red radiation. Another fundamental point is that all electromagnetic radiation, of whatever wavelength (or frequency), travels through space at the speed of light, which is $3 \times 10^8\,\mathrm{m\,s^{-1}}$.

Radiation in general is often described as either **ionizing** or **non-ionizing**, although in fact the distinction is not absolute. Ionizing radiation includes radioactive alpha and beta particles (neither are part of the electromagnetic spectrum), gamma and X-rays (both of which are transmitted as electromagnetic waves), and also some wavelengths of ultraviolet radiation. Ionizing radiation is potentially very dangerous to life. It may induce cancers, and in cases where an organism's genetic material is altered, the changes may be inherited. Fortunately, most of the energy of the ionizing radiation that reaches the Earth's atmosphere is 'consumed' in the ionization of atoms and molecules in the ionosphere which extends upwards from about 70 km into the lower atmosphere (Figure 43.3). These reactions dissipate the ionization potential of this radiation and protect life on Earth from its potentially harmful effects.

22.2 Solar radiation

22.2.1 Introduction

Some knowledge of solar radiation is essential for anyone interested in the physical environment. After all, it is the Sun's energy that is responsible for powering the major circulation systems of the Earth's atmosphere, oceans and biosphere. The amount of energy received by a surface perpendicular to the Sun's rays at the Earth's outer atmosphere is called the **solar constant**. Its average value is about 1370 $J\,m^{-2}\,s^{-1}$.

About 50 per cent of incoming radiation is lost within the atmosphere due to two processes: scattering (30 per cent) and absorption (20 per cent). Scattering involves the absorption of radiation by particles in the atmosphere and their immediate re-emission, often in a different direction. Scattering does not involve an exchange of energy. Much of the scattered component is accounted for by the reflection, principally from clouds, of radiation back into space. The remainder is scattered in other directions, and in consequence arrives at the Earth's surface from all directions. Another term for scattered radiation reaching the Earth's surface is **diffuse radiation**. The other key process is absorption. Unlike scattering, absorption involves energy exchange; it brings about an increase in heat within the absorbing medium. Absorption can result in the loss of an electron from an atom, ion or molecule, in which case it is known as **photoionization**.

Figure 22.2 shows the solar spectrum at the outer atmosphere and at the Earth's surface. The two curves show the relationship between the amount of energy received from the Sun and the wavelength. First, notice the range of wavelengths (horizontal axis), which is approximately 0.2–3 μm (200–3000 nm) at the outer atmosphere and slightly less at the surface. Wavelengths less than, and greater than, 0.8 μm (800 nm) are often referred to as shortwave and longwave radiation respectively. The shortwave component of the solar spectrum consists of two regions – the **ultraviolet** and the **visible** – while the longwave component is referred to as the **infra-red**. Second, note the general form of the curves. It is clear that the energy received is concentrated in the shorter wavelengths, whereas longer wavelengths possess comparatively little energy. Third, notice that the 'loss' of radiation in the atmosphere is not distributed evenly across the various wavelengths: the dips in the inner

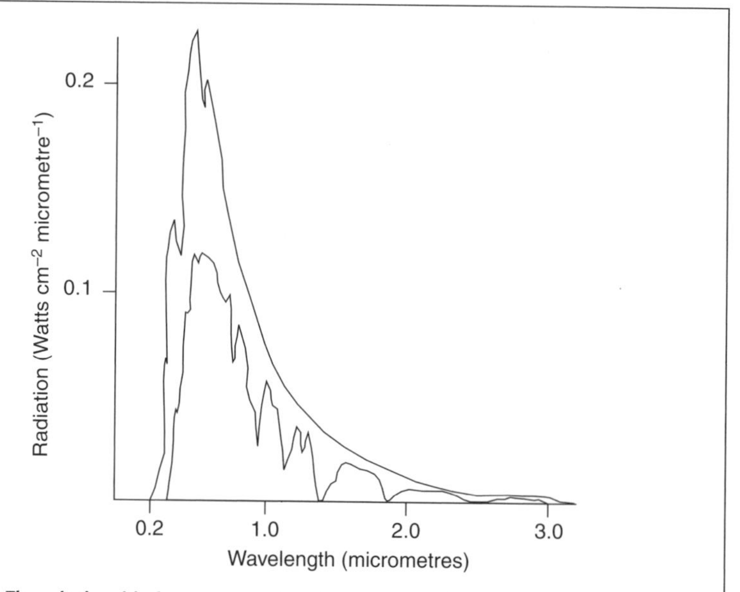

Figure 22.2 The relationship between wavelength and energy in the solar spectrum. The outer curve applies to the outer atmosphere, the inner curve to the Earth's surface. Notice the differential absorption of certain wavelengths.

curve show that some wavelengths are absorbed more effectively than others. This is because the various atmospheric constituents (for example water vapour, carbon dioxide, ozone) have their own absorption spectrum for electromagnetic wavelengths. Another relevant concept is that of the **action spectrum**, which relates the level of *activity* of a process to the wavelength of radiation. This is a particularly useful concept in the case of biological responses to electromagnetic radiation.

Now we turn to the principal named regions of the solar spectrum: the ultraviolet, visible and infra-red regions.

22.2.2 Ultraviolet radiation

About 7 per cent of the Sun's energy received by the Earth is in the UV wavelengths. Three components of UV radiation are recognized. Their designations, and wavelength ranges are, respectively, UV-C (0.2–0.28 μm), UV-B (0.28–0.32 μm) and UV-A (0.32–0.4 μm). Although the actual boundaries are arbitrary (and you will find they vary slightly in the literature), certain chemical and biological responses are associated with each UV band. UV-A, which lies just below the visible spectrum, is little absorbed by gases in the atmosphere, and is relatively inert biologically. Much of the solar UV-B radiation and practically all UV-C radiation are absorbed by the ozone present in the stratosphere (Figure 44.3). It is this ozone which makes life on land possible, because wavelengths of less than 0.3 μm are very active biologically. Nucleic acids, key molecules in development and reproduction (see **32.3**), absorb most strongly in the region 0.26–0.27 µm. Recent evidence for the depletion of stratospheric ozone therefore has considerable public health implications. Only tiny amounts of ozone, together with the oxygen in the atmosphere, are necessary to absorb nearly all UV-C wavelengths, but the UV-B range is seriously affected by ozone reductions.

Ozone is generated in the atmospheric by the action of UV radiation on oxygen. UV-C can split (or dissociate) molecular oxygen (O_2) into its two constituent atoms, and these are then free to combine with another oxygen molecule to form an ozone molecule (O_3). The UV-B radiation absorbed by the ozone molecules converts them to oxygen molecules and atoms. However, ozone is also fragmented by reactions with other chemical species including chlorine radicals (Cl^-). The principal source of these chlorine radicals is a group of chemicals called chlorofluorocarbons, which have been used extensively in refrigerators and as aerosol propellants. These chemicals are relatively stable as they are conveyed to the stratosphere, but once there they fragment under the influence of UV radiation, with subsequent release of their ozone-damaging components.

22.2.3 Visible radiation

The term 'visible' is used for radiation within the wavelength range 0.4–0.76 µm (400–760 nm) because these wavelengths are detected by the human eye. The other, more informal term is 'light'. Just under 50 per cent of the radiation reaching the Earth's surface is in the visible range. Wavelengths outside this range, including UV radiation, should not be referred to as light. Other animals see differently from humans though: certain insects see the longer UV wavelengths while many other types of animal see in the shorter infra-red wavelengths.

The different parts of the visible range of electromagnetic radiation correspond to colours (Figure 22.1). Each colour behaves in a unique way when a beam of white light strikes a glass prism. What happens is that wavelengths are 'bent' differentially causing the colours to separate out, thus revealing the colour composition of white light. (The 'bending' process is an example of **refraction**, which occurs when light passes from one medium to another.) Wavebands corresponding to violet and blue, i.e. the shorter wavelengths, are bent more than wavebands corresponding to the red part of the spectrum.

Other light sources differ from the Sun in the range of wavelengths they radiate. For example, mercury-arc lamps are characterized by a bluish light (plus invisible UV radiation). Fluorescent lamps consist of a glass tube containing mercury vapour. When an electric current is passed through this gas it emits UV radiation which is absorbed by substances layered on the inside of the tube. Different substances are used to emit radiation in different wavelengths, i.e. to generate light of particular colours. The fact that substances emit characteristic light spectra means that the presence and amounts of these substances can be detected. This is the basis of some invaluable scientific instruments, such as the flame photometer.

22.2.4 Infra-red radiation

The infra-red region of the solar spectrum occupies the wavelength range 0.76–3 μm (760–3000 nm), although the boundaries are somewhat arbitrary and marked variously. The infra-red region is also referred to as **thermal** or **heat radiation**, or **radiant heat**. The human eye is unable to discern infra-red radiation, but its warming effects are very apparent. Such incoming radiation is absorbed effectively, particularly by water vapour and carbon dioxide in the troposphere (Figure 22.2). This selective absorption by atmospheric gases applies also to radiation which is re-emitted by the Earth. It is the 'blanket' effect of these gases which makes our planet habitable. The term 'greenhouse effect' is used because the atmosphere is largely transparent, like glass, to incoming shortwave radiation, but is much less transparent to wavebands of the infra-red region and beyond, and it is this waveband range in which radiation is principally re-emitted.

Although the human eye cannot see infra-red radiation, instruments and certain photographic films are sensitive to such wavelengths. Films that are sensitive to infra-red radiation record temperature differences which are translated into photographic images. These have been used to observe nocturnal behaviour in animals and to distinguish buildings which emit heat from those that do not. Infra-red-sensitive instrumentation has also, unsurprisingly, found all sorts of military applications.

22.2.5 Radiation loss in water

Over 65 per cent of the Earth's surface is covered in water, which is a good absorber of incident radiation (low albedo – see **22.2.7**). As radiant energy passes through the water column, the energy declines and there is a shift in its spectral distribution. As with the atmosphere, it is scattering and absorption which are responsible. Red and infra-red wavelengths are absorbed by the water itself within the first half metre or so, reducing to about one-half the radiation received at the surface. In the very clearest waters it is the green and blue wavebands that penetrate furthest, sometimes to a depth of 400 metres or more. However, natural waters rarely permit the penetration of light to such depths; more typically, suspended material and organisms absorb much of the radiation, including the blue and green wavebands. The shift in spectral distribution with increased depth varies considerably between areas depending on the nature of suspended and dissolved substances and organisms in the water.

22.2.6 Terminology and units

The term used for the measurement of radiant energy (quality and quantity) is **radiometry**. The rate of flow of radiant energy, as electromagnetic radiation, is the **radiant flux**. It is usually expressed as joules per second ($J s^{-1}$). The **radiant flux density**, or **irradiance**, is the quantity of radiation received by a surface, and must include a measure of area as well as time. It is usually expressed as $J\,m^{-2}\,s^{-1}$. But 1 watt $= 1\,J\,s^{-1}$, so $J\,m^{-2}\,s^{-1} = W\,m^{-2}$. Formerly, energy values were often presented in calories, but the calorie is not an SI unit. Irradiance can also be expressed in terms of photons, as in $mol\,m^{-2}\,s^{-1}$.

The measurement of light (**photometry**) is not the same as the measurement of radiant energy. The quantity of light passing through a given area in one second is called the **luminous flux**. The derived SI unit of luminous flux is the **lumen** (lm). The amount of light per second which falls on a unit area of surface is the **illuminance**, and its derived SI unit is the **lux**, or lumen per square metre. The amount of light emitted per second from a point source, and in a given direction, is the **luminous intensity**. The appropriate SI unit is the **candela**. Previously, light levels were often expressed in foot candles, one foot candle being equal to one lumen per square foot.

Light measurements are important for many practical reasons because different light levels are appropriate for particular situations. However, measures of light should not be confused with measures of radiant energy. Under natural conditions, however, the solar spectrum remains pretty much the same during the daylight period, so there is a fairly good correlation between energy input and light level.

22.2.7 Reflected and absorbed radiation

Part of the incoming shortwave radiation that strikes the Earth's surface is reflected as shortwave radiation. The ratio between the radiation reflected and that received by a surface is called the **albedo**, and it is conventionally expressed as a percentage. Albedo values vary considerably between surface types, primarily because of their differences in colour. We know from experience that snow, ice and sand emit glare, while vegetation surfaces are dull in comparison. In fact the albedo of fresh snow can be 95 per cent while the albedo of sand dunes may be 65 per cent. Clouds, too, have high albedos, typically 60–90 per cent. In contrast, dense forests have values typically between 10 and 20 per cent; whole values for the ocean are usually in the 3–12 per cent range. Radiation that is not reflected by a surface is absorbed. So whereas snow might absorb a small proportion of incident radiation, most is absorbed by water and by dense vegetation.

22.3 X-rays

X-rays occupy the wavelength range (10^{-5}–10^{-3} μm) of the electromagnetic spectrum. They are emitted by atoms of all the elements when they are bombarded with electrons, and each element emits X-rays with a characteristic frequency spectrum. Energy is lost from the atom, as X-rays, because an inner electron is knocked from the atom's orbital by the incoming electron and is replaced by another, outer electron. The penetrating power of X-rays is used in a variety of medicinal and research situations.

22.4 Radio waves

Radio waves are electromagnetic waves in excess of 10^{-3} metres. They include wavelengths used in microwave ovens, and for television and radio transmission. The wavelengths of the longest radio waves are well in excess of a kilometre. The shortest radio waves, between approximately 10^{-3} and 0.03 m, are called microwaves. Heating of food in a microwave oven is brought about by the selective absorption by water molecules of radiation. The absorbed radiation causes excitation of electrons and the consequent heating of the water.

22.5 Heat and light

Some general aspects of radiant heat, introduced in **20.2.3**, have been implied in this section but now they can be put into context. A fundamental point is that all bodies (i.e. all objects) at a temperature above absolute zero emit radiant energy. (The reason the temperature of an object does not continue to decline is that objects are continuously receiving radiation from other objects.) Now, the temperature of the body influences both the rate at which radiation is emitted and the wavelength range of this energy. The rate of emission is also influenced by its colour: in general, matt black objects absorb and emit more efficiently than those of other colours. The rate of energy emission increases markedly with an increase in temperature. As temperature increases, the frequency distribution of wavelengths also changes: the mode shifts towards the shorter wavelengths. A good demonstration of this phenomenon occurs in a toaster or electric fire: when first switched on we sense a warming effect, but as the element reaches about 500 °C it begins to glow. This is because some of the radiant energy is now within the red wavebands, which are visible. At around double this temperature objects glow white; in other words they are emitting white light, which possesses all the wavebands (or colours) of the visible spectrum. (The expressions 'red hot' and 'white hot' therefore have a sound scientific basis.) The principal source of light on Earth is the Sun. What we see is the layer of ionized gases at the Sun's surface, known as the **photosphere**, which has a temperature of approximately 6000 °C (about 6000 K). The temperature has cooled markedly from that at the core, where it is estimated to be 15 million K. The temperatures of other stars can also be determined by their colour spectra. Heat and light are different manifestations of the same phenomenon: we distinguish the 'light' wavelengths simply because we can see them.

22.6 The particulate nature of radiant energy

So far we have been discussing electromagnetism in terms of waves. This was to facilitate understanding of some key environmental phenomena. The early development of electromagnetic theory during the eighteenth century was based primarily on observations of the behaviour of light, the one form of electromagnetic energy that is visible. And indeed it is observations of the way that light behaves that still provide the basis for the wave theory of electromagnetism. In the late seventeenth century, however, Sir Isaac Newton had postulated that light could be likened to a stream of particles, which moved in a straight line. This theory was resurrected in the early twentieth century and gained increasing support with the development of the unifying field of quantum mechanics. It is now widely accepted that the behaviour of electromagnetic radiation is consistent with both wave and particle theories. This topic is explored further in **29.2**.

The particle theory envisages the propagation of energy in discrete 'packets', called **quanta** (singular, quantum). A quantum of electromagnetic radiation is termed a **photon**. In practice, the two terms 'quanta' and 'photon' are used pretty much interchangeably. According to this model we can conceive of light, and other types of electromagnetic radiation, as a beam of photons.

23 Electricity and magnetism

23.1 Introduction

An introductory book on science for physical geography and environmental study would be incomplete without at least a cursory look at electricity and magnetism. In your studies of the environment you will not need to have a detailed understanding of any electromagnetic theory but it is worth noting that electrical terms are sometimes used in an analogous way in environmental systems. Also, you may well have to use battery or mains-powered instruments, or may use the magnetic characteristics of rock and sediments as an investigative tool. Therefore, it is important that you gain a reasonable knowledge of some key terminology and concepts. In addition, it is worth starting with the fundamental concepts, and then looking at the larger topic of electromagnetism.

23.2 The nature of electromagnetism

It was the ancient Greeks who found out that if you rubbed a piece of amber it could pick up small objects. This is because by rubbing the amber a charge is induced on its surface. Normally, objects are electrically neutral, which means they have equal amounts of negative and positive charge. By rubbing an object such as a piece of amber or a glass rod you upset this balance and the object is said to be charged. Objects with like charges oppose each other, and objects with unlike charges attract each other. The Greek word for amber, *elektron*, has given us the term 'electricity' in relation to the forces described above.

Since the end of the nineteenth century it has been known that electrical effects derive from an imbalance in the distribution of electrons. Thus, positively charged objects have a deficit of electrons, while negatively charged objects have a surplus. Before this time, however, a number of studies had shown that electricity and magnetism were linked. For example, if an electrical current is passed along a wire, a magnetic field is generated around that wire. Conversely, if a wire is passed through a magnetic field, then an electrical current is generated within the wire. A landmark advance in electrical theory was made in the late nineteenth century when it was shown that light, and other related types of radiation, were electromagnetic in character (see **22.1**). Electromagnetic radiation drives itself along by mutually supporting electric and magnetic fields. Fluctuating electrical fields generate fluctuating magnetic fields, a pattern which repeats itself indefinitely, carrying the wave along.

23.3 Insulators and conductors

Insulators are objects that can maintain charge differences because they do not conduct the charge away from the surface of the object. A glass rod, for example, is a good insulator. Conductors, in contrast, are objects that do not hold charge because they conduct the charge away from their surface and redistribute it. A metal rod makes a good conductor.

What makes a conductor conduct? Metals such as copper are good conductors. This is because some of the electrons are only weakly bound within the metallic structure, and are free to move around within the solid metal (see **37.3**). It is the movement of these free electrons that allows a metal to conduct electricity. Conversely, the metal cannot easily hold a charge, as this can readily be conducted away. The opposite is true for an insulator because it has very few free electrons: charge cannot be conducted away and will readily be retained.

23.4 Volts and amps

Volts and **amps** are the most commonly used electrical units. They are named after famous scientists who played a part in understanding electricity. You have probably heard of voltage, such as the amount of mains voltage or the voltage of a battery used in a car, torch or radio. But what does voltage mean? Perhaps the fact that voltage is also called a potential difference provides a clue. The state of charge can be viewed as a potential to give up charge. If two objects have the same potential there is no potential difference between them, and no net charge will flow. But if you put a conductor (a piece of wire) between the charged object and the ground (earth) there would be a potential difference between the charged object (with an arbitrary value of potential) and the ground, which we assume is earth or zero. Once the wire touches the charged object the charge held by the object is carried along the conducting wire to the ground. This is a potential difference and hence the flow of electricity will take place until the charge is exhausted.

Another way of thinking of voltage is that it is analogous to a hydrostatic head maintained by a reservoir dam. That is, you have a large gravitational potential difference between the top of the dam and the bottom of the dam. The potential difference, or voltage, can also be viewed as the 'pressure' of the electricity in the wire. The higher the voltage, the further the electricity can travel along transmission lines: hence the very high voltages – up to many hundreds of thousands of volts – used in long-distance power transmission. The mains electricity in your house is about 230 volts if in Europe, or about 115 volts if in the USA and Canada. Thus potential difference is measured in volts and it does have a strict definition, although this is not very useful in the present context.

The amp, or ampere to give its full name, is a measure of the current in an electrical circuit. The current is the flow of the charge through the conductor, and given our metal wire in the example above, this is the flow of electrons. Current may be thought of as the amount of charge moving along a conductor, and the voltage as the force driving the movement of the charge. Using the hydraulic analogy, the current would be the flow rate of water that was falling through our potential difference.

You may well have come across amp ratings on fuses used in a motor car or in electrical appliances or circuits in the home. The higher the energy usage of the circuit, the higher the amperage of the fuse. Circuits for electric lights generally have much lower amperage fuses than those for, say, an electrical stove or heater. The bigger the amperage of the fuse then generally the thicker the wire in the fuse. Old-fashioned household fuse boxes allowed you to rewire your own fuses: if they 'blew' you replaced them with a fuse wire of appropriate width (or amperage). A blown fuse is one that has melted under increased electrical loadings. The fuse is the weakest part of the circuit and its destruction preserves the rest of the circuit and other appliances.

23.5 AC and DC

AC stands for **alternating current** and DC for **direct current**. Mains electricity has an alternating current whereas power from a battery is direct current. An AC supply has a varying voltage and current that when traced out gives a sinusoidal signal (sinusoidal denotes a wave-like form – see **30.3** for a discussion of waves and sine functions). The frequency of this fluctuation in the mains supply depends on the country you live in, but 50 and 60 Hz are typical frequencies for mains supplies. The voltage varies about an average value, and it is this average value that is given. The value is normally given as a root mean square, or rms, value (see **26.9**). A DC circuit has a steady current and voltage passing through the load (for example a motor). AC is better for long-distance power-line transmission and it is used in the home. Initially, DC electricity supplies were used in the home, but for most places that was phased out a long time ago.

23.6 Power

We have seen in **23.4** that bigger electrical appliances need larger currents. This is simply because they require more power. In a DC circuit, power is calculated as the voltage multiplied by the current. A similar equation can be used in an AC circuit by using the rms values of voltage and current. However, the current and voltage are not normally in phase in AC (the voltage peaks before the current) and this phase difference needs to be accounted for in any calculation. Finally, a reminder that the unit of power is the watt (W), which is $J\,s^{-1}$.

23.7 Resistance

While conductors, by definition, conduct electricity, they also have a degree of **resistance** to the electrical flow. The resistance relates how much current will flow through a conductor (or electrical circuit) for a given potential difference or voltage that is applied. The unit of resistance is the ohm (Ω). The relationship between resistance voltage and current is called Ohm's law, and the resistance in a conductor can be defined as:

$$R = \frac{V}{i}$$

Here R is the resistance, V the voltage and i the current. Hence 1 ohm = 1 volt per ampere. The above equation can be easily rearranged (by multiplying both sides of the equation by i and dividing both sides by R) to show that:

$$i = \frac{V}{R}$$

You should now be able to see that as R (the resistance) increases for a given voltage the current carried in the conductor or circuit will decrease. Resistance can also be viewed in terms of the hydrological analogy. A narrowing of the channel leads to a local increase in the potential difference across the section.

The name **resistor** is given to the component with a given resistance in an electrical circuit. Resistors are used in electrical circuits to set voltages and currents. **Resistivity** is the resistance across a unit length of a given conductor. For copper it is 1.69×10^{-8} Ω.m (ohm metres).

23.8 Capacitance

A **capacitor** is a device that stores electrical charge. Capacitance is the amount of charge for a given voltage that a capacitor can store and it is measured in farads (F). A capacitor is in essence two facing plates set a small distance apart. The plates hold opposing charges and the space between is filled by a material to enhance the storage of charge. Such material is called the **dielectric**. Examples of dielectrics are air, glass and oiled paper. Water has good dielectric properties (see **41.2.4**), though it is not used in electrical devices. Capacitors have a large number of uses in electrical circuits.

23.9 Breakdown potential

Air is a poor conductor. Two objects of opposite charge separated by air can hold their charge. However, if the electrical field between the objects becomes strong enough (about $3 \times 10^6\,\mathrm{V\,m^{-1}}$) then an electrical discharge will occur (as in the case of lightning). This voltage, or difference in electrical potential, is termed the **breakdown potential**.

23.10 Conductivity

One application you may make of electricity is to measure the conductivity of some river water. The conductivity of river water depends on the concentration of ions and ionized solutes in the water. Conductivity shows how readily electricity passes through the water. Conductivity is defined as the reciprocal of resistivity, and has the units of $(\Omega.\mathrm{m})^{-1}$ or $\Omega^{-1}\,\mathrm{m}^{-1}$. This unit is called Siemens per metre ($\mathrm{S\,m^{-1}}$) or mhos per metre.

23.11 Magnetic fields

The space around a magnet where magnetic forces can be experienced is termed a **magnetic field**. The magnetic field you are probably most familiar with is that of the Earth with its north and south poles. A magnetic compass points towards the north magnetic pole. It is worth remembering that the magnetic north and south poles are not found in the same place as the geographical north and south poles and that the magnetic north is not the same as either grid north on a map or true north.

Magnetism and electricity are closely linked. Passing electricity through a wire generates a magnetic field around the wire. This also gives rise to electromagnetic radiation (see **28.4**). Electricity can be generated by rotating a coil of wire within a magnetic field (for example between the poles of a horseshoe magnet). A potential difference is generated across the two ends of the coiled wire as long as the coil rotates. An electric motor is in essence the same as the generator described above, but it operates in reverse. In a motor, a voltage is supplied to the coil causing it to rotate.

Magnetic fields are measured in tesla (T) although the former unit, the gauss, is still often used; 1 tesla $= 10^4$ gauss. The magnetic field at the Earth's surface is about 10^{-4} tesla.

24 Radioactivity

24.1 The nature of radioactivity and nuclear transformations

The basics of atomic structure (see **4**) should be appreciated in order to understand what is meant by **radioactivity**. Briefly, the nucleus of an atom, which contains positively charged protons and electrically neutral neutrons, is surrounded by negatively charged electrons. The number of protons in the nucleus (the atomic number) is fixed for each element, but most elements have atomic nuclei with a variable number of neutrons. The atomic mass of the atoms thus varies. The term **isotope** is used to refer to these different forms of the element. In other words, the isotopes of any named element are distinguished by the number of neutrons in the atomic nucleus. Elements typically have between two and seven isotopes, but some have more. For the majority of elements, one isotope is overwhelmingly predominant. To refer to an individual isotope, the mass number (i.e. proton number plus neutron number) is shown either as a superscript preceding the element symbol, as in ^{14}C, or after the element name, as in carbon-14. The reasons why each element has a fixed number of isotopes are to be found in the 'rules' governing the combinations of protons and neutrons that confer stability on a nucleus.

Most naturally occurring isotopes of the different elements are stable. However, there are fifty or so that are inherently unstable, meaning that they spontaneously disintegrate, or decay. In addition, a total of about eleven hundred unstable isotopes (or nuclides) have been produced artificially. Unstable isotopes, both natural and artificially created, are called **radioisotopes** or **radionuclides**, and are said to be **radioactive**. Just a few naturally occurring elements exist only as radioisotopes: one of these is uranium (atomic number 92). All the isotopes of elements with an atomic number of 84 or above, plus technetium and promethium (atomic numbers 43 and 61, respectively), are radioactive. (The distinction between isotope and nuclide is pointed out in the discussion of atomic structure in **4.2**; henceforth we use the terms 'nuclide' and 'radionuclide' unless referring to nuclides of a particular element.)

The presence of some fifty natural radionuclides means that there is always a **background** level of radioactivity in the environment. The sources of background radiation are the atmosphere, where radionuclides are continually being produced by cosmic rays (principally high-energy particles from deep space), and rocks. The continuous release of radioactivity from rocks is the source of heat in the Earth's interior. The level of radioactivity varies between rock types, being relatively high in granite, for example.

We turn next to what it is that is emitted when unstable atoms decay, and also to the products of the decay process. There are three types of emissions to consider here, termed **alpha particles** (α), **beta particles** (β) and **gamma rays** (γ). As their names suggest, the former two are particulate in nature while gamma rays are electromagnetic radiation of very short wavelength and high frequency (the electromagnetic spectrum is introduced in **22.1**). Most radionuclides emit either alpha or beta particles, and usually in association with gamma rays, but some emit both alpha and beta particles simultaneously. Just a few radionuclides emit only gamma rays.

24.1.1 Alpha particles

Alpha particles are relatively heavy and are associated with elements of high atomic numbers. They have little penetrating power (they may be stopped by a very thin sheet of aluminium or paper) but they are extremely damaging to living tissue. Each alpha particle consists of two protons and two neutrons.

Now, the element helium has an atomic number of 2. An alpha particle is therefore the same as a helium nucleus and can be represented as ($^4He^{2+}$). (There is another isotope of helium, mass number of 5, but it occurs in tiny concentrations.) So when an atom of an alpha-emitting radionuclide loses an alpha particle, its atomic number decreases by two and its mass number decreases by four. Consider the result of alpha emission by atoms of radium-226 (atomic number = 88; number of neutrons = 138). As the atomic number is reduced by 2, and the mass number is reduced by 4, what remains are nuclei with 86 protons and 136 neutrons, i.e. a mass number of 222. In other words a new element is produced. Inspection of the list of elements and their atomic numbers (see **3.1**) will reveal that the new atoms belong to the element radon (Rn). We can represent this process, called a **nuclear transformation** or **element transmutation**, as follows:

$$Ra \rightarrow He^{2+} + Rn$$

An important principle to be noted immediately is that one element can be transformed to another by radioactive decay. Before the discovery of radioactivity, while some speculated about the possibility of such transformations – particularly to produce valuable metals – scientists generally believed the elements to be immutable.

The general term for a product of radioactive decay is **daughter**. In the above example, radon is the daughter. You may have wondered why no reference was made to the electrons in the equation showing radium decay. As two protons are lost, the result should be radon *ions*, each carrying a double negative charge. However, the two outer electrons are rapidly lost, and combine with alpha particles to form helium atoms.

During alpha decay the combined mass of the products is slightly lower than in the disintegrating nucleus. This difference is accounted for by the kinetic energy of the alpha particles, and the energy of any gamma rays emitted during rearrangements within the nucleus. (The total energy is thus conserved, but not the rest mass, which is in accordance with the famous $e = mc^2$ equation.)

24.1.2 Beta particles

Beta particles are electrons: they thus carry a net negative charge, and they are represented by e^-, or just e. The loss of a beta particle thus has the effect of converting neutrons to protons. This increases the atomic number of the nucleus by 1. The artificial radionuclide phosphorus-32 (^{32}P) is a beta emitter which produces an isotope of sulphur on disintegration, thus:

$$^{32}_{15}\mathrm{P} \rightarrow \mathrm{e} + ^{32}_{16}\mathrm{S}$$

A variant of the beta particle is the **positron** which carries a positive charge. A positron is known as the **antiparticle** of an electron. In cases where positrons are emitted, protons are converted to neutrons. The atomic number of the daughter nucleus is thus lower than that of the parent nucleus. An example of positive beta decay is provided by the radionuclide zinc-65 (atomic number 30). By emitting positrons, this nuclide is converted to copper-65 (atomic number 29), a process represented as follows:

$$^{65}_{30}\mathrm{Zn} \rightarrow ^{0}_{1}\mathrm{e} + ^{65}_{29}\mathrm{Cu}$$

A beta particle has a mass approximately one-eighteen hundredth (1/1800) that of an alpha particle. But, as with alpha decay, the rest mass of the particles is reduced during beta decay. The velocity at which beta particles are emitted varies within and between nuclides but is always greater than the velocity of alpha particles. Also, beta particles are rather more penetrating than alpha particles although they can be stopped by a few centimetres of metal.

24.1.3 Gamma rays

Unlike alpha and beta radiation, gamma rays are not particulate in nature but are electromagnetic waves of very short length (wavelength range 10^{-14}–10^{-11} m). Gamma rays are emitted from nuclei of radionuclides as they lose energy – the 'excitation state' of the nucleus is lowered. Gamma rays are highly penetrating, and very hazardous to life.

24.2 Radioactive series

The result of alpha or beta emission from one element leads to the formation of daughter nuclides which are of another element. In the case of those few elements which emit both alpha and beta particles, nuclides of two different elements are produced. These daughters may have nuclei containing a stable combination of protons and neutrons, or they may be unstable. If unstable, the daughter will spontaneously decay to produce its own daughter nuclides, which again may be stable or unstable. A sequence of nuclides formed by such transformations constitutes a **radioactive series** or **family**. Each member is the decay product of the isotope preceding it in the decay chain, except of course the first nuclide in the series. The final nuclide in the series must clearly be stable. There are three natural radioactive series, and these contain many of the naturally occurring radionuclides. The series are named after the first nuclide in the series. These are, respectively, thorium-232, uranium-235 and uranium-238. Each of these series ends with a stable isotope of lead.

24.3 Rates of decay: half-lives

We have seen that radionuclides spontaneously decay with the loss of either, or combinations of, alpha particles, beta particles and gamma rays. Here we consider the *rate* of decay of individual radionuclides and see how this is expressed. Key points to remember are, first, that each radionuclide decays at a characteristic rate, and secondly, that there is enormous variation in rates of decay between radionuclides. For any given radionuclide it is not possible to determine when one of its *individual* atoms will disintegrate, but it is possible to calculate the proportion of atoms which will disintegrate in a given time. It is really the probability that an individual atom will disintegrate in a certain period of time that is being considered here. The **activity** of a radioactive sample is represented as the number of disintegrations per second. The SI unit of radioactivity is the **becquerel** (Bq), after the French physicist who, using uranium, played such an important role in the discovery of radioactivity in the 1890s.

The decay of radionuclides conforms to a particular pattern (Figure 24.1). For all radionuclides, when the value for radioactivity is plotted against time on an arithmetic scale, the resulting curve is a negative exponential function. This means that a logarithmic transformation of the activity values (or the use of a logarithmic scale) will give a straight-line relationship between activity and time, i.e. activity is a linear function of time. Remember, though, that each radionuclide decays at its own rate, and rates are hugely variable between nuclides.

The decay rate of each radionuclide is represented by its **half-life**. A half-life is defined as the time taken for the number of unstable atoms to be halved; in other words, the time taken for the activity of a sample of radionuclide to be halved. In Figure 24.1 you can see that in each successive half-life interval the

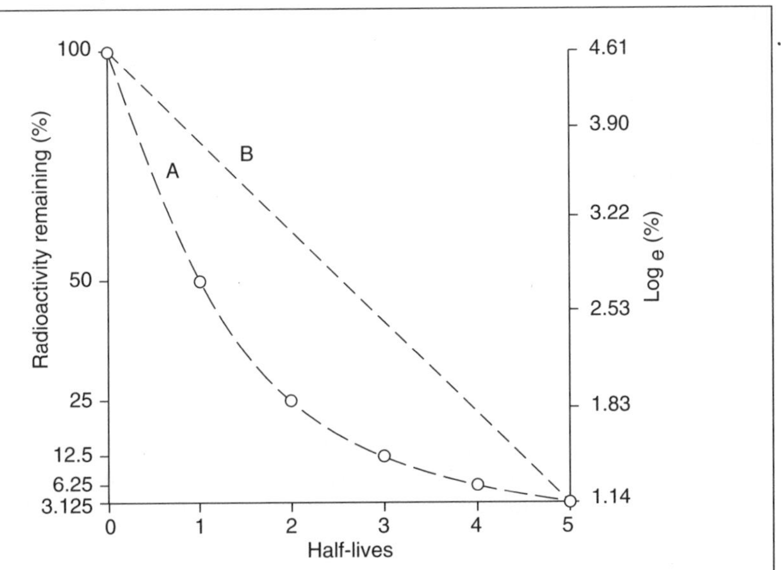

Figure 24.1 The pattern of decay of a radioisotope. The radioactivity remaining in a sample of a radionuclide is plotted against time. Curve A is an arithmetic plot (values are shown on the left-hand vertical axis; curve B is a logarithmic plot (values are shown on the right-hand axis). Note the percentage of radioactivity remaining after each successive half-life.

amount of the radioactivity is halved. Two half-lives are necessary to reduce the activity to one-quarter of its initial value, and three half-lives are necessary to reduce activity to one-eighth of its initial value. (Do not think of the half-life as the time needed for the sample of a radioisotope to become safe.)

As we have emphasized, half-life values vary enormously. For some artificial radionuclides, disintegration of the nucleus is virtually spontaneous, and half-life values are measured in milliseconds. At the other extreme, the half-lives of some natural radionuclides are many millions, even billions of years. Some examples are given in Table 24.1.

A little elementary mathematics is now presented to represent the decay process. In a sample of any given radionuclide, the number of atomic disintegrations per second is a constant proportion of the total number of unstable atoms in the sample. Expressed mathematically this is represented by:

$$\frac{dN}{dt} = -\lambda N \tag{1}$$

where N is the number of unstable atoms in the sample at time t, and λ is called the **decay constant**. Because of the great variation in rates of decay between the radionuclides, values of λ vary enormously. Manipulation of equation 1 allows us to calculate the activity of a sample at a particular time (N) if we know the activity of the sample initially (N_0), thus:

$$N = N_0\, e^{-\lambda t} \tag{2}$$

Table 24.1 Half-lives of some radioisotopes.

Radioisotope		*Half-life (approx.)*
Iodine-131	^{131}I	8 days
Phosphorus-32	^{32}P	14 days
Sulphur-35	^{35}S	87 days
Hydrogen-3 (tritium)	^{3}H	12.3 years
Carbon-14	^{14}C	5570 years
Chlorine-36	^{36}Cl	300 000 years
Uranium-238	^{238}U	4.47×10^{9} years

You may know that e (which is approximately 2.718) is the base in natural logarithms (see **26**). The loss of all radioactivity proceeds exponentially, i.e. according to a geometric series, although the exponent varies enormously between different radionuclides. From equation 2, the half-life (T) can be expressed as:

$$T = \frac{0.693}{\lambda} \tag{3}$$

The number 0.693 is the natural logarithm of 2.0, i.e. $\log_e 2 = 0.693$. From equations 2 and 3, and knowing the activity of a sample at any one time, it is possible to calculate its activity at any other time, in either the future or the past. This has a great number of useful applications (see **25**).

24.4 Creating nuclides

Of the twelve hundred or so radionuclides, only about fifty occur naturally. All the others are produced artificially, or they are the disintegration products of such radionuclides. The artificial production of nuclides involves bombarding atoms with particles in order to create new nuclear combinations of neutrons and protons. The first nuclide to be produced artificially was oxygen-17, which is stable, and which accounts for less than 0.04 per cent of natural molecular oxygen. This feat was announced in 1919 by New Zealand-born Ernest Rutherford who, by then, was working in the University of Manchester. It was accomplished by bombarding nitrogen-14 with alpha particles, thereby increasing the number of protons in the nucleus by one. Soon after, in the early 1930s, the first radionuclide was produced, which was nitrogen-13.

24.5 Nuclear fusion, nuclear fission

A **nuclear fusion** reaction involves the union of two light atomic nuclei. During the formation of a heavier nucleus an enormous amount of energy is released because the mass of the products is less than the mass of the original materials. Fusion reactions require temperatures of several millions of degrees. The most familiar manifestation of such reactions is the radiant energy emitted by the Sun and other stars. The ultimate source of such radiant energy is the fusion between hydrogen nuclei, with the formation of helium nuclei and positrons. Fusion reactions are also brought about artificially, for example during the detonation of hydrogen bombs.

Nuclear fission is a reaction involving the disintegration of an atomic nucleus (for example of uranium). As a result, lighter nuclides, neutrons and various forms of radioactivity are released, carrying enormous amounts of energy. The disintegration may occur spontaneously or as a result of the bombardment of nuclei by neutrons. Neutrons released during a fission reaction bombard other fissile nuclides, which in turn bombard other nuclides, thus forming a **chain reaction**. Fission reactions are responsible for atomic bombs, and, under more controlled conditions, for the generation of electricity.

25 Using isotopes in environmental study

25.1 Introduction

Here we look at some of the principles underlying the use of **isotopes** in studies of the environment. These require an appreciation of what isotopes are (see **4.2**), and the nature of radioactivity (see **24**). The key points are as follows. The different isotopes of a given element have the same number of protons (the defining feature of an element) but different numbers of neutrons; they consequently differ in atomic mass. Most chemical elements have more than one isotope, of which one is usually overwhelmingly predominant. Most naturally occurring isotopes are stable (i.e. *not* radioactive), but some are unstable, that is they are radioactive, and are referred to as radioisotopes or radionuclides. There are recognized standards for the proportion of each isotope in a sample of any given element, although in practice the relative abundance of isotopes varies depending on the source of the sample and its history. The relative amount of two isotopes in a sample is known as the **isotopic ratio**.

Chemically, the various isotopes of an element behave in the same way. However, the difference in mass between them means that they may differ quantitatively in their physical behaviour. Thus, when a sample of an element is subject to certain physical processes, a slight shift in the isotopic ratio occurs. Because of this, isotopic ratios are not constant in every sample of the element, although, as just stated, there are standards to which the ratio in a sample can be compared. Knowing what factors can bring about isotopic discrimination can provide important clues to the processes experienced by a particular chemical sample.

Isotopic ratios can be measured accurately in the laboratory using an instrument called a mass spectrometer. The predictable behaviour of isotopes, together with their precise measurement, provide the basis for a set of research techniques which are of great value for environmental scientists. Stable isotope analysis provides valuable insights to past environments, and also gives clues to the source and history of samples of rock, air, water and organic matter. One application of isotopic ratios in oxygen samples will allow us to elaborate on some principles and also to introduce some important techniques.

25.2 Oxygen isotopes

The isotopic ratios of oxygen, particularly within water, oxides and carbonates, are used extensively in environmental studies. By far the most abundant oxygen isotope is ^{16}O, but there are two heavier isotopes, ^{17}O and ^{18}O, both of which are stable. (Nuclei of oxygen atoms contain eight protons, so the three isotopes have eight, nine and ten neutrons respectively.) Now, evaporation of water results in oxygen isotope discrimination: specifically, there is discrimination against the heavier isotopes because the lightest isotope is more volatile. Thus, relative to the atmosphere and rainwater, the oceans are enriched with ^{18}O. The degree of discrimination during evaporation is affected by temperature, being greater at lower than at higher temperatures.

An understanding of the factors affecting isotopic discrimination can give vital clues to what environmental conditions were like in the past. This is because in some environments the oxygen isotope ratios have been preserved: thus their measurement today provides information about the conditions pertaining at the time. Such information is obtained from locations referred to as **depositional environments**, which are characterized by the progressive accumulation of materials over time. Important locations of this type are sediments on the sea- and lake-floor, peat in waterlogged bogs, and snow and ice at high altitudes. From such sites, cores of material can be extracted. An increase in distance from the top of the core represents increasing time since deposition. It is then necessary to establish the depth/age relationship of the core. In some situations, cores reveal banding which represents annual accumulations, much as in tree rings. Otherwise, radiometric dating (see **25.3**) is the favoured method of determining the age of particular layers. Cores therefore provide an opportunity to look at environmental conditions, and life at particular times in the past. Reading the record requires the identification of preserved life-forms and physical and chemical analysis of the deposited materials. This general approach is used routinely to study palaeoenvironments. Here our interest is focused on oxygen isotope ratios. These can be measured in the calcareous ('shelly') remains of marine organisms.

During the most recent 'ice age', which began around 2.4 million years ago, there have been several glacial advances and retreats as global climate cooled and warmed. Now recall our earlier comment about the degree of oxygen isotope discrimination being greater in cooler than in warmer conditions. We would therefore expect the $^{18}O/^{16}O$ ratio in seawater to be greater during cold intervals than during warm intervals. A record of the oscillations in oxygen isotope ratios in seawater is to be found in the carbonates incorporated by certain tiny marine organisms while they lived and which have been

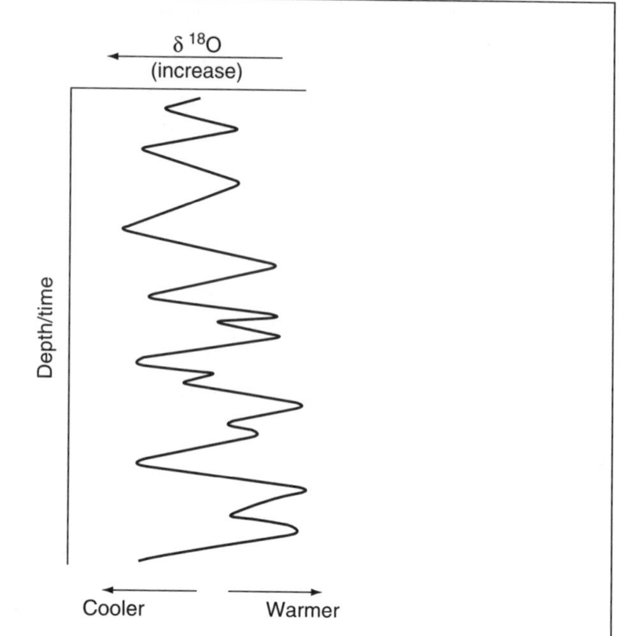

Figure 25.1 Hypothetical change in isotope ratio in preserved organisms in a core extracted from the sea-floor and the climatic changes indicated.

preserved since their death. The assumption here is that the oxygen isotope ratio in calcareous body parts is the same as that in the surrounding water. (We will note the fact that this is not always the case but overlook this potential source of error.) There is another factor at work here accentuating the temperature effect. The source of the glacial snow and ice is the oceans. Recall the general principle that evaporation discriminates against the heavier isotopes of oxygen in water. It follows, therefore, that during intervals of glacial advance the accumulating snow and ice become progressively enriched with the lighter isotope, and the relative abundance of the same isotope declines in the oceans. In non-glacial times there would simply be an annual oscillation in oxygen isotope ratios as melting snow and ice return the ^{16}O-enriched water to the oceans.

A plot of oxygen isotope ratio against time (core depth) therefore provides information about the climatic shifts during the last major 'ice age'. Figure 25.1 illustrates such a plot. Isotopic composition is usually denoted by the symbol δ, as in the case of $\delta^{18}O$. Values for δ are conventionally expressed as deviations from an agreed standard, and expressed as permils (parts per thousand), as here:

$$\delta^{18}O = 1000\frac{[^{18}O/^{16}O]_{sample} - [^{18}O/^{16}O]_{standard}}{[^{18}O/^{16}O]_{standard}}$$

As noted earlier, for nearly all elements one isotope is overwhelmingly predominant. In the case of a 'standard' sample of oxygen, only about 0.2 per cent is accounted for by ^{18}O. Furthermore, under natural conditions, discrimination shifts the isotopic composition only very slightly. Thus it is more convenient to express the deviations of isotope ratios in terms of per thousands than to use the absolute values or even percentages. Also, note that values can be either positive or negative. A negative value indicates that the $^{18}O/^{16}O$ ratio of the sample is lower than that of the standard. A large number of samples would need to be analysed in order to construct the sort of time sequence of oxygen isotope ratios as shown in Figure 25.1.

25.3 Radioactive isotopes

Radioactive isotopes (or radionuclides) are defined by their inherent instability (see **24**). Each radioisotope decays at a unique, and known, rate with the emission of particles and/or rays. The decay rate of a radionuclide is expressed as its half-life, and values vary enormously. Two properties of radionuclides make them invaluable tools for environmental research. First, the fact that the presence and the amounts of such substances can be determined means they are very useful as 'tracers'. For example, we could supply the roots of a plant with some radioactive phosphorus in order to determine the movements of this key nutrient at different times of the year. The second major use of radionuclides is in dating materials. In outline the principles are fairly simple to understand, but it is important to appreciate that there are complex issues involved, and possible sources of error. Failure to appreciate these has led some researchers to erroneous conclusions about the age of materials they are working with. The technique, known as **radiometric dating**, is based on a knowledge of decay rates of radionuclides and the nature of the products (daughters) of the decay process. In essence, as radionuclides decay, their abundance declines while that of daughters increases. Thus the ratio of parent to daughter shifts with time in a predictable way.

The method assumes that when the material which is being dated was formed, it contained no decay products, and also that neither parent nor daughter has leaked out of the material since the time of formation. Leakage can, however, be a problem for some parent/daughter systems. In addition, extreme heat and pressure can alter rocks and minerals (metamorphosis), which may cause radioactive clocks to be reset. The usefulness of a particular parent/daughter system depends largely on the half-life of the parent. The most commonly used parent/daughter systems are shown in Table 25.1.

The very long half-lives of these parent isotopes, together with possible sources of error, make them unsuitable for dating materials from the recent past. There is one nuclide, however, that is commonly used to date materials less than about 40 000 years of age. It is carbon-14 (^{14}C), which has a half-life of 5568 years. **Radiocarbon dating** is more-or-less confined to organic materials, for example wood and peat. However, if there is good evidence that local non-biotic materials are contemporary with the dated biotic material then this permits an estimation of their age also.

Table 25.1 Half-lives of some long-lived radioactive isotopes used in rock dating.

Parent isotope		*Half-life (years)*	*Daughter isotope*	
Uranium-238	^{238}U	4.5 billion	Lead-206	^{206}Pb
Uranium-235	^{235}U	0.7 billion	Lead-207	^{207}Pb
Rubidium-87	^{87}Rb	48 billion	Strontium-87	^{87}Sr
Thorium-232	^{232}Th	14 billion	Lead-208	^{208}Pb
Potassium-40	^{40}K	1.3 billion	Argon-40	^{40}Ar

Naturally, carbon contains a tiny proportion of the radioisotope ^{14}C. (This is continuously being produced in the upper atmosphere by the bombardment of nitrogen by cosmic radiation.) Carbon-14 is incorporated by plants, together with the more abundant, stable isotopes of carbon, in the process of photosynthesis (see **34.3.1**). It is assumed, first, that all living organisms acquire carbon in more-or-less the same isotopic ratios throughout their lives, either directly from photosynthesis or by consuming organic matter, and secondly, that this ratio has not changed appreciably over time. As carbon is being continually taken up from the environment, the ratio of ^{14}C to the other carbon isotopes should not change during an organism's life. When an organism dies, however, the ^{14}C begins to decay (back to nitrogen-14), with the result that the ratio of ^{14}C to the other carbon isotopes declines. The greater the age of the organic substance, the lower the ratio of ^{14}C to the most abundant carbon isotope, ^{12}C. The known half-life of ^{14}C permits the date of the death of the organism to be estimated.

Unlike the parent isotopes used in the parent/daughter systems mentioned earlier, ^{14}C is continually being replenished. Carbon dating also differs from the other radiometric techniques in that it is the *ratio* of ^{14}C to other carbon isotopes which is measured, and not the ratio of parent to daughter isotopes.

Radionuclides are also used to date samples from depositional environments. We have just noted that such environments are characterized by the progressive accumulation of material which, according to location, may be peat, sediments, or snow and ice. The example of sea-floor sediments was used above when describing oxygen ratios. Accumulating material traps within it organisms and substances such as charcoal and airborne pollutants, but also radionuclides. The depth at which artificial radionuclides are first encountered in any quantity in a core from a depositional environment probably marks the onset of nuclear testing in the 1950s, while more subtle changes in relative abundance in the core may well mark other nuclear events. Evidence of this sort is very useful in dating different parts of a core from a depositional environment, and thus in providing evidence of rates of accumulation in such environments. It also permits other events that have left their signatures in the sampled material to be dated.

26 Working with numbers

26.1 Use of numbers

A basic but sound numeracy is essential if you are going to quantify the processes occurring within an environmental system. During the introductory years of a geography or environmental science degree it is unlikely that you are going to come across many mathematical concepts that are completely new to you. The difficulty is that many people forget the mathematics they have learnt or have not had the need to apply what they know to real problems. You must always remember that mathematics is just a shorthand and concise way of explaining something that could be written out in full using plain words. The best way to overcome any problems with mathematics is to practise: it is very easy to forget even the basics when you are not using them on a day-to-day basis. Here we introduce some of the basic rules of mathematics, and some useful tips.

26.2 Operators

The commonly used operations in both algebra and arithmetic are multiplication, division, addition and subtraction. The symbols used to indicate that an operation should be carried out are termed **operators**. Thus, the operator $+$ is used when the operation of addition is required (i.e. $2 + 2 = 4$). The common operators are thus: $+$, $-$, $\times$ and $\div$. However, the last two of these are rarely represented in this form in algebraic expressions (see **26.4**). The multiplication operator ($\times$) is commonly left out, or assumed. Thus, $a \times b$ is usually written ab. Clearly, this is not done in arithmetic, as 2×2 is not the same as 22. You will, however, see $50a$ written instead of $50 \times a$. Sometimes a full stop (period) is used as a substitute for the multiplication operator ($a.b = a \times b = ab$). The division operator, $\div$, is commonly replaced by the symbol '/'. Thus, $a \div b$ is usually expressed a/b. It is also expressed $\frac{a}{b}$.

26.3 Order of arithmetical operations

One of the most common mistakes when evaluating numerical expressions is to carry out the arithmetical operations (addition, subtraction, etc.) in the wrong order. To help you, just remember BIDMAS (brackets, indices, division, multiplication, addition and subtraction). This is the correct order in which to calculate your equation. First, calculate anything within brackets, then calculate all indices, division, multiplication, etc. Note that this is not the order in which the operators (subtraction, addition, etc.) appear in an equation. One word of warning here: computer programs or applications do not always follow BIDMAS exactly. If you are using a spreadsheet and are unsure in which order the spreadsheet carries out the equation, then split it into parts or use extra brackets to force the order of the calculation. You should always carefully check any results from a computer program or spreadsheet.

Here are some examples:

$$3 + 4 \times 10/2 = ?$$

$$10/2 = 5 \text{ and } 5 \times 4 = 20$$

$$\text{and } 20 + 3 = 23$$

The following is evaluated using the following order: division, multiplication, addition:

$$3 + 10/2 \times 4 = 23$$

This is exactly the same calculation as shown above. It demonstrates that the order across the page does not matter. But if you carry out the calculations in the written order (i.e. $3 + 10 = 13$; $13/2 = 6.5$; $6.5 \times 4 = 26$) the answer is incorrect.

If we use the same numbers and operations as before, but this time add some brackets, the order of calculation changes, and so does the answer:

$$(3 + 4) \times 10/2 = ?$$

$$(3 + 4) = 7 \text{ and } 10/2 = 5$$

$$\text{and } 7 \times 5 = 35$$

The reason is that the calculation within the bracket must be carried out first, i.e. $3 + 4$, followed by the division of 10 by 2 and then the two resulting values are multiplied together.

So far we have used only whole numbers in our examples; such numbers are called **integers**.

26.4 Algebra and algebraic rules

Before we move on, let us return to an earlier example but this time replace the numbers with letters. This is another part of mathematics that causes many people needless confusion. Our previous example was $(3 + 4) \times 10/2 = 35$.

Now we simply substitute letters for numbers. So, let $A = 3$, $B = 4$, $C = 10$, $D = 2$ and $E = 35$. Now we can write:

$$(A + B) \times C/D = E$$

All we have done is to replace the numerical values by letters. The great advantage of this is that we can now use the same expression to apply to different values. If, for example, we need to know the value of A given the value of the other letters, we can simply rearrange the expression. We can likewise further rearrange it to find the value of any one of the letters. To do this we must progressively move letters, remembering to treat both sides of the expression identically. Before looking at an example, let us consider the standard algebraic rules.

Some operations (i.e. multiplication, subtraction, etc.) are termed **associative**. Such operations give a result that is independent of the grouping (by brackets) of the numbers. Multiplication is associative but division is not. (Remember to calculate the terms in the brackets first.) Thus these two expressions have the value of 48:

$$2 \times (6 \times 4) \text{ and } (2 \times 6) \times 4$$

However, if we replace the × symbol by the / symbol, the expressions do not have the same value, thus:

$$2/(6/4) = 1.333 \text{ and } (2/6)/4 = 0.0833$$

Another class of operations is termed **commutative**. A commutative operation is one in which the result is independent of the order of the numbers or operators (+, ×, etc.). So, $5 + 6$ gives the same result as $6 + 5$, but $5 - 6$ is different from $6 - 5$. Thus, we can see that addition is commutative but subtraction is not.

A third class of operation is termed **distributive**. In this case a number can be distributed between the items within a group (i.e. in brackets). Multiplication is distributive. Take the expression $3 \times (4 + 5)$. The 3 can be distributed to the 4 and 5 in the brackets:

$$(3 \times 4) + (3 \times 5)$$

The two expressions have the same value, which is 27. Division is also distributive: $(4 + 5)/3$ has the same value as $4/3 + 5/3$: the answer is 3. However, $3 + (4 \times 5)$ is not equal to $(3 + 4) \times (3 + 5)$. The values are 23 and 56, respectively. So addition (and subtraction) are not distributive.

Now let us turn back to our earlier example, and rearrange the expression $(A + B) \times C/D = E$ in such a way that we can find the value of A. First, as both multiplication and division are distributive, we can deal with these two first. (The order in which this is done does not matter.) Multiply both sides by D:

$$D \times (A + B) \times C/D = D \times E$$

$$\text{or } (A + B) \times C = D \times E$$

Then divide by C:

$$(A + B) \times C/C = D \times E/C$$

$$\text{or } (A + B) = D \times E/C$$

Finally, now that nothing is outside the brackets on the left-hand side (and thus we have no problem with the 'distributiveness' of the operation), we can subtract B:

$$A + B - B = D \times E/C - B$$

$$\text{or } A = D \times E/C - B$$

A similar approach can be taken to rearrange the expression in terms of the other letters.

26.5 Indices, powers and exponents

Remember from BIDMAS that after dealing with operations within brackets you deal with the indices. Indices are those superscripted numbers (or terms) that follow a number or algebraic symbol, which is the base. An example is x^y. Here, x is the **base** and y is the **index**. Another term for index is **exponent**. The quantity of the index (or exponent) shows the number of times that the base is multiplied by itself. Put another way, we say that a number multiplied by itself, say y times, is raised to the **power** of y. Index terms can be negative as well as positive and are not confined to whole numbers. They are simply the number indicating the power of a quantity.

To provide some simple numerical examples, the expression 2^2 is shorthand for 2×2, or 'two squared', which is 4, and the expression 2^3 means $2 \times 2 \times 2$, or 'two cubed', which is 8. (On many calculators it is the x^y button which evaluates a number raised to the power of another.)

You might like to evaluate the expression:

$$3^4 + 5^6 \times 2^5 = ?$$

First, deal with the indices:

$$3^4 = 3 \times 3 \times 3 \times 3 = 81$$

$$5^6 = 5 \times 5 \times 5 \times 5 \times 5 \times 5 = 15\,625$$

$$2^5 = 2 \times 2 \times 2 \times 2 \times 2 = 32$$

Then carry out the multiplication:

$$15\,625 \times 32 = 500\,000$$

Finally, do the addition:

$$81 + 500\,000 = 500\,081$$

When multiplying powers of the same quantity (for example $x^a \times x^b$) the indices are added, thus:

$$x^a \times x^b = x^{(a+b)}$$

$$5^2 \times 5^3 = 5^5$$

And when dividing powers of the same quantity the index of the divisor is subtracted from the index of the numerator, thus:

$$x^a \div x^b = x^{(a-b)}$$

$$5^8 \div 5^3 = 5^5$$

When we see a **negative index** to a quantity (for example x^{-2}) it simply means the reciprocal of the quantity, so:

$$x^{-2} = 1/x^2$$

$$3x^{-4} = 3/x^4$$

$$1/x^{-a} = x^a$$

If you encounter an expression like nx^y it is only x, and not n or nx, that is raised to the power of y. However, with an expression such as $(3x)^2$, 3 is raised to the power of 2, which is 9, and x is raised to the power of 2. This becomes $9x^2$.

A power of a power, as in $(x^y)^n$ is evaluated as x^{yn}.

You will sometimes encounter a term or quantity that is raised to a fractional power, for example $x^{1/2}$. This is evaluated as $\sqrt{x^1}$. Here, the numerator in the power, i.e. 2, means that the square root is taken (i.e. two numbers with the same value are multiplied together) and x^1 is simply x. The same principle applies to other fractional powers. Thus:

$$x^{4/3} = \sqrt[3]{(x \times x \times x \times x)}$$

The expression $\sqrt[3]{}$ is called the cube root.

26.6 Logarithms

Logarithms and indices are related. The written definition of a logarithm which follows may appear a bit vague at first glance but the idea is quite straightforward if you appreciate the terms 'base' and 'power' introduced above. A logarithm is a power to which a number (the base) is raised to give another number. So in the expression:

$$x = y^z$$

z is the logarithm and y is the base. So if $x = 100$ and $y = 10$, then $z = 2$. The logarithm is thus 2. We can rewrite the equation as:

$$\log_y x = z$$

Substituting numbers for letters, we get:

$$\log_{10} 100 = 2$$

In words, this says 'the logarithm, to the base ten, of one hundred is two'. We have deliberately set y (the base) at 10 because logarithms to the base 10 are in such common use. In fact we refer to them as **common logarithms**. No doubt you already know that the common logarithms of 10 (10^1), 1000 (10^3), 10 000

(10^4) and 1 000 000 (10^6) are 1, 3, 4 and 6, respectively. Recalling what we learnt earlier about manipulating power terms, and knowing that we use the decimal system for quantities, we can see one very useful application of logarithms. Note that where the type of logarithm is not specified it is always assumed to be to the base 10.

Now, if we changed the value of y (the base) in the above equation $x = y^z$, the power, i.e. the logarithm, would no longer be 2. In theory we could choose any base, but in mathematics there is one other base of great value and utility. It has the approximate value of 2.718 and is referred to as 'base e'. Logarithms to the base e are called **natural**, or **Napierian logarithms** (after the Scottish mathematician John Napier), and they are written as $\log_e$ or ln. You might like to verify on your calculator that the natural logarithm of 100 is 4.605.

So far we have considered just positive numbers and positive logarithms. A key point is that the logarithm of 1, whether it be the common or natural logarithm, is zero ($\log 1 = 0$). Thus the logarithm of a number less than 1 is always negative.

Logarithms are useful in all sorts of ways. They are useful in mathematical manipulations. For any base the following rules hold for logarithms:

$$\log(a \times b) = \log a + \log b$$

$$\log(a/b) = \log a - \log b$$

$$\log(a^b) = b \log a$$

Logarithmic transformations are commonly carried out before plotting data and conducting statistical analysis. Sometimes, untransformed data are spread over many orders of magnitude (i.e. powers of 10). Such data points can be plotted much more easily on 'log paper', or alternatively, the logarithmic transformed numbers can be plotted on an arithmetic scale. Some environmental phenomena are expressed as $\log_{10}$ transformed values, including decibels (sound), pH (acidity) and Richter numbers (magnitude of earthquakes).

26.7 Exponentials

In the case of logarithms, the index (see **26.5**) is derived from the number of interest. In the case of exponentials, the index *is* the number of interest. Any base can be used. However, the natural number e and 10 are most commonly encountered. The particular merit of exponentials based on e is that they mimic many naturally occurring processes. For example, the reduction in radioactivity of a radionuclide through decay (see **24**) is exactly modelled by the expression:

$$N = N_0 \times e^{-\lambda t}$$

where N is the number of nuclei remaining at time t, N_0 is the initial number of nuclei, and λ is the decay constant.

26.8 Trigonometry

Trigonometry is a branch of mathematics that is concerned with solving problems that are defined in terms of right-angled triangles (Figure 26.1). In practical terms this means calculating the unknown values for sides and angles of a triangle when other values are known. Trigonometric functions are therefore very useful for solving problems you may come across in environmental study, such as when carrying out simple surveying.

The right-angle triangle in Figure 26.1 shows the labelling of the sides in relation to the angle θ (called theta, see **2.7.1**). The longest side, which is always opposite the right angle, is termed the **hypotenuse** (h). The side opposite angle θ is called the **opposite** (o), and the remaining side is termed the **adjacent** (a) (as it is adjacent to the angle θ, but is not the hypotenuse). A link between the angle θ and the length ratio of pairs of these sides can be established by transforming the angle into another form. This transformation is achieved using a mathematical tool termed a **function**. There are three main functions of relevance here, called **sine**, **cosine** and **tangent**, thus:

$$\text{sine}\,\theta = \sin\theta = \frac{\text{o}}{\text{h}}$$

$$\text{cosine}\,\theta = \cos\theta = \frac{\text{a}}{\text{h}}$$

$$\text{tangent}\,\theta = \tan\theta = \frac{\text{o}}{\text{a}}$$

These functions are interrelated as follows:

$$\tan\theta = \frac{\sin\theta}{\cos\theta}$$

To appreciate how useful these functions are, consider this example. We find that the angle from the ground to the top of a tree is 30°, and we are standing 50 m away. How tall is the tree? Making use of the triangle in

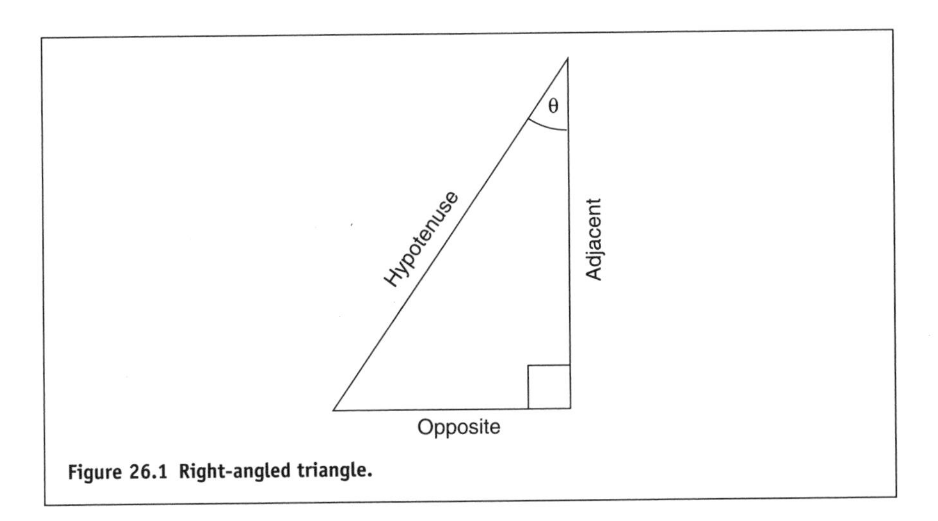

Figure 26.1 Right-angled triangle.

Figure 26.1, we know the length of the adjacent side (from observer to tree: 50 m), and we know the angle θ (30°) and we need to know the length of the opposite. From the three expressions listed above it is clear that we need to use the tangent. We can find the tan of θ using a calculator or spreadsheet, and we can rearrange the expression (see **26.4**) to find the value of o (which is the height of the tree):

$$\tan\theta = \mathrm{o}/\mathrm{a}$$

which rearranges to

$$\mathrm{o} = \tan\theta \times \mathrm{a}$$

(Note: $\tan\theta$ is a single entity, as if it had a bracket around it. Thus, we are talking of $(\tan\theta) \times \mathrm{a}$, not $\tan(\mathrm{a} \times \theta)$, which has quite a different value.) To complete the example:

$$\begin{aligned} \mathrm{o} &= 50 \times \tan 30^\circ \\ &= 28.9 \end{aligned}$$

Therefore the tree is about 29 m high.

Now try a few calculations for yourself. If you are unsure of your calculations, you can carefully draw scaled distances and the angles on graph paper. Then you can read off the unknowns to check your calculations.

26.9 Root mean square

The **root mean square** (**rms**) method is a way of measuring an average. Normally, to obtain an average (or mean) we simply add together all the relevant values and divide the total by the number of individual values. The rms method is similar, but involves squaring the individual values before summing them, and then dividing this value by the number of measured values. Finally we obtain the square root of this number.

$$\mathrm{rms} = \sqrt{\frac{a^2 + b^2 \ldots z^2}{n}}$$

where a, b, etc. are the individual values, and n is the number of values. The advantage of this method is that it gives extra importance to high values. You will quickly appreciate this point by looking at this example:

$2 + 3 + 4 + 5 + 28 = 42$ mean $= 8.4$

$2^2 + 3^2 + 4^2 + 5^2 + 28^2 = 838$ (mean $= 167.6$) rms $= 12.95$

Root mean square is often used when discussing the nature of molecular movement in gases.

27 Use of data and graphs

27.1 The nature of graphs

An experiment is often designed in such a way that the researcher must collect values for a number of variables. Regardless of whether they vary in time (such as the NO_2 concentration in air varying through the day) or in space (such as the increased sea salt deposition as you near the sea), these variables fall into three principal classes. These may be termed responses, causes and interference. Strictly, the latter two are little different from each other: interferences are simply causes that our experiment does not wish to consider. In the analysis stage of the experiment, we want to explore the relationship between the cause and response variables. An ideal starting point is the graph.

On a scatter plot, or XY graph, we have two variables plotted against each other. By convention, the causal variable is plotted along the horizontal axis (or x-axis) of the graph, and is termed the **independent variable**. The response variable is plotted on the vertical axis (or y-axis), and is termed the **dependent variable** simply because it is assumed that values of this variable are dependent upon those of the independent variable. However, there are many cases where two variables appear to be related, or associated, but we are not sure which is dependent and which is independent, or indeed whether there is another causal variable on which both are dependent.

The first job you must always do with data you have collected is to plot it out. With the widespread use of spreadsheets this is very easy to do. By plotting your data you are taking the first step between your raw observations and any governing physical/mathematical relationships (or models). The graphical representation of data is a very powerful tool in understanding an environmental system. Often, two variables will be shown to have a characteristic relationship between them. Sometimes this is unvarying, indicating that this relationship is governed by physical laws. Some areas of study have developed specific and specialist graph types. Also, it is a very good way of spotting potential problems such as instrumentation problems or even the simple misreading of instruments.

Table 27.1 Data for velocity calculation.

x (m)	*t (s)*	$\Delta x/\Delta t$ ($m\,s^{-1}$)	$\Delta v/\Delta t$ ($m\,s^{-2}$)
0	0	0	0
10	10	1	0.1
20	15	2	0.2
30	20	2	0
40	30	1	−0.1
50	40	1	0
60	50	1	0
70	60	1	0
80	65	2	0.2
90	70	2	0
100	80	1	−0.1

A simple example is based on someone walking down a 100 metre track. The first column in Table 27.1 shows the distance travelled from the starting point (at 10 metre intervals) and the second column shows the time taken to reach each point. We can calculate the average velocity for the entire 100 metre walk using this information. As we see, it has taken 80 seconds to complete the 100 metres. The average velocity is thus $100/80 = 1.25$ metres per second, or 4.5 kilometres per hour, which is a reasonable walking pace.

To relate distance to time over the duration of the walk, we can use a graph. Time, because it is the independent variable, is represented on the horizontal (x) axis, and distance on the vertical (y) axis (Figure 27.1). Note that we have

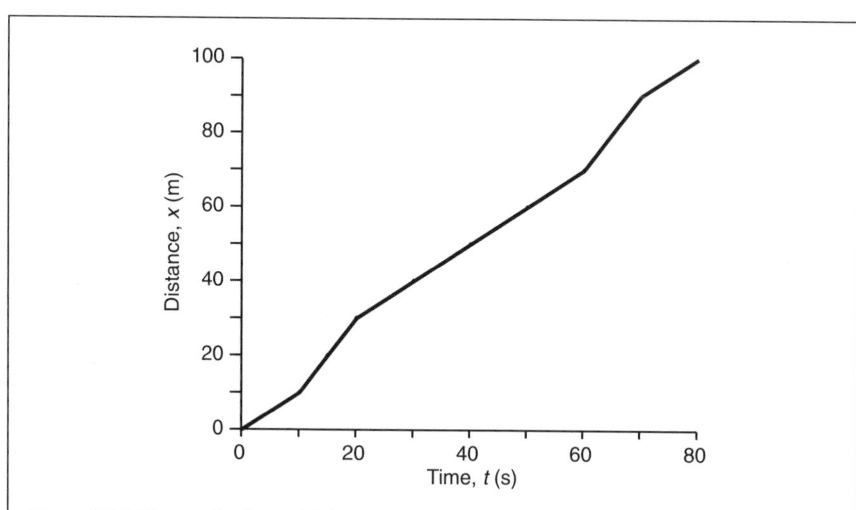

Figure 27.1 Time on horizontal (x) axis, and distance on the vertical (y) axis, for person walking.

measurements only at certain points in time. On the graph, however, the points have been joined up. This is standard practice, but makes *assumptions* about what is happening between the measured points. We say we are **interpolating** between the points.

The third column shows the velocity over each section of the walk. For this calculation we have used the time taken to cover the previous 10 metres. The first 10 metres were covered in 10 seconds, hence an average velocity of 1 metre per second. The second 10 metres took only 5 seconds to complete, i.e. an average velocity of 2 metres per second. What we are doing here is comparing the change in distance with the change in time for small intervals of distance. We have used the equation $\Delta x/\Delta t$ (pronounced 'delta x by delta t') to calculate these velocities. The Greek capital letter delta (Δ) is used to show that we are looking at changes in a quantity. Indeed, it can be helpful to read Δx as 'change in x'.

In calculating the velocities, we have used quite large distance intervals (10 metres), i.e. Δx is large. The method is said to use **finite differences**, and it is for these that the delta symbol is reserved. The method gives us average velocities for periods of time. If we want to know the exact velocity for any one point in time, we must shorten the time interval until it becomes infinitely small (or infinitesimal). Such a velocity is termed the **instantaneous velocity** and is represented by the expression dx/dt (pronounced 'dee-x by dee-t'). This is the subject of calculus. For many practical purposes with observed data, the method of finite differences provides an adequate approximation, and is certainly easier to compute.

The fourth column in Table 27.1 shows the finite difference acceleration ($\Delta v/\Delta t$). This shows the rate of change of velocity over the indicated time periods. Acceleration has the units of m s^{-2} (metres per second per second, i.e. velocity per second). Again, the calculations are based on the previous time period for each distance mark. The person is stationary at the start, and initially there is acceleration. Later, there are periods with zero acceleration, or even deceleration (negative acceleration). A good exercise to reinforce these ideas would be to plot the velocity vs. time and acceleration vs. time graphs.

27.2 Lines of best fit

We often want to establish whether any relationship exists between two variables, and if so, the form of this relationship. This is particularly the case when we carry out experiments in which we alter values of one variable and measure the effect of this change on another variable. For the distance/time graph shown in Figure 27.1 you could plot a line of best fit, either graphically by hand or using the option in the spreadsheet (which will probably employ what is known as a least squares regression technique to fit the line). In either case we can, if we choose, make the line start at zero on both axes (the point 0,0 is termed the **origin**). In our example, this is appropriate as no distance is covered at time zero. (In other situations this need not be the case.) The equation for our line is:

$$y = 1.24x$$

This means that y, in this case the distance measured in metres, will be 1.24 times the elapsed time measured in seconds. If you plot this line on your graph you will see that at certain times our walker is ahead of the predicted line (above in this case) and at times below the line. This is because the line that is fitted through the points is a line of best fit, i.e. an 'average' line.

Whenever a line of best fit can be drawn, it has an angle with respect to the x-axis. Just as with an actual hillside, this is referred to as a **gradient** or **slope**. What does this slope tell us? In the example we are looking at, it is really the average velocity. The calculated value of the slope is 1.24, which is almost the same as the average velocity that we calculated above.

The general equation of a straight line is:

$$y = mx + c$$

(which is often written in statistics texts as $y = a + bx$). The variable m (or b) is the slope of the line, i.e. the difference or interval between two points on the line on the y-axis divided by the corresponding difference, or interval, between two points on the x-axis. Now, in our example the slope is obviously positive – more time leads to a greater distance travelled – but in many relationships between two variables the slope value may be negative, meaning that higher values of one variable result in lower values of a response variable.

The letter c (or a) is the **intercept**, defined as the point at which the line of best fit reaches the y-axis. In our example, c is zero because we forced the intercept to be at zero as we knew this to be true. In most cases, though, you will not be able to say that the intercept is zero and will have to draw or calculate (for example using a spreadsheet) the intercept of the line. In experiments, the intercept value is often a key value needed to solve the equation. Note that the intercept can also be a positive or negative number. You should also note that if you derive a relationship between two variables this will hold true only within the observed range; extrapolating lines of best fit beyond the observed range may lead to erroneous conclusions.

28 The fundamentals of energy

We are all familiar with the term **energy**. We know that energy can be derived from water bodies, solar radiation, fossil fuels or nuclear fission. We know it can be transported by media such as electricity, and put to use far from its source. However, when we consider how the term 'energy' is used in the sciences, we find a concept which is far from easy to grasp. In this section we consider the basic concepts which allow us to understand what is meant by 'energy', what the related terms 'power' and 'work' mean in a scientific context, and what the fundamental characteristics of energy are. It is also worth noting that while a great deal is known about the properties of energy, exactly what energy is remains one of the most fundamental mysteries of science.

In the dictionary, energy is defined as the ability of matter or radiation to do work. To understand what is meant by 'energy', we must first consider what is meant by work.

28.1 Work and energy

Work, in a scientific sense, can be defined as the product of a force and the distance moved in the direction of the force (Figure 28.1). It is instructive to consider the properties shared by a number of different mechanical systems. Think of the following examples as thought experiments. Let us define three systems: a ruler fixed at one end, a coil spring mounted on the ground, and an object resting on the ground (Figure 28.2). By bending the ruler, or compressing the spring, or lifting the object from the ground, we are applying a force over a distance. After application of this force, the systems have changed. Each is capable of returning most of the force over the same distance: the ruler can straighten, the spring decompress and the object falls back to ground level. In other words, each system is now capable of giving the work back. Looking back to our initial definition of energy, it is clear

A definition for mechanical work

Imagine that a pulley is used to lift a weight from the ground up to a platform.

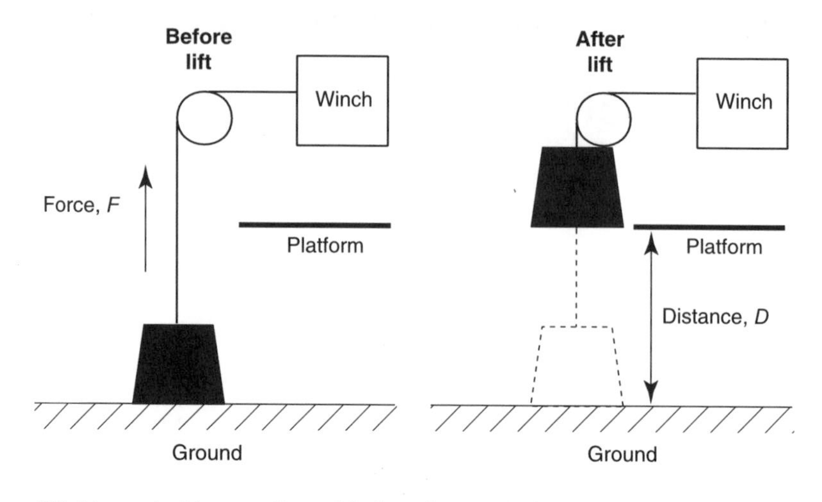

Effort is required to move the weight from the state before lifting, to that after. This effort is termed *work*, and can be quantified by multiplying the distance that the weight has moved, by the average force needed to move it. This can be summarized by the expression:

$$Work = Force \times Distance$$

Figure 28.1 A definition of mechanical work.

that in doing work upon our three systems, we have given each the potential to give the work back. Thus in doing work we are giving the system *capability* of doing mechanical work. It is this capability which is energy.

How much energy do we get back from our systems? In an ideal, frictionless world we would get it all back: the amount of work done is equal to the amount of energy stored by the system. In the real world, however, some of the energy would be lost from our simple system (as is discussed later, the *loss* of energy described, mostly as heat, is only lost in the sense that it cannot be used to do mechanical work – the *total* energy, if we allow for the lost heat, is conserved).

From the above it is clear that in doing work we are transferring energy to the system. Furthermore, energy is stored by the system, which is then made more capable of doing work itself. Yet energy is defined as the ability to do work. No surprise then that work and energy are measured in the same units.

Energy and work are clearly very similar, but what exactly is the difference? The formal definition of work, as given above, owes its origin to the development of mechanics and thermodynamics as the theory of engines. Engineers were interested in how energy, of whatever kind, was converted into mechanical work: in other words, how engines could apply forces over

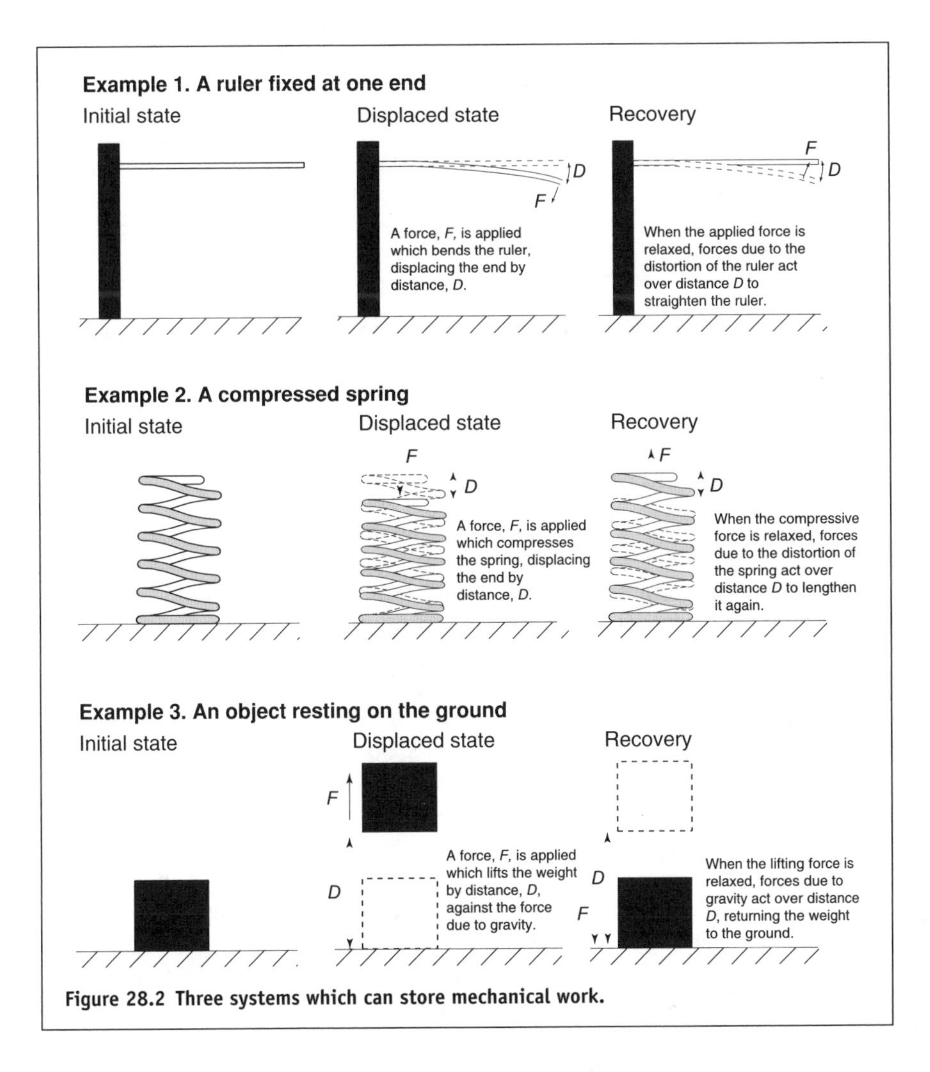

Figure 28.2 Three systems which can store mechanical work.

distances. However, there is no fundamental difference between mechanical work (where energy is used to move, lift or distort an object), and, for example, chemical work (where energy is used to change the configuration of atoms or molecules). In a general sense, work might be defined simply in terms of the transfer of energy between systems of any kind.

Before considering the different forms of energy, it is useful to consider the units by which energy is measured. The modern standard unit is the joule (J). As energy is defined in terms of mechanical work, it is clear that the joule must have the dimensions force times distance:

$$\text{Energy (J)} = \text{Force (N)} \times \text{Distance (m)}$$

In older literature, the calorie is frequently encountered as a unit of energy. This is the amount of energy needed to raise the temperature of one gram of water by one degree Celsius (1 calorie = 4.187 joules). The unit **kilocalorie** was also widely used, but was also called the **Calorie**. Confusion was widespread. In the study of atomic scale systems there is another widely used unit of energy: the electron volt (eV).

$$1\,\mathrm{eV} = 1.6 \times 10^{-19}\,\mathrm{J}$$

28.2 Forms of energy

We are all familiar with a number of different forms of energy. Here we run through the common ones, consider their basic properties and characteristics, and then see what they have in common.

28.2.1 Potential energy

If the energy of a system is due to the displacement or distortion of a part of that system, which is resisted by internal forces, then it is known as **potential energy**. There are numerous wide-ranging examples of potential energy. The best known of these is **gravitational potential energy**.

A force of attraction exists between separated masses; we call this force **gravity**. The properties of this force are very well known, even though its origin remains a mystery. The magnitude of the attracting force is directly proportional to the combined mass of the bodies, and decreases with the square of their separation. Because of this force, energy is required to separate the bodies, and the energy conferred upon the system by this separation is termed gravitational potential energy. For any pair of bodies, the gravitational potential energy is maximum when the separation is greatest, and minimum when they are in contact. Considering the Earth's surface, separation means simply distance above ground. Hence, lifting an object against the force due to gravity gives it potential energy. Strictly, an object lying on the floor still has a great deal of gravitational potential energy, because it lies at a great distance from the Earth's centre. However, for practical purposes it may be treated as having zero potential energy, if that is the lowest amount it can have in the system under consideration. This idea ties in closely with the geomorphological concept of a baseline. It is a common property of potential energy systems that absolute scales are inconvenient. In most, zero energy is defined arbitrarily at some level which can be conveniently benchmarked.

Returning to Earth, and to gravitational potential energy, how do objects acquire potential energy? There are numerous ways, and these can be broadly separated into forces due to the interior of the Earth, and forces due to solar energy (extraterrestrial objects such as meteors are too rare for consideration). Mountain formation and volcanoes provide gravitational potential energy to rock through slow steady uplift. On the other hand, solar radiation drives wind, waves and currents, which are then capable of lifting objects. However, the most important source of potential energy from a geomorphological perspective is evaporation of water, driven by solar energy. By mixing with the atmosphere, and recondensing at altitude, potential energy is given to water, which then drives most geomorphological processes.

While gravitational potential energy is the best known, potential energy can exist wherever forces can be maintained between parts of a system. We have considered the example of a spring. The same arguments can be applied to any elastic object, that is any object in which deformation leads to a proportionate opposing force. Large rock masses behave elastically: a fact which is particularly well demonstrated during earthquakes. Crustal movements, driven by forces from within the Earth, deform the elastic lithosphere (Figure 43.1). The potential energy due to the deformation is very great. When rupture occurs, the potential energy is converted to wave energy and is transmitted rapidly away from the point of origin.

28.2.2 Chemical energy

Gravity is not the only force that can act between objects. Electrical forces can also attract or repel, and thus support potential energy. Electrical forces are trivial at the scale over which gravity operates, leading to the impression that they are weak. However, the opposite is true; it is the very strength of the electrical force that compels large masses to be electrically neutral. At molecular or atomic scales, the separation of charges across atoms leads to very great forces. Indeed, if the electrical force between two electrons is compared with the gravitational forces between them, the former is greater by a factor of 4.2×10^{42}.

Where we have a force, we can also have potential energy. It is electrical potential energy (termed, in this case, chemical potential energy) that is the source of heat derived from chemical reactions. Different chemical forms contain different amounts of chemical potential energy. If two substances are brought into contact, they will not react with each other unless that reaction leads to a product with less chemical energy than the individual reactants.

If a reaction takes place as shown in Figure 28.3, then the difference between the chemical energy of the reactants (here H_2 and Cl_2), and that of the products (HCl) is released as heat (thermal energy). A reaction involving the release of heat is termed **exothermic**.

For humans, the most important exothermic chemical reactions are **respiration** and **combustion**: in both processes, molecules based on carbon react with molecular oxygen, liberating energy in the form of heat (see **34.3.3**). In respiration, the body carbohydrates (such as sugar) react with oxygen to liberate carbon dioxide, water and heat:

$$\text{Sugar} + O_2 \rightarrow CO_2 + H_2O + \text{Heat}$$

The essential part of this reaction, which applies also to combustion, is:

$$C + O_2 \rightarrow CO_2 + \text{Heat}$$

For human society, combustion is the most important source of energy. It must be remembered, however, that this energy ultimately derives from the Sun (so ultimately is nuclear energy). This is because organic matter and

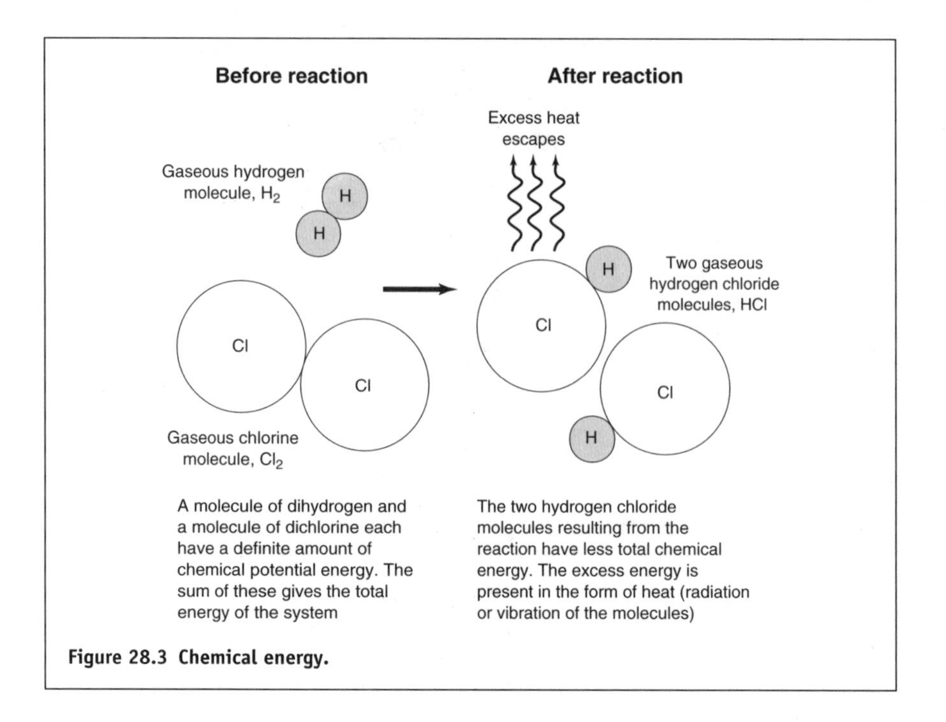

Figure 28.3 Chemical energy.

molecular oxygen are present only because of photosynthesis (see **34.3.1**) which converts the Sun's radiant energy to chemical energy. Photosynthesis is thus the most important example of an **endothermic** reaction, i.e. one in which energy is taken from the surroundings (in this case solar radiation). Following an endothermic reaction, the reaction products have more chemical energy than the reactants. This is a good illustration of work in a chemical sense. The solar radiation works on the chemical system, transferring energy to that system in the form of elevated chemical potential energy. It is this energy which provides the power for practically all life forms, and, via their dead remains, powers human society as fossil fuel.

28.2.3 Nuclear energy

If we move to a smaller scale than the atom, we start to encounter stronger forces still; these are the strong nuclear forces which hold together the component parts of the nucleus. These opposing forces within the nucleus lead to nuclear potential energy. It is well known that huge amounts of energy are stored this way. The total amount of nuclear potential energy in a gram of coal is 2.7×10^9 times the chemical energy. On the other hand, most of that nuclear potential energy is firmly held within the various subatomic particles and is not available to do work. We are well aware of the enormous amount of thermal energy released by the small fraction of nuclear energy freed during nuclear fission or fusion reactions.

As an aside here, we remind you of the world's best-known equation: $E = mc^2$ (E is energy in joules, m is mass in kg, and c is speed of light in m s^{-1}). But what

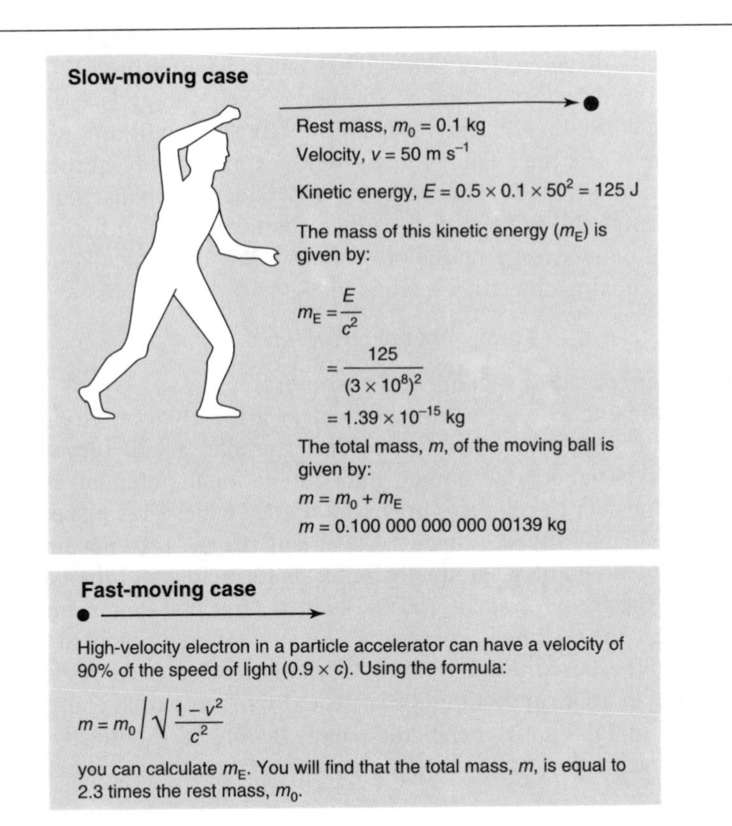

Figure 28.4 Energy and mass.

does it mean? The c^2 part is a constant, equal to the square of the speed of light. The equation is thus commonly regarded as stating that mass is proportional to energy. This is misleading, however. It would be better to say that grams are proportional to joules. In other words, we can use the joule as a unit of mass, and we can use the gram as a unit of energy. This is true at all scales, but it is only at the nuclear scale where the effects become obvious. If you throw a ball, its mass increases due to the kinetic energy you have given it, but the increase is negligible (Figure 28.4). However, electrons in particle accelerators can have their mass changed by very significant amounts. Further, the amount of thermal energy released by nuclear fusion is such that the resulting atoms have several per cent less mass than the reactants.

28.2.4 Kinetic energy

The other major class of energy is that due to the motion of a body. According to Newton's first law of motion (see **21.5**), bodies remain at rest until a force is applied. When a force is applied, the body starts to move, and accelerates while the force remains (Newton's second law). If we look at our definition of work given above, it is clear that work has been done; force has been applied over a distance. Thus, energy has been given to the ball. This energy has been stored in the motion of the body. Such energy is termed **kinetic energy** (the word 'kinetic' derives from the Greek for motion).

The amount of kinetic energy possessed by a moving object increases with the square of the velocity. Thus, if we have two identical balls, one of which is moving at twice the velocity of the other, then the faster of the two has four times the kinetic energy of the slower one. We can calculate the kinetic energy of a moving object if we know its velocity and its mass:

$$\text{Kinetic energy } (E) = 0.5 \times m \times v^2$$

where E is in joules, m is in kg, and v is in m s^{-1}.

The kinetic energy of a moving body gives it the ability to do work, i.e. to transfer its energy into other forms. For example, a ball thrown upwards converts its initial kinetic energy into gravitational potential energy. It is helpful to consider the changes in energy storage which take place during this event. Immediately after it is thrown, the kinetic energy is at a maximum. Thereafter, the potential energy gradually builds as the velocity of the ball declines. Eventually the throw is spent and the ball has reached its maximum height. Now the ball has no kinetic energy, but the gravitational potential energy is at a maximum (because the ball is now at maximum distance from the Earth). At that instant, in an ideal frictionless world, the ball's potential energy is exactly equal to its initial kinetic energy, the energy having been perfectly conserved. On falling back to its starting height, the energy transfers are reversed.

28.2.5 Vibration and waves

Exchange of energy between one storage form and another is commonly encountered in the study of natural systems. Different parts of a system are capable of doing work on other parts, leading to a transfer of energy. This is exemplified by vibrations and waves, two systems which are characterized by cyclical exchange of energy between them.

Consider again a ruler fixed at one end (it was one of our earlier examples, in **28.1**). If we bend and then release the ruler, an oscillating or vibrating state is established (Figure 28.5). Consider the energy storages during this vibration. Just before release, the bent ruler has maximum potential energy (**potential energy of distortion**). Following release, the potential energy falls and is converted into kinetic energy. When the ruler has reached its centre line, it has no potential energy and maximum kinetic energy. Beyond this point, the kinetic energy is gradually used to store potential energy in the system again, and after a certain time the ruler is fully bent in the opposite direction, and has maximum potential energy once again. If we look at a graph showing both the kinetic and potential energy of the system over time, we see a sinusoidal form, resembling waves that are out of phase with each other. A graph like this can be drawn for any vibrating system, and for other oscillating systems such as pendulums and waves. Indeed, the graph nicely illustrates the similarity between vibrations and waves (Figure 28.5).

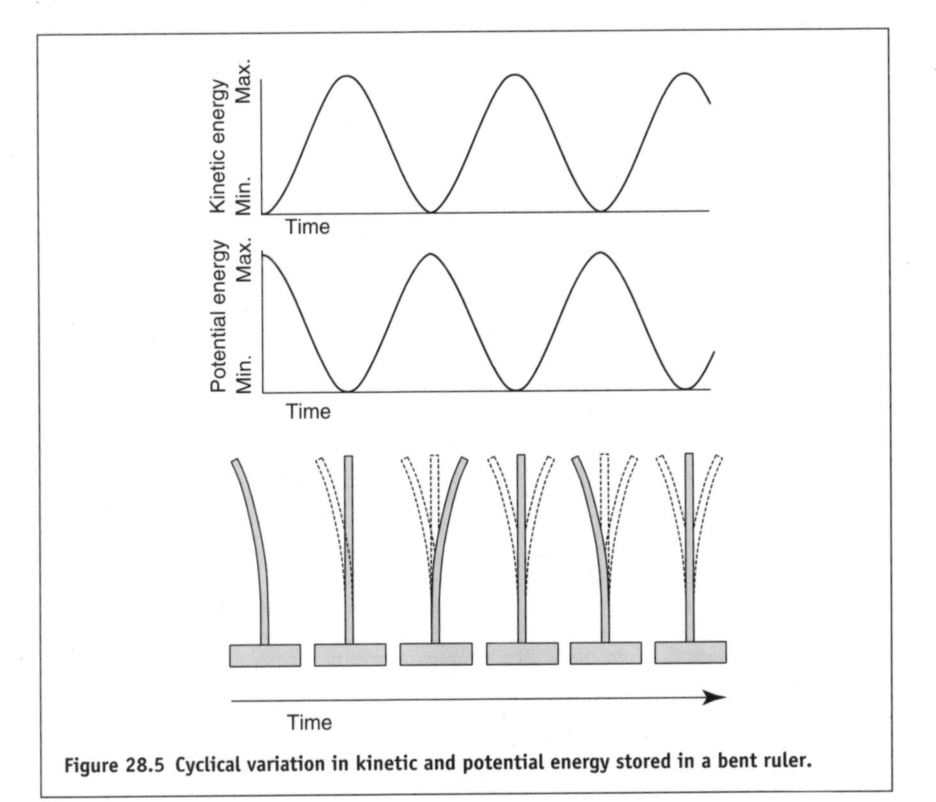

Figure 28.5 Cyclical variation in kinetic and potential energy stored in a bent ruler.

Waves also involve a cyclical transfer of energy between competing energy stores. In all common wave types except electromagnetic waves, the two stores are kinetic and potential energy. At fluid surfaces this is gravitational potential energy and vertical kinetic energy. In sound waves it is kinetic energy and potential energy of compression (of the gas, water, or whatever other material is carrying the wave). What distinguishes waves from vibration is that in vibration the energy storage is centred on a fixed point. For waves the energy moves at a uniform velocity in a specific direction, the velocity being a function of the physical properties of the medium carrying the wave.

The two common types of wave carried by solid media are described as transverse and longitudinal. In transverse waves the potential/kinetic energy exchange takes place in a direction perpendicular to the motion of the wave (Figure 28.6). In longitudinal waves the kinetic and potential energy are stored in motion that is parallel to the direction of the wave. In liquids, only longitudinal waves are possible. This is because liquids do not support shear stresses, and thus cannot store potential energy of shear deformation. Instead, they can support the temporary compressions needed to carry longitudinal waves.

Notice that all of these wave types need a medium to carry them. This led scientists during the nineteenth century to assume that all waves needed such a medium. Thus, when electromagnetic waves were discovered it led

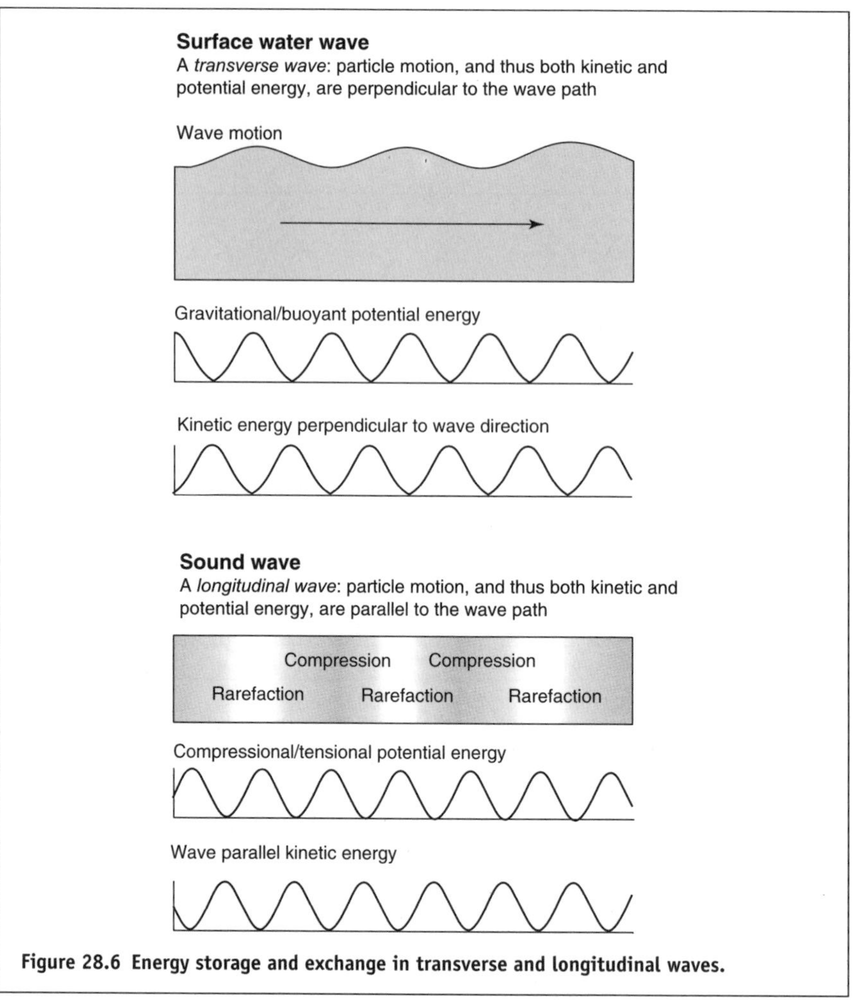

Figure 28.6 Energy storage and exchange in transverse and longitudinal waves.

to a false search for a supporting medium, which was termed the ether. Later, it was realized that, as described below, electromagnetic waves need no medium, but instead are self-supporting (see **23.2**).

There is no need here to go into detail about the many remarkable properties of waves. However, it is important to understand how energy is transformed into waves and how this energy is then dissipated.

Disturbance causes waves. For example, brittle failure of stressed rock causes vibration at the site of the failure. The potential energy that had been stored in the stressed rock is converted to vibrational energy. In our ruler example above, such vibration is isolated, and in the absence of friction would continue indefinitely. However, the vibrating rock is in direct contact with surrounding stationary rock. The vibrational energy is rapidly transmitted away from the point of origin, through to this rock, as waves carry the energy outwards in all directions. If we picture a pulse of vibration expanding away from a single point, it is easy to see that we have a spherical wavefront. As time goes by, each spherical wavefront expands. Now, the total amount of energy carried by the wavefront cannot increase; indeed, in the real world it would decrease because of energy lost to heat. Therefore, as the wavefront expands, the amount of energy at any one point on the front decreases. An examination of the geometry of this decrease shows that the energy at any point decreases with the square of the distance from the point of origin. This phenomenon is a characteristic of waves expanding through three-dimensional space, and it is described by the **inverse square law**.

28.2.6 Thermal energy (heat)

Heat is simply kinetic energy. If one body is hotter than another, it means that its component atoms or molecules are vibrating more energetically than in the other body: in other words they have greater kinetic energy.

In practice, heat differs from other forms of energy in three main ways. First, if we store energy in a coiled spring it stays put. However, if we store energy as heat in an object, then thermal conduction allows the energy to drain away. Second, there is a link between vibrations of atoms (and therefore of electrons) and electromagnetic radiation. Thus, heat can be transmitted from one body to another either by conduction or by radiation. Third, above the atomic scale, mechanical processes in the real world are subject to friction. It is friction that causes mechanical energy to be converted to heat. While much of this heat is turned back into mechanical energy, for example by convection in the atmosphere, some of it is lost via radiation and escapes the Earth. Such 'loss' is particularly obvious in the case of machines: hence the human view of heat generation as energy loss.

28.3 Heat and molecular motion: a simple demonstration

Let us look at heat in more detail, first considering the simplest of all systems, a cylinder of ideal gas. (An **ideal gas** is one in which the separate molecules are small and elastic; at normal pressures most gases behave like ideal gases.) If we add some heat to the container, the temperature and pressure of the gas increase. Furthermore, as we add more heat, the pressure and temperature rise in strict proportion. Given that our cylinder contains only gas molecules, and that the number of molecules is fixed, it is clear that the change in both temperature and pressure are due to something that has happened to the gas molecules. So what can we tell from correlation between pressure and temperature?

First, we know two things about the conditions:

- We can deduce that the pressure is due to momentum exchange as the gas molecules bounce off the cylinder walls.
- We know that the momentum of a moving object is proportional to its velocity.

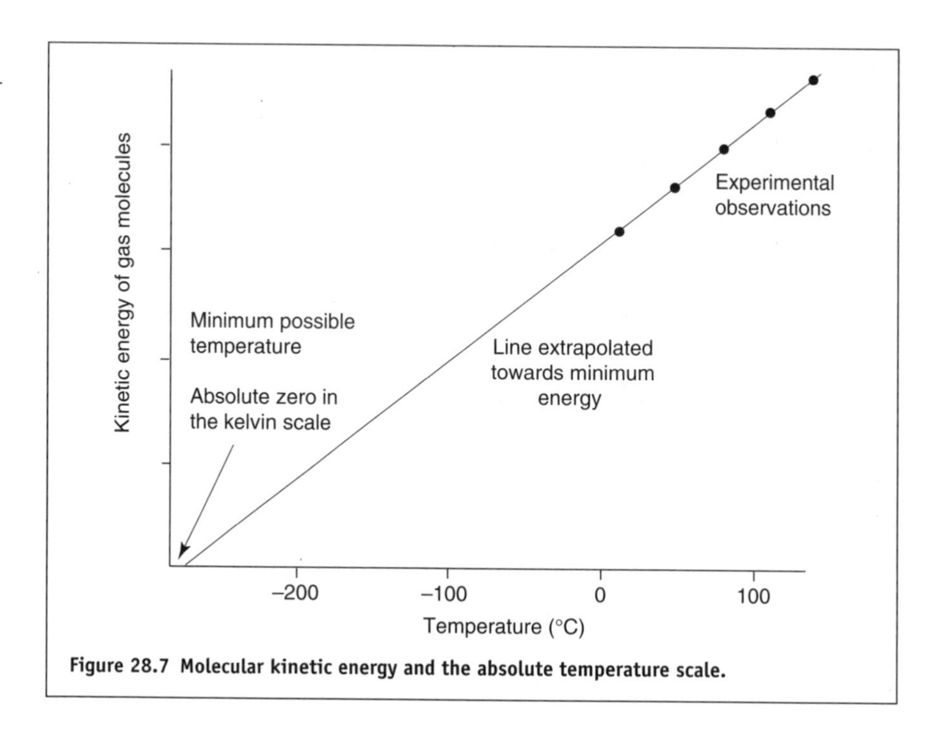

Figure 28.7 Molecular kinetic energy and the absolute temperature scale.

From this information we can tell that any increase in pressure must be due to the molecules moving faster: it is the only way in which their total momentum, and thus the pressure, can increase. We can conclude therefore that *if you heat a gas its molecules move faster*.

Can we find out how much faster they are moving? If we add enough heat to double the pressure in the cylinder, it is relatively easy to show that the speed of each molecule has increased by $\sqrt{2}$. This is because an increase by this much would increase the momentum change for each collision of a molecule with the cylinder wall by $\sqrt{2}$. It would also increase the rate of collisions by $\sqrt{2}$ because the accelerated gas molecules are crossing the cylinder in less time, thereby reducing the time between collisions.

Now, if the speed of each gas molecule has increased by $\sqrt{2}$, then its kinetic energy has doubled (because kinetic energy is proportional to the square of the velocity). Therefore, we can also conclude that *doubling the pressure in the gas corresponds with a doubling of the kinetic energy of the gas molecules*.

How does all of this relate to temperature? Let us plot the measured temperature of the gas (°C) against molecular kinetic energy in the gas. What we see is a straight line (Figure 28.7). This is more curious than it appears. There is nothing strange to us about the idea that a gas molecule is able to have zero kinetic energy, it would simply be at rest. But, a straight-line relationship with temperature implies not just that it would be at rest but that it would have a temperature of zero. This observation underlies the concept of an absolute temperature scale. Absolute temperature is measured in **kelvin**, the scale of which starts at zero. We can find the Celsius value for

0 kelvin (K) by extrapolating the straight line back to zero kinetic energy; it gives us a value of approximately −273 °C (Figure 28.7).

What we have shown in this experiment is that the temperature of a gas corresponds to the kinetic energy of the molecules. An analogous situation is found with liquids and solids, although there it is kinetic energy of vibration rather than motion in a straight line.

28.4 Heat and radiation

Most forms of energy can be stored in a permanent way. Heat is special in that it is hard to contain. Because heat is stored as molecular vibrations, it is easy for the kinetic energy to be transferred to neighbouring molecules, and thereby dispersed. This is known as thermal conduction. However, conduction is only one of the ways by which heat can be lost (see **20**). In a body at any particular temperature, the molecules are vibrating with a range of energies, up to a specific maximum energy defined by the temperature. These vibrations mean that electrical charges on the molecules are also oscillating. Oscillation results in repeated acceleration of those charges, which generates electromagnetic fields. Occasionally, those fields lead to the emission of a photon of electromagnetic radiation (see Figure 28.8). The energy of the emitted photon is equal to the excess vibrational energy of the molecule it came from. Thus, a whole range of different energy photons is emitted up to the energy specific to the temperature. This is the basis of black body radiation.

What happens to the molecule which gave rise to the photon? It vibrates less energetically, by an amount equal to the energy of the emitted photon; in other words it gets colder. Of course, it quickly warms up again because of conduction from the surrounding vibrating molecules, but everything has cooled off a bit. Thus, all bodies lose heat by radiation.

Conversely, matter can also gain heat by radiation. Whether heat is gained or lost depends on the number and energy (temperature) of the incoming and outgoing photons. In the end it is temperature difference that governs this. Bodies at the same temperature have no net flow of heat between them. This is in fact a definition of temperature.

At molecular or atomic scales, vibration stores thermal energy. There is no friction, but energy can be lost through two mechanisms: conduction and photon emission. In conduction, heat (and thus vibrational energy) is transferred to neighbouring molecules either by induction (forces arising from electric/magnetic fields), or in the case of metals by the advection (physical displacement) of electrons. (Advection of electrons transfers heat extremely rapidly, which explains why metals are such good conductors of heat.) Conduction causes heat to be gradually lost from molecules (assuming that the neighbouring molecules are colder). In the case of photon emission there is nothing gradual about it. At the instant that the photon is generated, the heat of the molecule has dropped by the amount carried by the photon.

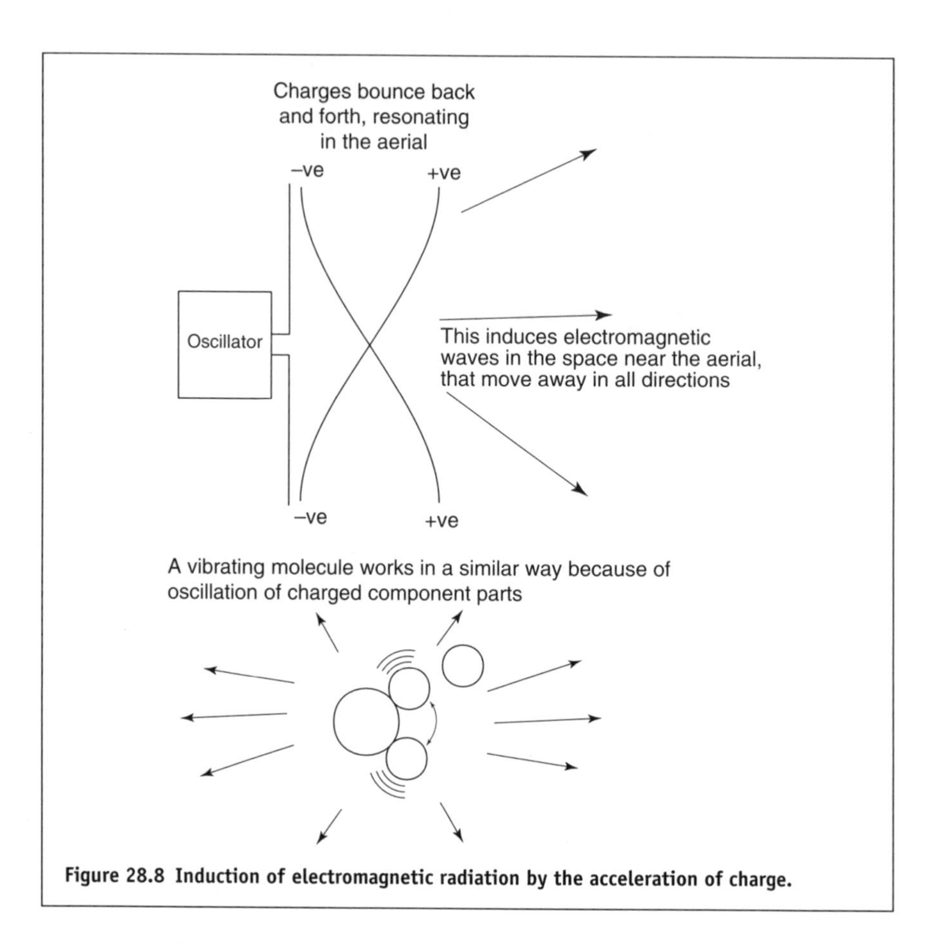

Figure 28.8 Induction of electromagnetic radiation by the acceleration of charge.

28.5 Energy conservation

In the middle of the nineteenth century, experimental results led to the idea that energy is conserved. If we account for energy exchange, both within the system, and with the surroundings, we find that the total is constant. In all the years since this principle was first stated, no exceptions have been found. Indeed, it turns out that conservation of energy lies at the heart of our understanding of the world at all scales from the subatomic to the astronomical.

Conservation of energy does not mean, however, that energy will not be lost from our particular system: transfers of energy between systems are a normal part of the world's operation. Nevertheless, if we could accurately measure the fluxes of energy in and out of our system, then we would find the total to balance.

28.6 Power

Power and energy are closely related concepts (see **21**). Whereas energy is the ability to do work, power is the rate at which that energy can be delivered. Two cars with similar fuel tanks but with different engines may have the same amount of usable energy (though this will vary with the **efficiency** of the engine), but the power may differ greatly: a car with higher power will be able to utilize that energy at a faster rate, and will have greater acceleration (given that the cars have the same mass).

29 A closer look at matter

29.1 Some fundamentals

An introduction to matter and changes in state of matter is found elsewhere (**18** and **19**). Here we consider the basic properties of matter in more detail. The topic is divided into three sections corresponding to the three main states of matter: gas, liquid and solid. Each section starts at the smallest scale, looking at molecular or atomic explanations for the observed properties. The sections then go on to consider large-scale properties.

Air, water, rock, organic tissue: a list of very different materials; however, they are all made of matter. In other words, they are made of something which occupies space and has mass. To better define matter we need to consider some of its properties. From an early period in the history of science, it became apparent that matter is conservative. (We now know that this is not entirely true: nuclear reactions, either the spontaneous nuclear decay of unstable isotopes or the forced nuclear interactions caused by artificial bombardment, cause formation or destruction of matter; however, this effect can be ignored in most situations.) In practice, being conservative means that if we isolate and contain some matter – say, a given amount of gas – then that amount does not change over time. For most practical purposes the gas's mass does not change either (strictly it increases slightly with both temperature and kinetic energy, but this is normally negligible – see **28.2.3**). Again, for the past two to three centuries it has been understood that the reason for this shared conservation across different kinds of matter is that all of it is made of similar particles, i.e. atoms.

Most of the matter which is of significance to geographers is dealt with in very large quantities. So why do we complicate things by considering atoms? The reason is simplification. Far from complicating things, atomic theory is one of the greatest of all human simplifications. Nature employs countless chemical substances, yet all of these are composed of atoms, and most are dominated by only a few kinds of atom. The great strength of the atomic theory is that so much complexity can be reduced to a relatively simple problem.

Another question arises if we are considering matter at the atomic scale. How far do we need to understand the implications of modern quantum mechanics? For geographers, the answer is very little. Some aspects of radioactivity and chemistry require an explanation which has its roots in quantum mechanics. However, most aspects of matter relevant to geographers operate at a scale where the pre-quantum, **classical mechanics** are adequate. In other words, atoms can be treated as minute solid spheres. This is fine when dealing with large quantities of matter; it is only when matter must be considered at the atomic scale that we have difficulties. Having said that much, it is worth taking a quick look at the small, weird end of things, if only to see why it does not matter to most of us.

29.2 The nature of matter

Instead of reviewing the theory, let us simply note some of the facts. Before doing so, we must start by understanding some of the key properties of waves, as this will make things simpler later on. To do this, we use the example of waves on water. Consider a series of experiments in which sets of linear waves are moving across an experimental tank. In this tank, we will assume that no waves are reflected back from the sides. The wavefronts are shown as lines in Figure 29.1.

What happens when the waves encounter a physical barrier, parallel to the wavefront, that has a gap in it (Figure 29.1a)? There are three outcomes. Waves that hit the barrier are simply reflected back. Waves that completely miss the barrier carry on uninterrupted. More interestingly, waves that pass close to the edge of the gap are **refracted**, that is they form circular wavefronts with the same wavelength as the original waves. If the gap is narrow enough, then we have only these circular waves beyond the barrier (Figure 29.1b).

Now consider what happens if we have two gaps. The two sets of circular wavefronts impinge upon each other, and this leads to **interference**. Where wavefronts cross we get **constructive interference**: the waves add to each other to form an especially high peak. Where the troughs (midway between wavefronts) coincide, we also get constructive interference, but this time leading to especially low troughs. Where wavefronts meet troughs we get **destructive interference**, and we have no wave at all. The net effect is a complex, dimpled **interference pattern** (Figure 29.1c). It turns out that interference patterns are a fundamental characteristic of all waves. It was Thomas Young's demonstration in 1801 of the interference pattern for light that showed that light comprised waves. He demonstrated that when light illuminated narrow slits, interference of the two beams of light passing through the slits caused a striped pattern to be seen (Figure 29.2). He concluded from this observation that light must be in the form of waves.

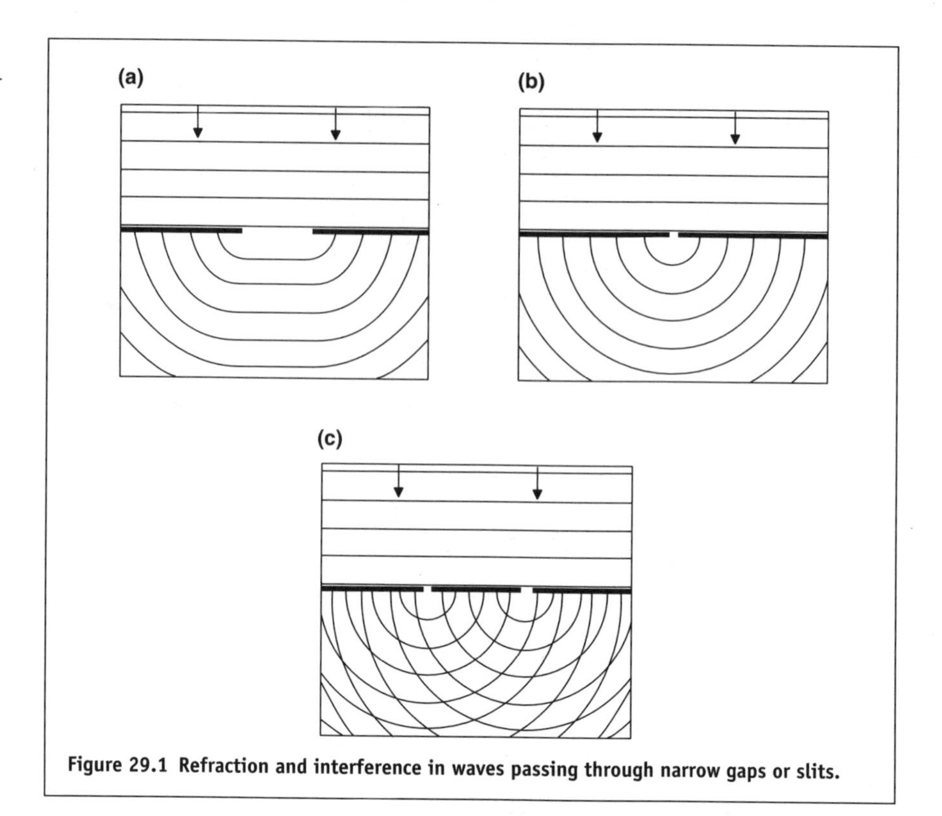

Figure 29.1 Refraction and interference in waves passing through narrow gaps or slits.

This view of the nature of light held sway for about a hundred years. However, in the first years of the twentieth century, experimental physicists found some peculiar results. They discovered that electrons, which had only recently been discovered, were ejected from some materials, for example zinc, when exposed to light. The odd thing was that the speed with which the electrons were ejected was independent of the amount of light (and thus energy applied), and was instead strictly governed by the wavelength of the light. The only possible explanation, first suggested in 1905 by Albert Einstein, was that light must comprise discrete packets of energy, called **quanta**, the magnitude of which depended on the wavelength.

So we have two experimental findings: one showing that light must be a wave, and the other showing that it can behave as an object. This led to the concept of duality (i.e. there are two ways to look at electromagnetic waves), and marks the start of quantum mechanics. It removed some of the certainty from physics. It meant that it was no longer possible to have a single model for electromagnetic radiation. Instead, it became necessary to treat it as particulate in some cases and as a wave in others. The truth is probably that light is neither a particle nor a wave. We can speak of a photon (the name for a quantum packet of light), and by this mean a precise amount of energy with a precise wavelength, but in a physical sense a photon is only a poorly defined fuzzy centre of energy moving at the speed of light.

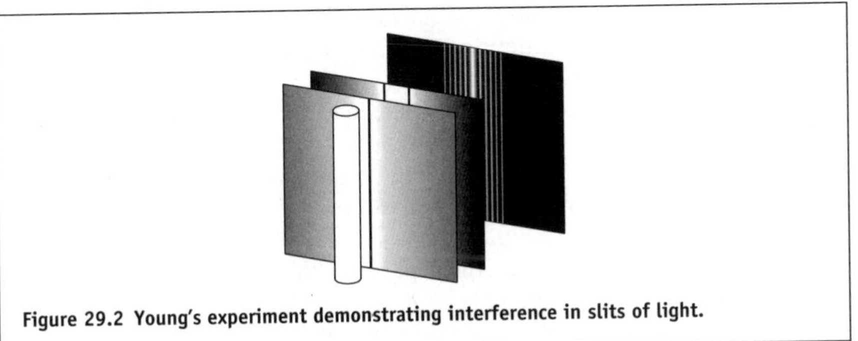

Figure 29.2 Young's experiment demonstrating interference in slits of light.

What about matter? If photons have a precise amount of energy, and thus mass, are they a form of matter? Before addressing that question, let us look at the properties of those entities which are taken to be matter. The electron is a good starting point. This is the smallest of the 'normal' subatomic particles, with very little mass compared with a proton or neutron (1/1840). In the 1920s it was discovered that electrons too show diffraction effects, and therefore must behave as waves. Thus, duality is a feature of matter as well as of electromagnetic waves. Indeed, the wave property of the electron is at the heart of quantum mechanical theory, because it explains why only a few energy levels are available to electrons in atoms.

One more experimental finding needs to be described. If electrons are sent through a pair of holes, analogous to the slit experiment described earlier for light, then we get, as stated above, a diffraction interference pattern whose shape can be accurately predicted by atomic wave theory. This means that the electrons must be passing through both holes, diffracting and interfering. However, we know that electrons are particles, even if they do have wave-like properties. What happens if we slow down the rate at which the electrons arrive at the holes? Common sense tells us that when the rate falls low enough, there can be no interference. Each electron will simply go through one hole or the other, and we will encounter no interference. However, this is not the case. *Experiments show interference even when only a single electron is involved*. In some way, that single electron passes through both holes simultaneously and interferes with itself. It is because of such results that in quantum mechanics we cannot treat electrons as minute solid spheres. Not only can we not know where the particle is, but we cannot even say that it is in only one place at one time. Now that is duality. These effects decrease with the mass of particles. They can be demonstrated for alpha particles (two protons and two neutrons), but rapidly become negligible for large atoms and molecules.

So, what we see by experiment is that electromagnetic radiation has the properties of both particles and waves, and so do atomic-scale particles. Does this mean that there is no difference between photons and particles? No: there are some important differences. A particle, such as an electron, can move at velocities from stationary up to just short of the speed of light. When the particle is stationary its mass is at a minimum, termed the **rest mass**. If we cause it to move, its mass increases with its kinetic energy according to Einstein's famous equation, $m = E/c^2$, where m is the enhanced mass in kilograms, E is the kinetic energy in joules, and c is the velocity of light in a vacuum, in metres per second.

In contrast, a photon can have only a single velocity (in a vacuum). To say it has no rest mass is misleading; it simply cannot exist at rest. The energy of the photon has a single fixed part, depending upon its wavelength. The energy of an electron has two parts: a fixed rest mass, and kinetic energy relating to its velocity.

Normally, by 'matter' we mean those particles that have rest mass. The mass of matter is clearly not constant, as it varies with velocity. It is also not conserved during nuclear disintegration, although the total mass, including that of any emitted photons, and the mass of the kinetic energy of ejected particles, is conserved.

Clearly, the concept of matter is not simple when addressed at the smallest scales. However, the good news is that from the molecular scale upwards we can neglect the more confusing aspects of quantum mechanics and we can visualize chemically the behaviour of atoms and molecules using the classical (non-quantum) model without significant loss of accuracy.

29.3 Gases

29.3.1 Gas laws and the kinetic theory of gases

Early experimental chemists showed that gases tended to have simple physical properties: for a given mass of gas, the values of temperature, pressure and volume were linked in a simple way. If, while holding the temperature constant, you compress a parcel of gas to half its volume V, then its pressure P doubles (approximately), i.e. $PV = k$, a constant (called Boyle's law). If, instead of compressing the gas, you heat it to double its absolute temperature T (kelvin) while maintaining a constant volume, its pressure also doubles (approximately), i.e. $P = kT$ (this is Charles' law). Linking these two laws eventually led to the universal gas law:

$$PV = RT$$

where R is the gas constant, which equals approximately $8.3\,\mathrm{J\,K^{-1}\,m^{-1}}$. It was recognized that this expression worked better for some gases than for others, and that it worked perfectly only for a hypothetical ideal gas. The gases for which it worked best were those with small molecules (for example H_2). By the mid-nineteenth century a number of physicists had developed these ideas into the kinetic theory of gases.

Consider a box filled with separate, freely moving, gas molecules (Figure 29.3). Imagine that each of these molecules is perfectly elastic (perfect elasticity means that kinetic energy is conserved during collisions – see **28.5**). The gas molecules bounce off the sides of the box, transferring momentum as they do so. This exerts a force on the wall. Two factors should influence the amount of force exerted. First, the greater velocity of the molecule, the greater the momentum transfer. As momentum is linearly dependent upon velocity (see

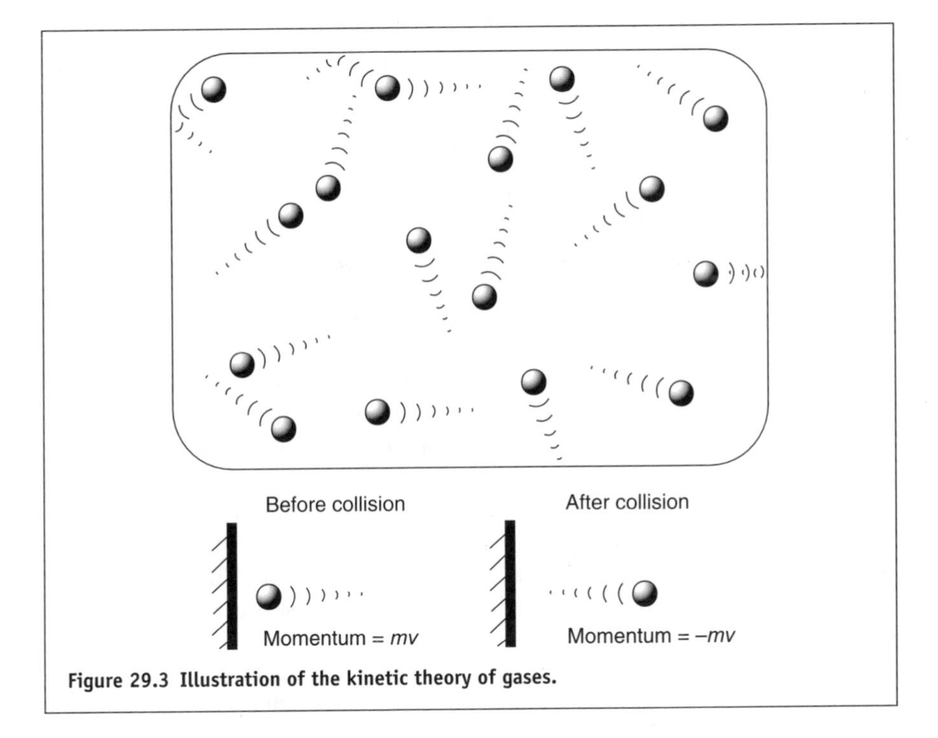

Figure 29.3 Illustration of the kinetic theory of gases.

21.5), so the force exerted on the wall is linearly related to velocity. Secondly, the more molecules there are, the more force is exerted. Again, there is a linear relationship between the force and the number of molecules. Thus, from these two associations:

$$\text{Force exerted} \propto \text{Velocity}$$

and

$$\text{Force exerted} \propto \text{Number of molecules}$$

The number of molecules in our experiment is determined by us; it depends only on how much we put in. But, what controls the velocity of the molecules? We know that velocity and kinetic energy are linked (see **28.2.4**), such that: kinetic energy $\propto v^2$. Further, from thermodynamics (see **28.3**) we find that the kinetic energy of a molecule is proportional to its temperature. Inverting the previous expression we find:

$$v \propto \sqrt{\text{Temperature}}$$

From this we can see that the force exerted on the wall of the box by any one gas molecule is proportional to the square root of its absolute temperature. So, doubling the temperature would increase the velocity by $\sqrt{2}$. At first sight, this appears to contradict the gas law, which states that doubling the temperature should double the force (pressure is simply force divided by the area over which that force is applied). However, this is because we have failed to consider the role of collision rate. We noticed that the force exerted increases with the number of molecules in the box; this is because with more molecules we get more impacts with the wall. But, the rate of impacts is also affected by the length of time it takes for any one molecule to bounce back from the far side of the box and undergo a further collision. Clearly, this depends on how fast the molecule is moving. If we double its temperature, the molecule's velocity has increased by $\sqrt{2}$. This means that the number of times the molecule collides with the wall has increased by $\sqrt{2}$. The combined effect of increasing by $\sqrt{2}$ both the number of collisions and the force exerted by each collision is to double ($\sqrt{2} \times \sqrt{2} = 2$) the total force exerted.

What have we achieved by this story? We have assumed three simple things: (1) gases are made of molecules; (2) the kinetic energy equation; (3) the velocity of a molecule is proportional to its absolute temperature. From these assumptions we have been able to deduce Charles' law $P = kT$.

What about Boyle's law? Simple: if we compress a gas to half its original volume, and then allow it to return to its original temperature, then we have not changed the velocity of (and hence force exerted by) each molecule, but we have halved the distance travelled by each molecule. This has the effect of doubling the collision rate. Halving the volume must therefore double the pressure; thus $PV = k$.

Therefore, from first principles, or as near as you can get to them in physics, we have deduced the gas laws. Needless to say, the model is a simplification. For example, not all of the molecules are moving at the same velocity. Energy, and thus velocity, is being continually passed around between the different molecules. The simplification works because the bulk properties of the gas depend on the average velocity of its molecules. Another simplification is the assumption that the collisions are elastic. This is not actually the case, but the model works because collisions are relatively infrequent. This is why the laws work better for gases at low density, where the likelihood of molecules colliding with each other is reduced. They also work better for molecules which do not readily interact with each other, i.e. the noble gases. In fact, deviation of gases from ideal gas behaviour is a useful way of studying general interactions between their molecules. This can be illustrated by considering the subject of vapour.

29.3.2 Vapour

We all know that there is water vapour in the atmosphere, but why do we not call it water gas? If the water molecules are separate, what makes it different from a gas? In a sense there is no difference. The water vapour is subject to the same diffusive and turbulent properties as a gas. To illustrate the contrast with gases, it is informative to go back to the gas law experiment above. If we take a volume of helium at room temperature, and reduce its temperature by 150 K, the pressure halves as the gas law predicts. If we do the same with water vapour, the pressure rapidly becomes undetectable, as the water first condenses to liquid, and then freezes. Instead of getting the change in gaseous behaviour we were looking for, we get a change in state.

Table 29.1 Critical temperatures for some common gases.

Gas	*Temperature (°C)*
Helium	−268
Hydrogen	−240
Oxygen	−118
Air	−140
Carbon dioxide	31
Sulphur dioxide	155
Water	374

What about other gases? It turns out that for any gas there is a critical temperature above which it cannot be liquefied whatever the pressure. The critical temperatures of some common gases are shown in Table 29.1. Above the critical temperature the gas will conform (approximately) to the gas laws regardless of the pressure. Below the critical temperature, there is always the possibility that the gas will condense. From this it is clear that a vapour is a gas which is potentially subject to a change in state (chiefly condensation) *under the conditions of the experiment.* Water vapour behaves as a gas at a high enough temperature, just as oxygen can be considered a vapour at −150 °C. From the point of view of the natural environment, water is the most important of the vapours. Subtle changes in atmospheric conditions have a profound impact on the state of the water it contains.

29.3.3 Motion of gases: diffusion

Gases are in constant flux. We know that molecules in a gas are moving at high velocities in all directions. If we tag an individual gas molecule, how fast does it actually move? The instantaneous velocity can be calculated easily for any given temperature (refer again to Figure 20.1). However, the molecule does not get far before hitting another one: the mean free path of an air molecule at normal temperatures and pressures is only about 1×10^{-7} m. Following collision, the mean speed is unchanged (the actual speed can be different depending on the exchange of momentum), but the direction will be different. The net effect is that the gas molecules wander, or diffuse, only slowly through the bulk of the gas.

An illustration of how slowly diffusion takes place is provided by the cold layer of air found in contact with the glass of an outside window in winter. Heat loss through the window leads to cooling of the adjacent air, and even though the air molecules are individually moving at hundreds of metres per second, this does not lead to enough mixing with the warm air of the room to disperse the cold air layer. What we can feel instead is a steady, cold down-draught close to the window, revealing mixing of the cold air with the warm air of the room. This is an example of **advection**, which is the bulk movement of a fluid material such as air. Wherever pressure gradients exist, advection is the most effective way of transporting gas molecules. For example, while gaseous diffusion is important in soils, the pressure change due to the passing of a weather front causes air–soil advective fluxes that briefly greatly exceed molecular diffusion.

Before moving on to advective movement of air, we need to consider the factors that influence the rate of molecular diffusion. Diffusion is due to the motions of individual molecules, and is therefore controlled by molecular velocity. From this, two important properties of gaseous diffusion can be deduced. First, the relative rate of diffusion for any gas type is inversely proportional to the square root of its molecular weight. Therefore, O_2 (molecular weight 32) diffuses slightly faster than CO_2 (molecular weight 44), but both are far slower than H_2 (molecular weight 2). Secondly, the rate of diffusion for any gas increases in proportion to its absolute temperature.

Brownian motion is a related phenomenon and leads to the dispersion of particles in still air. Particles suspended in a gas are subject to repeated collisions with gas molecules. If the particles are small enough (for example fine dust or smoke), each collision may cause a detectable movement of the particle. Because the collisions are caused by molecules coming from random directions, the particle experiences a jerky, erratic movement termed Brownian motion. Because the particles come from all directions, one might reasonably suppose that there would be no net movement. However, this is not the case: the motion is chaotic, and the chance of the particle returning to its starting point is negligible. Such motion, popularly known as a 'drunken walk', can easily be simulated in a computer spreadsheet, allowing you to get a better feel for this effect.

Simulating Brownian motion in a spreadsheet

In cell A1 of the spreadsheet insert the formula = 0.5 − rand( ). Copy and paste this to cell B1. This will place a number lying between −0.5 and 0.5 in the top cell of each column. Set cell A2 to = A1 + 0.5 − rand( ). Copy this cell, and paste it to cell B2. Finally, paste the formula to the block A3...B200. What we have done is to simulate the x and y coordinates of the cumulative random motion of a particle in two dimensions subject to randomly orientated forces. To view your findings, insert into the spreadsheet a scatter plot (showing markers joined by a line). Each time you press the function key F9 a new path is calculated. You can see that the particle moves erratically, but usually wanders away from the starting point. Now you know what Brownian motion looks like.

29.3.4 Motion of gases: advection and turbulence

Advection is the physical movement of a mass of substance from one place to another. In gases, such movement is caused by differences in pressure. Pressure gradients can be caused by large-scale phenomena, for example the differences in heat budgets between the equator and the poles, giving rise to the Earth's large-scale air circulation. Alternatively, they can be caused by smaller-scale phenomena, such as the interaction of topography with the larger circulation, or by local differential heating. At these smaller scales, a special form of advection becomes important. This is termed **convection**.

Convection involves vertical mixing of air masses, and can be further subdivided into two types according to the principal driving mechanism. **Free convection** is caused by the generation of low-density air beneath higher-density air. This is normally caused by solar heating near the ground. Once such low-density air is formed it becomes buoyant following Archimedes' principle, and it rises until its excess energy is dispersed. This contrasts with **forced convection**, which is caused by the passage of air masses over hills and mountains. In such cases, the air is forced to rise whether buoyant or not.

Advecting fluids (including both liquids and gases) may move in an efficient ordered way known as **laminar flow**, or they may move in an inefficient chaotic manner known as **turbulent flow**. Movement of liquid water exhibits both kinds of flow (see below), while all flow in the lower atmosphere is turbulent. What does this actually mean? In laminar flow, all molecules in the fluid are moving in the same direction, though they need not all move at the same velocity. Therefore, all the kinetic energy of the flow is held by the linear motion parallel to the flow direction. In turbulent flow, while the average movement is the same, any individual molecule is being spun around in **eddies**. At any one moment it might actually be moving in the opposite direction to the flow. Thus, kinetic energy in turbulent flow has two distinct forms. There is linear kinetic energy as before, due to the average motion of the molecule, and there is also turbulent kinetic energy stored in the eddies, and not directed with the flow. Much of the turbulent kinetic energy is dispersed as heat, making this a very inefficient way of advecting the fluid. These eddies are of variable size and duration. Their impact on any one gas molecule is chaotic and unpredictable.

The phenomenon of turbulence is extremely complex and remains one of the frontier areas of science. The real problem lies in predicting the path of a particle subject to turbulent motion. Fortunately, for many purposes it is not necessary to know anything about individual particles, but rather about the motion of the *average* particle, and this is less problematical.

29.4 Liquids

29.4.1 Atomic theory and liquids

Like gases, liquids are made of molecules. Both are described as fluids because external forces applied to them cause indefinite deformation. However, two fundamental properties distinguish these two states of matter from each other. First, whereas gas molecules can be compressed in accordance with the gas laws, liquids compress very little when subjected to similar pressures. Secondly, gases, when given more space, expand to occupy the space available; liquids, in contrast, do not. If a mass of liquid is not subjected to an external force it naturally forms a spherical mass. On the surface of the

Earth, however, because of the force due to gravity, the liquid instead spreads laterally until contained.

These properties tell us that liquids comprise molecules between which strong forces of attraction resist any tendency to drift apart. Unlike solids, however, liquids do not possess rigidity. That is to say, they cannot resist shear stresses (except at scales small enough to allow surface tension to be significant). This tells us that the strong intermolecular forces do not fix a molecule of the liquid to any other *particular* molecule. Instead, individual molecules continually exchange neighbours, wandering through the mass of the liquid. The pathway taken by any one molecule through the liquid can be pictured in just the same way as with Brownian motion (for a computer simulation, see **29.3.3**).

If thermal energy in gases is expressed as lateral motion of the constituent molecules, what about liquids? Once again, kinetic energy is the key. However, unlike gas molecules in which the motion is predominantly from place to place, in a liquid, the motion is mainly in the form of vibration. In fact, it is this vibration which prevents liquid from being rigid. The vibration is just sufficient to stop the molecules from forming stable lattices (and thereby becoming solid), but not so much that they break away from each other completely and become vapour. Again, it is important to remember that all this is true only for the *average* molecule. In fact, there is a very uneven distribution of energy, which is continually being passed around between the molecules. Consequently, there are some clusters of molecules which are briefly 'colder' than the others, and temporarily form crystal-like structures. Likewise, there are other molecules which become 'hotter' than the rest, and manage to escape the liquid to form vapour.

This latter phenomenon explains why evaporation from a liquid causes cooling. The only molecules capable of escaping from the liquid are those which have thermal energies (vibrations) that are far greater than the average. By escaping, these hot molecules leave the remaining liquid with slightly less average energy than before, and therefore slightly cooler. Conversely, the vapour molecules with the best chance of condensing on to a liquid surface are those with the least thermal energy, which explains why condensation leads to warming in the vapour above a liquid.

29.4.2 Compression of liquids and thermal expansion

Gases compress easily because confinement is resisted only by the combined momentum of the gas molecules. Liquids, in contrast, resist compression because the molecules are in close contact, and strongly resist closer contact. This effect is illustrated in Figure 29.4. Molecules which are close to each other experience strong mutually attractive forces due to electrostatic effects (van der Waals' force). At great distances these forces are attractive but negligible; hence gases behave ideally. However, as two molecules get to within a nanometre (10^{-9} m) or so, the positive forces become significant: they become attached to each other, and settle at a distance which leaves on average a zero net force between them. Any attempt to pull the molecules

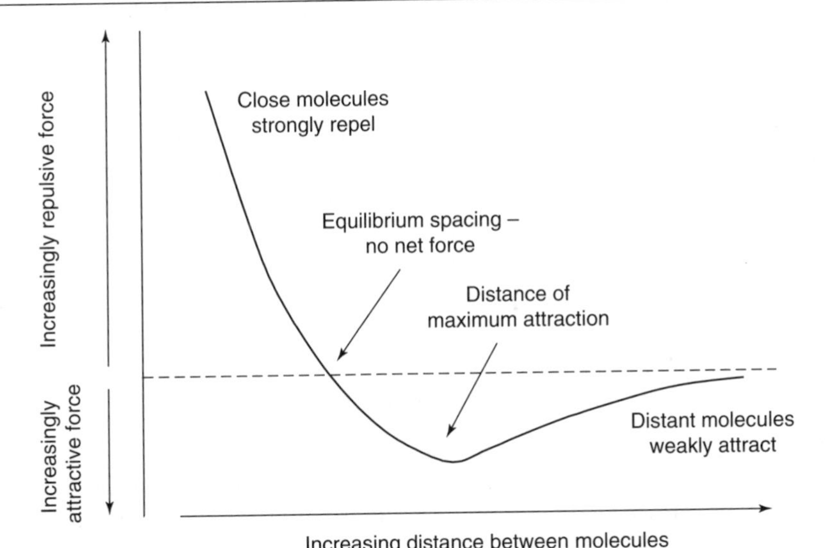

Figure 29.4 Forces exist between two adjacent atoms in a molecule, crystal, compound or liquid. When too close, this force is repulsive, while when further separated it is attractive. There is an optimum spacing at which the attractive and repulsive forces are in balance.

apart will be resisted by the electrostatic forces of attraction. On the other hand, if they are pushed towards each other, their electron shells start to impinge, and the repulsive effect of the locally intense fields is extremely strong. The molecules resist compression far more strongly than they resist tension.

While liquids resist compression far more than gases, they do nonetheless compress slightly in response to pressure increase. Strain of this kind is termed **bulk strain** (a reduction in volume) and is expressed as:

$$\text{Bulk strain} = \frac{\text{Change in volume}}{\text{Original volume}}$$

Note that as the expression is simply the ratio of two volumes, there are no units. The compression of a liquid is approximately elastic, which means that bulk strain is proportional to bulk stress (force per unit area, or pressure). The proportionality constant in this expression is termed the **bulk modulus**:

$$\text{Bulk modulus} = \frac{\text{Bulk stress}}{\text{Bulk strain}}$$

The units of bulk modulus are the same as of pressure, usually $\mathrm{N\,m^{-2}}$ – see **21.2**. The bulk modulus of water between 1 and 25 atmospheres is $2 \times 10^{-9}\,\mathrm{N\,m^{-2}}$. Thus, the water at the bottom of a 250 m deep lake (i.e. subject to 25 atmospheres of pressure) is compressed by approximately 0.5 per cent compared with water at atmospheric pressure.

If the molecules of a liquid are held at a temperature close to absolute zero, they vibrate very little and would maintain a nearly constant (subject to quantum constraints) ideal intermolecular spacing. However, if the temperature is raised, the molecules gain thermal (kinetic) energy in addition to their molecular potential energy. They vibrate about the ideal intermolecular distance. They do not, however, move an equal distance to and from the minimum energy position. It is easier for them to move apart than to come together (Figure 29.4). Consequently, they vibrate asymmetrically: they move further, and more slowly, on the separation part of the cycle than on the compression part. Consequently, their *average* spacing increases with increasing thermal energy. A greater average spacing for all of the molecules in a liquid, means that the liquid must expand as the thermal energy increases. This is the molecular explanation for thermal expansion.

29.4.3 Cohesion, adhesion and surface tension

The force between identical molecules, such as two molecules in a liquid, is termed the force of **cohesion**. As stated above, cohesive forces can be very great indeed. However, liquids are never in isolation, and forces of attraction or repulsion will exist between the liquid and the surrounding materials. The attractive forces are termed forces of **adhesion**. Such forces exert a powerful influence over a number of properties of liquids, and are most important at liquid surfaces. A familiar example is the **surface tension** effect of water, which is now explained.

As stated previously, the molecules of a liquid have strong forces of attraction between them. Within a mass of liquid, each molecule is surrounded by approximately eight other molecules, and electrostatic interactions exist with each of these. Within the body of the water, the forces between the molecules are minimal, unless an external action tries to interfere. The spacing between each molecule is maintained such as to minimize force, and hence, potential energy.

Now, imagine what happens at the surface of the liquid. The molecules right at the surface have interactions only with those below, as illustrated in Figure 29.5. Each surface molecule is in contact with fewer other molecules. For a molecule to get to the surface of the liquid, it must break contact with two or three other molecules. This means that the surface molecules have higher potential energies than molecules within the body of the liquid; in effect there are forces which pull molecules away from the liquid surface into the body of the liquid. This results in a *slight* tendency for liquid molecules to avoid the surface, causing a slight deficit at the surface. The slight deficit leads in turn to a slightly greater than ideal molecular spacing, which generates attractive forces. Thus, the whole surface of the liquid is in tension, giving rise to **surface tension**. This tension, and the reluctance of molecules in the liquid body to exchange with the surface of the liquid, leads the surface to behave as if covered by a weak skin.

There are a number of consequences of this tension. First, while liquids ideally have no rigidity, at small scales they can oppose stresses. A small insect can sit on the surface of the water because the surface tension effect is strong enough to resist the small forces due to the weight of the insect.

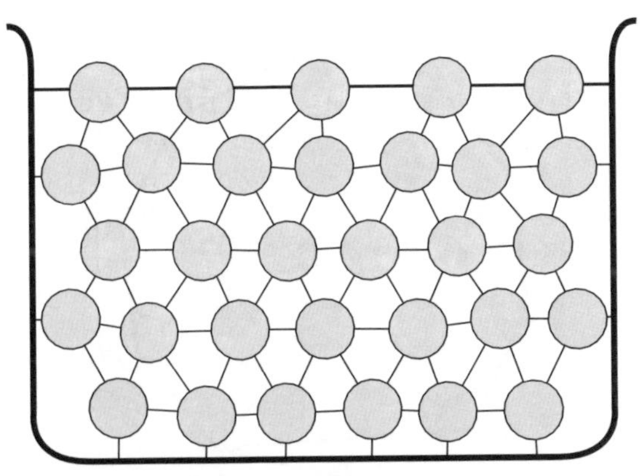

Figure 29.5 Illustration of surface tension. Imagine water molecules in a container. Within the body of the water, each molecule has forces of attraction in all directions with other water molecules. At the surface of the water, each water molecule is attracted only to others at its side or below. This leads to a net force away from the water surface, and thus to a slight deficit in water molecules in the surface layer. The consequent increased separation leads to increased forces of attraction between surface molecules (see Figure 29.4), and hence to surface tension.

Thus, from the point of view of an insect, water exhibits rigid properties, and indeed behaves elastically up to the yield point of the water surface.

Another consequence of surface tension is the enhanced pressure which is encountered within the body of a liquid. All masses of liquid are bounded by surfaces, and these are always in tension. Because the surface goes right around the liquid, and is a closed surface, the forces act on average towards the centre of the liquid. Thus, any body of liquid is at a higher pressure than its surroundings. For large bodies the effect is negligible. However, for small masses of liquid the effect can be highly significant. A typical raindrop, approximately 0.3 mm across, is held to a very close approximation of a sphere because of the surface tension (the potential energy due to the surface tension is minimized for a sphere). The pressure inside that droplet is 1.006 times higher than the air around it. This has very significant practical effects, as it allows the water of the droplet to evaporate, because the higher total pressure leads to a higher water vapour pressure within the droplet.

The pressure increase is proportional to the curvature of the surface, which in turn increases with reduced radius (Figure 29.6). Thus, the effect is greater for cloud droplets than for rain droplets: in a typical cloud droplet (10 μm radius) the pressure factor is 1.21. In cloud droplets this effect is sufficiently great that water vapour moves from small to large droplets, which contributes to the generation of precipitation from clouds.

The surface tension effect described above is due to **cohesion** within the liquid. Typically, however, liquids are contained by other materials, so we must consider the effect of **adhesion** of water to surrounding materials. Consider what happens when our raindrop lands on a surface (assume that it remains intact). The shape that it adopts depends upon the nature of the

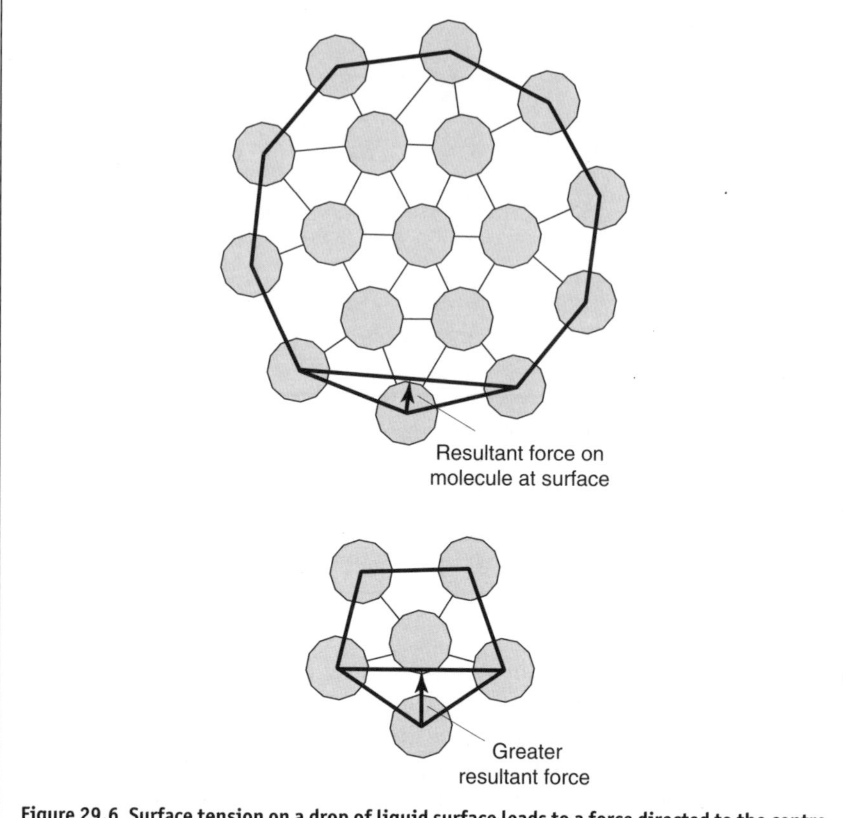

Figure 29.6 Surface tension on a drop of liquid surface leads to a force directed to the centre of the drop. This increases the pressure in the drop compared with the external atmospheric pressure. This effect is greater for more tightly curved (and thus smaller) drops.

surface it lands on. On waxed glass, the water would form a bead (Figure 29.7). This would not be completely spherical as it would be distorted by the effects of gravity, but the droplet would remain pretty much as before. On dirty glass, the effect would be quite different. The drop would spread across the glass, forming a dome. The difference between these two cases is the level of adhesion. Adhesion of water to wax is much weaker than cohesion of water to itself. The drop is thus affected by the surface only to the extent that the surface is a physical barrier. However, adhesion of water to glass is greater than water cohesion, so the drop attaches to the glass as well as to other water molecules.

29.4.4 Capillary effects

A further important effect of cohesion and adhesion is the **capillary effect**. This is of fundamental importance in the movement of water in soils and plants. Liquids confined within narrow tubes experience forces due to adhesion. Water in a clean glass tube clings to the glass, and will climb up

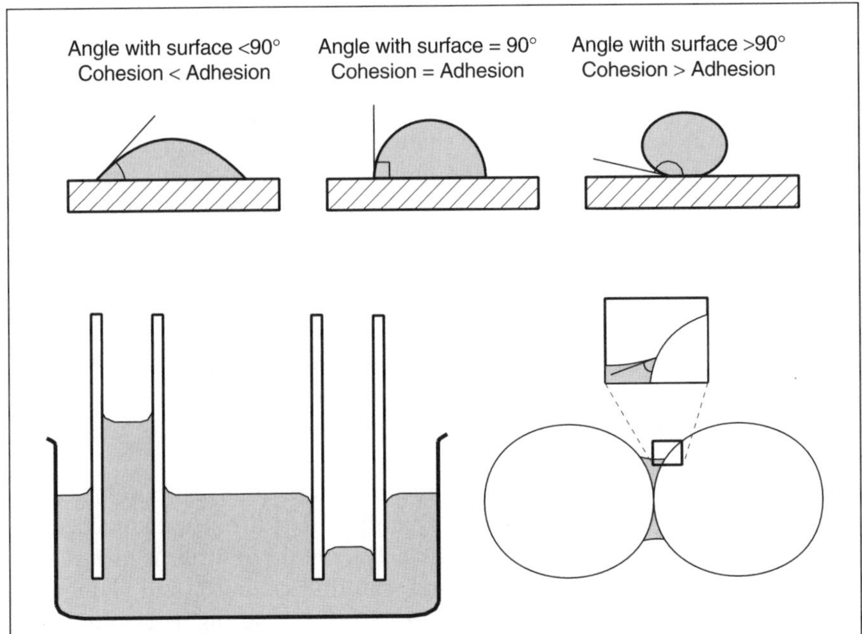

Figure 29.7 The capillary effect is due to the relative strength of the adhesion of the water to a surface compared with internal cohesive forces within the water. The 'neck' between two sand grains behaves as a capillary.

the tube. The angle of contact is the same as in the example above. Thus, water in a wax capillary will actually fall (Figure 29.7).

The shape of the liquid surface in the capillary, together with our knowledge that the liquid surface is in tension, tells us the direction of the forces involved. Thus, in Figure 29.7, we can tell that the forces must be upwards in the case of pure glass. The surface is concave upwards: thus the net effect of the tensional forces in the surface must be directed upwards. Once again, the magnitude of the force depends upon the curvature, which in turn depends upon the diameter of the tube.

This effect can be illustrated by the capillary 'neck' which exists at the contact between two sand grains (Figure 29.7). Although such a contact is not a tube, the behaviour and physical principles are identical. As the water drop decreases in size (evaporation or physical removal), the water surface contracts, the curvature increases, and the under-pressure, or tension, in the water droplet increases. This capillary effect in soils, at water contents well below saturation, is what leads to matrix suction (water pressure far below atmospheric), and explains why matrix suction increases as a soil dries.

29.4.5 Viscosity, inertia and intermolecular forces

If a force is applied to a solid, it deforms (becomes strained), and as the molecules are attached to each other, this strain leads to opposing forces, which resist the applied force. Liquids, on the other hand, are not rigid (neglecting the surface tension effect). When a force is applied to a liquid it also deforms, or strains, in response to the force. However, its molecules are not attached: they simply move around each other. Indeed, a defining feature of fluids is that they change shape indefinitely in response to even a small applied force. Does this mean that liquids do not offer resistance to an applied force? Experience tells us otherwise. A person belly-flopping into a swimming pool certainly experiences forces of resistance. But where do these forces come from if water is not rigid?

There are two mechanisms which permit liquids to oppose applied forces. One is simple inertia, and the other is due to the work which must be done to move liquid molecules past each other. **Viscosity**, by which we measure the ability of liquids to resist forces, is dominated by molecular effects when the movements under consideration are slow (and therefore laminar as opposed to turbulent).

Inertial effects are particularly significant where changes in velocity are occurring. The belly-flopper discovers painfully that it takes a considerable force to get a large mass of water moving. The inertia of water is also significant in free-fall events (rain-splash and waterfall erosion), and changes in direction of flowing water (a bend in a river). It is also highly significant in turbulence, where eddies created by the turbulence take up considerable amounts of inertia and kinetic energy. However, in many situations considered in geography, such changes can be considered negligible (for example, any case involving laminar flow), and forces of resistance are due largely to molecular effects.

Viscosity is controlled by the detailed structure of the liquid at the molecular scale. If a liquid comprises identical, perfectly spherical molecules, then viscosity is minimal, and resistance to external force is dominated by inertial effects. However, in practice, molecules tend to have irregular shapes, which means that the strength of attachment to surrounding molecules varies. In addition, different molecules in the liquid have different, and fluctuating, thermal energies. Thus, temporary clustering of molecules into crystal-like objects takes place. These effects mean that it takes work to physically move molecules around each other, and that allows the liquid to oppose applied forces. Clearly, these effects decrease with temperature, as the thermal energy of each molecule weakens its link to the rest. Thus, viscosity decreases with temperature.

An object falling through water, driven by the force due to gravity, is resisted by the effect of viscosity. There is no net inertial effect as long as the flow around the falling body is laminar. In other words, the body is giving kinetic energy to water as it is parted by the bow-wave, but that energy is (mostly) given back as the water closes in behind the object. If, however, turbulent flow is initiated, then the inertia given to the water is not given back, but is dispersed through the water as eddies, and the object will fall more slowly. In practice, the main factor causing turbulent flow around a falling sphere is velocity and, other things being equal, velocity increases with particle size. Therefore, for any given temperature and particle density, there is a critical

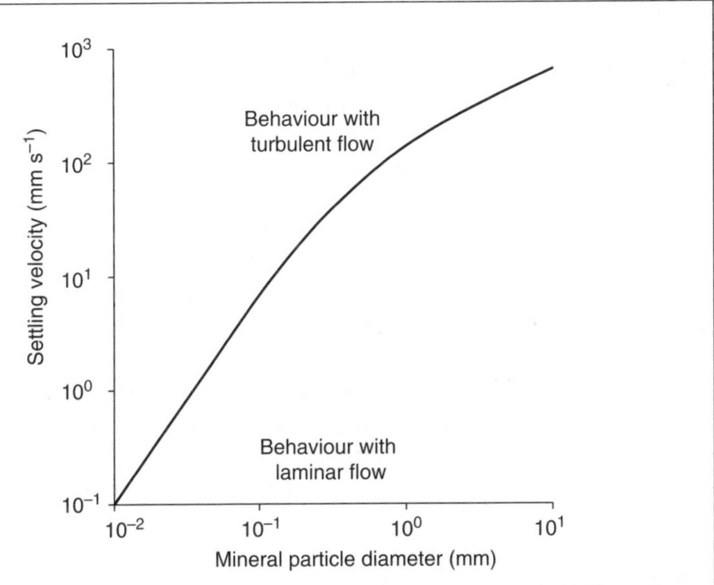

Figure 29.8 The simple linear relationship between particle size and settling velocity in a liquid breaks down when particles are so large that they lead to turbulent flow.

particle size which leads to turbulent flow. (This explains why the settling velocity method for particle size analysis does not work for particles much greater than 0.1 mm in diameter – Figure 29.8.) A net effect of this is that bodies which cause turbulence experience far greater drag than those that do not. A boat goes faster if the hull can be designed to maximize laminar flow.

29.4.6 Rate of strain effects

The work done in moving molecules around each other is not significantly affected by the rate of movement. Thus, the viscosity of a liquid due to molecular effects is not influenced by the rate of movement. This means that the force experienced (due to molecular effects) by an object moving through a liquid is proportional to its velocity. On the other hand, the inertial effects increase much more rapidly than the velocity, as the eddy kinetic energy is proportional to the square of the velocities. Thus, the force experienced by a sphere which is being forced through a liquid at high velocity is far greater than predicted by the molecular viscosity. At high enough rates of strain, liquids behave still differently, and elastic shock fronts (sound waves) are very readily transmitted by liquids.

29.4.7 Laminar flow, turbulence and friction

The fact that forces associated with moving liquids are critically dependent on whether flow is laminar or turbulent means that we need to know what governs the switch between the two states. Whether we are interested in hydraulic conductivity in a river channel or aquifer, or erosion of sediments on a river bed, we need to know what controls the onset of turbulence.

In the 1880s the physicist Osborne Reynolds carried out a series of famous experiments which involved passing coloured liquids through glass tubes. He showed that, below a critical velocity, friction in the tube increased linearly with flow velocity. Above that velocity, however, he showed that resistance increased more rapidly, and with approximately the square of the velocity. Using another experimental set-up, he showed that this change in behaviour was due to the onset of turbulence at higher velocities. By studying different fluids and different tube sizes, he developed a universal dimensionless number, the **Reynolds number**, which can be used to predict the onset of turbulence in tubes.

It is popular to think of the Reynolds number as measuring the relative importance of viscous and inertial forces. Viscous forces resist break-up of the fluid, and thus oppose turbulence. On the other hand, at higher velocities, and in large systems, the liquid is weaker in relation to its viscosity.

29.5 Solids

29.5.1 Atomic theory

Solids, like liquids, comprise molecules between which strong forces of attraction resist any tendency for them to drift apart. Unlike liquids, however, solids are rigid: they have the ability to resist both compressional and shear stresses. This tells us that the molecules are held together in fixed structures which do not allow the molecules to exchange positions. Once a solid is formed, each molecule retains its neighbours.

Just as with liquids, thermal energy in solids is mainly in the form of vibration. However, in solids the level of vibration is insufficient to allow the molecules to break free of their lattices. As with liquids, there is an uneven distribution of energy between the molecules. Molecules within the solid body do not escape. However, those at the surface are capable of escaping to form vapour if chance fluctuations give them enough thermal energy. This process of 'evaporation' from solids is termed **sublimation**. In principle, all solids are subject to sublimation to a degree that increases with temperature. However, for most common solids at room temperature the vapour pressures are vanishingly small (that is why they are solid), and sublimation is in effect zero. Any sublimation that does occur has the same effect as evaporation in cooling the object.

29.5.2 Types of solid: a molecular view

The properties of pure solid materials owe a great deal to their molecular structure. **Ionic solids** comprise rigid lattices of positively and negatively charged ions (see **37.6**), held together by electrical attraction. The structure is extremely rigid, but the attractive forces weaken rapidly on separation. If an applied stress is sufficient to divide the lattice in two, then the lattice is broken. In other words, ionic solids may be either weak or strong (depending on the strength of the attractive forces) but they are always brittle.

Metallic solids comprise rigid lattices of positively charged ions which are held together by clouds of electrons which are shared between the ions. Because the negatively charged component is not rigidly attached to the lattice, the attractive forces do not fall as rapidly with separation as is the case for ionic lattices. This reduces the rigidity of the lattice, but also makes breakage more difficult. Tensile stress leads to internal reordering of the lattice rather than breakage (within limits). Consequently, metallic solids have high tensile strength and flexibility.

Covalent solids are enormously variable. Many members of the vast family of silicate minerals (see **38**) are predominantly covalent, as are most organic substances. What they all have in common is that bonds are specific between the atoms, rather than the general electrical attraction in the ionic lattice. The silicate minerals tend to form hard, brittle solids. The organic solids, on the other hand, vary with molecule type. Long-chain materials can be very strong and flexible, while others can be hard and brittle.

29.5.3 Effect of stress on solids

Compression and thermal expansion effects are in principle the same as for liquids (above, **29.4.2**), and have the same molecular explanation. However, the rigid character of solids allows these phenomena to operate differently. A heated solid can lengthen, while a liquid can only increase its volume.

Under compression, the behaviour of a solid depends upon whether it is confined or not. If confined, then it behaves very much like a liquid, and only the bulk modulus (see **29.4.2**) is relevant. If it is unconfined then the effect depends on shape and internal structures. A column of solid loaded from above will initially shorten and become fatter. But what happens next depends upon how **ductile** the material is. If a material is ductile then it will permanently deform in response to strain. If it is **brittle** it will break. A column under load will experience shear stresses at *circa* 30° to the direction of loading. As the stress increases beyond a critical limit, a brittle substance will fail, breaking along the line of the shear stress (Figure 29.9). On the other hand, a ductile material, which will fail more quickly, will buckle.

Natural materials are more complex. Perfect columns rarely exist in nature; though a 'calving' glacier can exhibit 30° shear surface failure at its front. A more common circumstance is a sloping surface. Whether the slope is stable or not depends upon the angle of the slope and the strength of the material. Engineering geologists have studied this problem in detail and have developed equations to estimate the shear strength of materials. The strength of Earth-surface materials varies greatly with water content. If saturated with water, the normal stress is reduced, which weakens the material. If failure occurs, it takes place along fractures that angle towards the slope. Blocks may rotate rigidly, coherently (Figure 29.9), or may break up entirely.

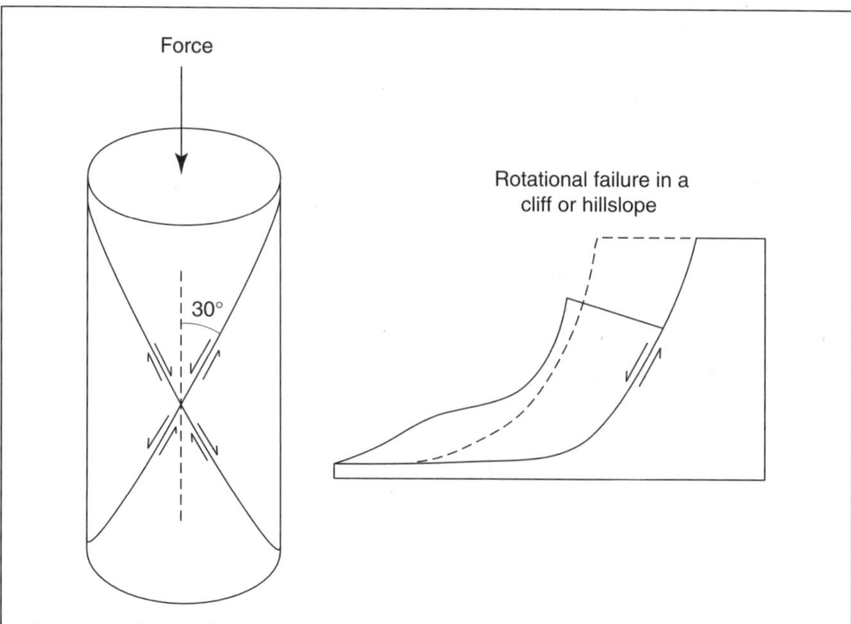

Figure 29.9 Shear failure in a column of homogeneous solid material leads to a characteristic angle of *c.* 30°. On steep slopes, failure commonly takes place along fractures angled towards the slope. The block which breaks away can rotate rigidly, or deform but remain coherent during movement as shown here.

Whereas the properties of pure materials are well known, most natural materials are mixtures, and hence more complicated. They are described as complex solids. The properties of a solid mixture are a complex function of the individual components, and the boundaries between those components. Secondly, mixtures are usually highly heterogeneous (for example, layers of different rock or soil types), and contain planes of weakness (faults, joints and bedding planes). This makes it very difficult to make general predictions about the physical properties of Earth materials. They are also subject to internal disruptive forces which modify the properties, for example gradual strain linked to plate movements, volcanic activity, burial compaction in rocks, root activity and burrowing.

An example of the complicating effect of mixtures is liquefaction during earthquakes. An otherwise rigid, saturated sand can form a non-Newtonian liquid when high shear is applied. In such a material, the viscosity varies strongly with the level of shear. Materials of this kind are termed **thixotropic**.

A large-scale effect of the complexity of rock materials is the dependence of their properties on strain rates. The Earth's mantle is clearly a rigid elastic solid, as it can carry transverse shock waves (that is how we can measure earthquakes which happen thousands of kilometres away). On the other hand, it is clearly a liquid because it flows owing to convection (the measured heat fluxes in the crust are too high to be due to conduction alone). The belief concerning the liquid nature of the mantle is also supported by the shape of the Earth: given the forces due to gravity and to rotation, the Earth's flattened ellipsoid shape is exactly that expected for a liquid. This paradox is explained by strain rates. At high strain rates (shock waves for example) it is a solid, comprising solid crystalline materials, but at slow strain rates the convection cells move at a few millimetres per year. Restructuring within crystals and creep along crystal boundaries are sufficient to allow the whole mass to behave as a highly viscous fluid. There are other materials familiar to us all that exhibit similar properties. Glass is demonstrably solid we might think, yet it is strictly a highly viscous liquid; ancient glass windows are measurably thicker at the bottom than at the top. Silly putty can be bounced or even shattered if hit with a hammer, yet if left on a flat surface for a few days will flatten and spread laterally.

Another example of this kind of behaviour, but with a very different cause, is the movement of soil or colluvium on hillslopes. Physical examination will show that these are solids (except in rare mass movement events), yet, over periods of decades, they creep slowly down-slope. This is due not to slow mineral reorganization, but to biological activity. Burrowing animals – rabbits, gophers, worms, ants, etc. – bring deep soil to the surface. On flat ground this leads simply to mixing, but on a slope, gravity ensures that on average the displaced material moves down-slope. Root activity pushes soil aside. Once again, the slope means that downhill movements are easier than uphill movements; thus, on average, root activity on slopes leads to movement down-slope.

29.5.4 Surfaces of particles

We learnt earlier how the strange properties of liquid molecules at the liquid surface lead to surface tension. But what about solids? Tension is clearly not an issue, and molecules cannot migrate to the mineral surface: but what if we break a mineral in half? What will we find at the exposed surface? If we picture a framework silicate like quartz (see **38.5**), where a lattice of oxygen and silicon atoms is interlinked by covalent bonds, it is clear that breaking the crystal means breaking the bonds. The exposed surface will have much higher potential energy than the rest of the mineral, and will have a net electrical charge. This has very important effects on the chemical properties of solid particles in water (see **13**).

30 Waves

30.1 Introduction

Waves are found in all parts of the natural world. You have seen waves at the sea's surface, and may have seen evidence of waves in the atmosphere (wave-like cloud forms). Earthquakes cause shock waves to pass through the Earth's interior. Light, other forms of electromagnetic radiation (for example X-rays and radio waves), and sound all travel in waves. While these wave types are very different from each other in some respects, they share a number of common characteristics. They carry energy, transporting this energy from place to place; and they do this by cycling energy between pairs of internal energy storages (see **28.2.5**).

Waves are so important in the natural environment that we feel it is useful to describe some of their key features. We will look at some different wave types, see how waves can be described mathematically, and demonstrate how you can apply mathematics to waves. We will also look at two additional facets of waves: the Doppler effect, and the relationship between wave energy and loudness in sound.

30.2 Describing waves

Waves can be categorized in a number of ways. The two main types are termed **mechanical** and **electromagnetic**. Mechanical waves, such as sound, ocean waves and seismic waves, need a medium to travel through. The medium allows temporary storage of the wave energy, permitting the wave to move (**propagate**) through the medium (see **28.2.5**). Electromagnetic waves, in contrast, need no medium. They store energy within self-sustaining electrical and magnetic fields, and can thus travel through a vacuum. (We deal more fully with electromagnetic radiation in **22.1**.)

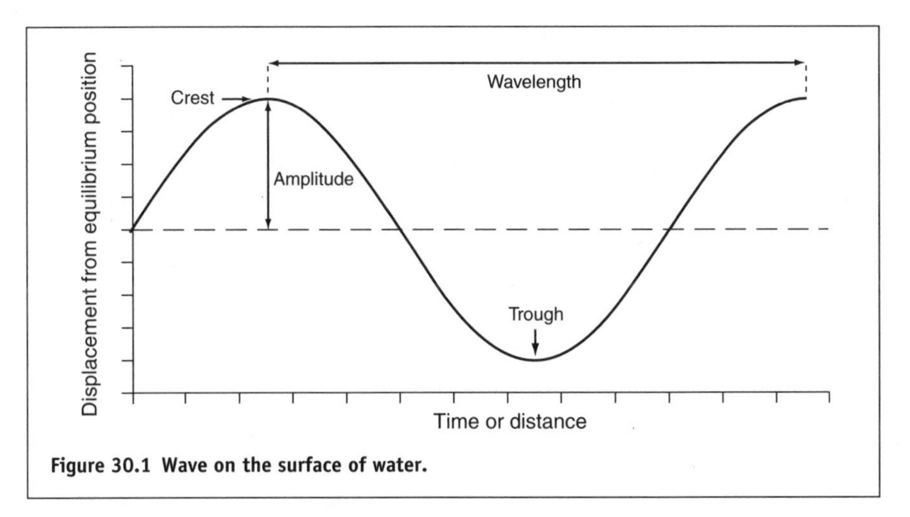

Figure 30.1 Wave on the surface of water.

Mechanical waves can be further categorized according to the direction of particle motion associated with the wave (this is discussed in more detail in **28.2.5**). In a **transverse wave**, the particles of the medium move up and down at right angles (transverse) to the direction in which the wave is travelling. Ripples on a pond and electromagnetic waves are examples of transverse waves. In a **longitudinal wave**, the motion of the particle is backwards and forwards in the direction in which the wave is travelling. Examples of longitudinal waves are sound waves and compression waves in a spring. If you hold the ends of a long, flexible spring in your hands and push one end you should see a compression wave move along the spring. Note, in all mechanical waves, whether transverse or longitudinal, it is only energy that moves with the wave. After the wave has passed through, the particles that carried the wave remain in their original positions.

Before going further, we need to introduce the structure and some of the terminology of waves. In Figure 30.1 we see an example of a water-surface wave. What we see is a snapshot of the wave frozen in time – in reality the wave would be moving. In this example the vertical dimension of the diagram represents height displacement. However, it could equally represent either kinetic or potential energy stored by the medium (water surface), as explained in **28.2.5**. This is important because, while not all waves can be represented on a diagram using height displacement, they could all be represented by variations in energy. The top and bottom of the wave (whether measured in height or energy) are called respectively the **crest** and the **trough**. The crest and trough show displacement of the water surface with respect to the average state. The water surface is said to be **oscillating** about the midpoint, or equilibrium position. The distance between successive wave crests (and successive wave troughs) is called a **wavelength**. The wavelength is normally denoted by the Greek symbol λ (lambda). The amplitude of the wave (a) is the maximum displacement of the water surface from its equilibrium position, which is the distance from the midpoint to the crest (or midpoint to the trough).

The descriptors above can describe the snapshot of the wave. However, we also want to describe the wave in motion. There are two ways of looking at wave motion: as it changes through time and as it changes through distance along its path. However, wave motion in terms of distance and time are closely related, and many of the descriptors can be used for both. Thus, the **period** (T) of the wave is both the time taken for the wave to complete a cycle and also the time taken for the wave to travel one wavelength. Conversely, the **frequency** (f) is the number of oscillations, or **cycles**, completed in one second, and also the number of wavelengths travelled in that time (the unit for frequency is the **hertz**, Hz). Note that frequency and period are simply opposite ways of looking at the same thing, i.e. they are the inverse of each other, thus:

$$T = \frac{1}{f} \quad \text{and} \quad f = \frac{1}{T}$$

The wavefront moves with a specific velocity, usually given the Greek character ν (nu). Velocity, wavelength and frequency are clearly interrelated. For any given wavelength, a higher frequency (i.e. more wavelength travelled per second) requires that the wavefronts travel faster. Conversely, for any given frequency, an increase in wavelength requires the wavefront to travel further in the same time interval, and thus to move faster. This interrelationship can be expressed mathematically:

$$\text{Velocity} = \text{Frequency} \times \text{Wavelength}$$

that is:

$$\nu = f\lambda$$

Because of the relationship between T and f described above, velocity can also be expressed in terms of wavelength and period:

$$\text{Velocity} = \frac{\text{Wavelength}}{\text{Period}}$$

that is:

$$\nu = \frac{\lambda}{T}$$

These expressions can be used to convert units of measurement of waves. For example, if we know the frequency of a radio signal, then given that we know the velocity of the wave (the speed of light, usually denoted c, is $3 \times 10^8\,\text{m}\,\text{s}^{-1}$), we can easily convert from frequency to wavelength. So, if we have a radio station that transmits at 99 MHz (megahertz), or, using more conventional units, 99×10^6 Hz, we can perform the calculation as follows. From the expressions above we know that $c = f\lambda$. This can be rearranged to $\lambda = c/f$. Putting in the known values for c and f gives:

$$\lambda = \frac{3 \times 10^8}{99 \times 10^6}$$

Multiplying out these numbers gives an answer of 3 m. Thus, a radio signal at 99 MHz has a wavelength of 3 m.

30.3 Waves and mathematics

In this section we use some trigonometry, so it may be useful to refer to **26.8** if these concepts seem unfamiliar. There are two kinds of motion that must be described if we are to apply mathematics to waves. First, there is the motion of the medium, whether this is transverse or longitudinal movement. Secondly, there is the motion of the wavefront through both time and distance. It is easier to start with the motion of the medium because this does not move with the wave, which means that we do not need to allow for distance along the wave path. When we have looked at this case, we can extend the approach to looking at motion of the wave along its path.

When considering the effects of a wave on its medium, it is conventional to think of the motion of an arbitrary particle within that medium. Consider a particle that has a position on the equilibrium line (see Figure 30.1). As waves pass the point in the medium that contains this particle, the particle moves back and forth (either parallel or perpendicular to the wave path depending on wave type). Its behaviour can be described in the same way as a vibration. A graph of the motion has the same shape as the wave itself, and in mathematical terms is said to be sinusoidal (meaning that it is shaped like a sine function – see **26.8**). If we know the frequency (f) and amplitude (a) of the vibration, then the following expression tells us the displacement of the particle from the equilibrium line at time t (written y_t) after a given starting point:

$$y_t = a \sin(2\pi ft)$$

As an example, let us assume we have a wave with an amplitude of 2 m and a frequency of 0.1 Hz (i.e. a period of 10 seconds). Let us also assume that at time 0 (seconds) the displacement is 0 m. We can find out what the displacement will be 5 seconds later by putting the appropriate values in the expression above and calculating the result. Thus, for $t = 5$ seconds:

$$\begin{aligned} y_5 &= 2 \times \sin(2 \times 3.1417 \times 0.1 \times 5) \\ &= 0.1 \end{aligned}$$

A useful exercise would be for you to plot this expression (or function, as it is usually termed in mathematics) on a spreadsheet. In column A of the spreadsheet, place a number of increasing values for *t*, from say 0 to 30 in one-second intervals. In column B write a formula for the expression above (in EXCEL this would be = 2* sin(2*PI()*0.1*A1), if you started the columns in row number 1). You can easily modify this approach to allow for different values of the amplitude and frequency (by using absolute cell references to values for a and f, rather than put actual values in the formula). You will notice that the pattern you have plotted repeats itself after a full wavelength. Thus, at 10 seconds (one period), you will be back to zero and have passed through one peak and one trough. If you create a new column on your spreadsheet to look at the values of $2\pi ft$ you will see that it cycles through multiples of 2π for each wavelength. This is because there are 2π radians in a circle.

The approach just outlined allows us to predict displacement of a specific particle. As said above, this is the same as treating the particle as if it were vibrating, rather than moving with a wave. If we want to consider the passage of the wavefront, we need to add a new dimension; this is the distance along the wave path. We are now no longer considering only one particle, but rather a line of particles spread along the wave path. The equation looks like this:

$$y_{x,t} = a \sin(kx + 2\pi ft)$$

The symbols f, t and a have the same meaning as in the first expression, while $y_{x,t}$ is slightly different. In the previous expression, y (the symbol for displacement) had only one subscript, t, because it meant the displacement of the particle at time t. Now we have two subscripts (x and t) indicating the displacement at time t for a particle at distance x along the wave path. Note, the part of the expression $2\pi f$ is sometimes termed the angular frequency and given the symbol ω (omega). We have one new term in the equation: k. This is the angular wave number, and has units of radians per metre (rad m^{-1}). Its value is related to the wavelength, and can be calculated using the expression:

$$k = \frac{2\pi}{\lambda}$$

To put the whole expression into a spreadsheet you can follow the same procedure as above. Use the same example (values of a, f and t), but include (as is now needed) a wavelength of 20 m. The expression can be used in a number of different ways. If, for example, you fix x at zero (i.e. you find displacement through time of a particle at x, or $y_{t,0}$), then you will get the same result as with the first expression. Indeed, if x is set at zero, then the term kx becomes zero, and the two expressions are identical. Alternatively, you could choose a time t of, say, 7 seconds, and plot displacement at time 7 seconds ($t_{7,x}$) for all values of x up to, say, three wavelengths.

If you try out these two methods you will gain a feel for the basic mathematical treatment of wave.

30.4 Doppler shift

No doubt you have experienced the lowering in apparent frequency of a police or ambulance siren as the vehicle passes you. This effect is called the **Doppler shift**. It is worth considering the mechanism that leads to this effect. Imagine that a vehicle is moving towards you, sounding a siren. The sound of the siren comprises a succession of wavefronts, each emitted at intervals of period T. However, each successive wave crest is delivered from a nearer point than the last, because the vehicle is moving towards you. As a consequence, each successive wavefront has less distance to travel to get to you, and thus takes less time to arrive. The overall effect is that the sound frequency is greater (i.e. more wave crests per second) at your position than at the source. The pitch of a note is related to its frequency, thus the pitch of the siren is higher where you are standing than it is on the vehicle. As the vehicle draws level with you, you briefly hear the siren's pitch exactly as it is delivered. When the vehicle is moving away from you, the initial situation is reversed, and the apparent pitch of the siren is lowered.

Doppler shift has a number of practical applications. Doppler rainfall radar gives us the direction of movement in rainstorms, while the shifts in the frequency of the electromagnetic spectrum from certain stars (their redshift) tells us their velocity with respect to the Earth, and tells us that the universe is rapidly expanding.

30.5 Sound level and decibels

We have mentioned that a key parameter in relation to waves is energy transfer. We know, for example, that sound waves carry energy. A loud noise carries considerable energy and is capable of moving an object it encounters. Indeed, we can all feel loud noises as well as hear them. Because of this association of loudness and effect, it would be easy to imagine that loudness and energy were proportional for sound waves. However, this is not the case. Rather, doubling the energy carried by a sound wave does not double the noise as we perceive it. This fact is recognized in the decibel scale, which is based on the logarithm of the energy (measured as sound intensity). Using a log scale allows a wide range of sound intensity values to be expressed using a manageable range of numbers. The equation for the decibel sound level (S) is:

$$S\,(\mathrm{dB}) = 10\log I/I_0$$

where I_0 is a standard reference sound intensity, which is set at $10^{-12}\,\mathrm{W\,m^{-2}}$ (which is close to the human threshold of hearing), and I is the intensity of the sound which you are measuring. Notice that sound intensity is measured in units of power (i.e. energy per unit time) per unit area, and is thus directly related to the energy carried by the sound. This system gives a value close to 1 dB for a barely audible sound. A whisper is about 20 dB, a normal conversation is about 60 dB, and the threshold of pain is about 120 dB. To see how convenient this is, you can use the expression above to convert these decibel values into sound intensity units. We find that the pain threshold is at a sound intensity of about $1\,\mathrm{W\,m^{-2}}$, compared with the lower limit of hearing at $10^{-12}\,\mathrm{W\,m^{-2}}$. Try working out the sound intensity for a whisper. How many times greater is the intensity at the threshold of pain compared with the intensity of a whisper?

Notice that by using a logarithmic scale to quantify loudness, a small range of numbers can be used to cover a range of twelve orders of magnitude in sound intensity.

31 Life

31.1 Introduction

There is general agreement that life-forms have been present on Earth for nearly 4 billion years, in other words for most of our planet's 4.6 billion-year history. In this section we introduce some of the fundamental properties of life. Our justification for this section is, first, that organisms are key components of the environment and, secondly, that life processes have a major influence on the state of the physical environment. It follows that most environmental problems require an understanding of both physical and biotic processes and the interactions between them. A discussion of the environment that omitted the living component would therefore be incomplete.

A useful and commonly used term is **biosphere**. Usually, this term is used to refer to the zone in which life occurs, and often appears to mean the living organisms as well. At other times though, the term is used to refer to all life on Earth. Because the suffix '-sphere' suggests a zone, it is probably best to use biosphere in the former sense, particularly as the organisms themselves can conveniently be referred to using the term **biota**.

The preceding remarks beg the question: what is it that distinguishes living from non-living entities? Surprisingly perhaps, it is not easy to provide a simple, close definition of life. However, we recognize living entities by certain characteristics, some, but not all, of which are possessed by some non-living objects. Not all life-forms are considered to be organisms in the conventional sense. The most notable exception are **viruses**, which consist of 'naked' molecules of nucleic acids (see **32.3.5**) and proteins (**32.3.4**). Viruses meet the criteria for being alive but they function only within living organisms, whose chemical machinery they use. They can also bring about severe disruption of their host organism, which may severely weaken it, or even lead to its death. Viruses thus share some, but not all, of the features of organisms.

All organisms have a high degree of both structural and functional organization, all have an ordered and reasonably predictable pattern of development, and all process matter and energy. Another key feature of all organisms is that they respond to environmental stimuli. The stimuli vary, as does the nature of the response, but all respond to temperature (a consequence of all chemical reactions being temperature dependent), and moisture (as all biochemical processes occur in water). Organisms also have the potential to replicate themselves, even though many types of organisms require two individuals of different sex for the process. In addition to these general features, all life-forms share certain aspects of their chemistry, most notably two chemical substances: the proteins and the nucleic acids. The chemistry of life is introduced in **32**.

31.2 Cells and cell types

A defining feature of all true organisms is their cellular construction. **Cells**, in essence microscopic entities, can be considered the building-blocks, or the modules, of which all organisms are constructed. Some organisms, bacteria and amoebae for example, consist of a single cell. The number of cells in multicellular organisms varies by several orders of magnitude: for a large mammal it is hundreds of billions. The growth of multicellular organisms occurs owing partly to the increase in the size of cells, but much more importantly because of cell multiplication in the process formally known as **cell division**.

All organisms are assigned to one of two groups based on cell type. This is one of the most fundamental categorizations in biology. In one group, which consists of *all* multicellular, and many unicellular organisms, the cells contain discrete, membrane-bound structures (**organelles**). Examples of organelles include the **nucleus**, in which the chromosomes and genes are located (**32.3.5**), and the **mitochondrion**, where key energy reactions take place (**34.3.3**). Cells, and organisms, of this type are called **eukaryotic** (from the Greek *eu*, meaning 'true', and *karyon*, meaning kernel – used here to represent the nucleus). In contrast, cells of the second major group of organisms do not contain organelles. Such cells (and organisms) are described as **prokaryotic** (from the Greek *pro* meaning 'before' to denote their origin before eukaryotic cells). It is easy to remember which organisms belong to which major group because it is *only* bacteria, and *all* bacteria, which are prokaryotes. (Note that it is now standard to divide the bacteria into two groups, the archaebacteria and the eubacteria on the basis of differences in chemistry, and these terms are becoming increasingly common.) Bacteria, which are essentially unicellular organisms, are exceedingly important members of the biota. Although they lack subcellular structures, collectively, they perform all the fundamental metabolic processes associated with eukaryotic organisms (**34.3**).

A key difference between prokaryotes and eukaryotes is that the former do not reproduce sexually; instead, the single cell of a prokaryote simply divides in two in the process known as **binary fission**. Prokaryotic organisms were the only life-forms present on Earth from around 3.8 billion years ago until the eukaryotic cell evolved. The timing of this momentous event is not known for sure: it had certainly occurred by 1.5 billion years ago and some claim it occurred up to a billion years earlier. Either way, prokaryotes were the only life-forms present for a significant part of the biosphere's history.

32 Life's chemistry

32.1 The chemical elements of life

Organisms have made use of around one-quarter of the ninety or so naturally occurring chemical elements on the planet. Most of these elements are required by *all* organisms, but a few of them are required by some organisms and not by others. The elements, which are often called the **nutrient elements**, **inorganic nutrients** or **bioelements**, are shown in Table 32.1, together with their chemical symbols.

The list of elements in Table 3.1 reveals that all but four of the bioelements have an atomic number of 30 or less and comparatively low relative atomic masses. It thus appears that the size and mass of nuclei have been important factors in their appropriation for life processes. The chemical analysis of living tissue usually reveals the presence of a few other elements in addition to those listed. In the case of a plant, they may have been taken up by the roots, and in the case of an animal, consumed with the food. Some such elements, lead and cadmium for example, are toxic and may harm the organisms concerned, or those other organisms which feed upon them. Also, note that some of the essential elements, zinc and copper for example, are harmful if present in relatively high concentrations.

32.2 Carbon

The chemistry of life is intimately associated with the chemistry of carbon compounds. (However, it should not be forgotten that some of the many other elements which organisms require are not necessarily always bound to organic molecules.) Carbon accounts for approximately 45 per cent of the dry mass of the Earth's biota, although the percentage varies between different types of organism. Because carbon is so fundamental to life, the term 'organic chemistry' was coined early by chemists to distinguish this

Table 32.1 The elements essential for life.

Name	*Symbol*	*Name*	*Symbol*
Carbon	C	Cobalt	Co
Hydrogen	H	Iodine	I
Oxygen	O	Molybdenum	Mo
Nitrogen	N	Manganese	Mn
Sulphur	S	Zinc	Zn
Phosphorus	P	Selenium	Se
Potassium	K	Fluorine	F
Calcium	Ca	Silicon	Si
Magnesium	Mg	Chromium	Cr
Sodium	Na	Vanadium	V
Iron	Fe	Tin	Sn
Copper	Cu		

branch of the discipline from inorganic chemistry, which deals with all other compounds. However, large compounds containing carbon are produced synthetically, so organic chemistry is no longer solely concerned with the chemistry of life. We describe in more detail some general properties of carbon in **16**. Key points are that carbon atoms (atomic number 6), with four electrons in the valence shell, form stable bonds with each other, and these may be single, double or even triple bonds. The molecules so formed may be long chains (which may be branched or unbranched), they may be rings, or they may be more complex structures. Carbon also forms stable bonds with certain other elements, particularly oxygen (single and double bonds), hydrogen (single bonds), nitrogen (double and triple bonds) and sulphur (single bonds). The size of carbon compounds varies enormously. Some of the main ones that we consider contain just a few carbon atoms but others contain many thousands. In illustrating some of the more common compounds we use the conventions introduced when we discussed organic chemistry (see **16.3**).

Together, carbon, hydrogen and oxygen make up the greater part of the dry mass of an organism. For plants, and therefore for the biota as a whole, it is typically between 95 and 98 per cent. If we included water, then hydrogen and oxygen would account for a much higher proportion of the total mass. Water, although essential for life, is not really part of the organic structure of organisms, and its content is very variable. Thus it is often better to consider amounts on a dry mass basis.

The other elements in the list, i.e. from nitrogen onwards, make up a small proportion of the dry mass of organisms. These elements are often referred to as the mineral elements. Nitrogen, and to some extent sulphur, differ from the other elements in this part of our list in having gaseous phases at the Earth's surface. The nutrient elements are often categorized as **macronutrients** or **micronutrients** (or trace elements) on the basis of the concentrations in which they are required. The concentrations in which micronutrients are found are extremely tiny, often less than 0.001 per cent of dry mass. However, concentration should not be equated with importance: if an element is required even in very tiny concentrations then its absence will impair normal metabolism. The mineral elements may be covalently bonded to the atoms of other elements in organic compounds or they may exist in ionic form in solution.

32.3 The substances of life: biomolecules

Collectively, living organisms synthesize millions of different compounds (the exact number is unknown because all the species named so far have not been thoroughly investigated, and anyway these comprise only a proportion – some think quite a small proportion – of the number of species on Earth). Despite this enormous biochemical diversity, most of the dry weight of any organism is accounted for by a handful of different classes of organic compounds. These compounds have a 'backbone' of carbon atoms to which are attached atoms of hydrogen and oxygen, and to a lesser extent nitrogen and sulphur. A little information about each major group will enhance your understanding of living processes. These groups, whose names are in everyday use, are now dealt with in turn.

32.3.1 Carbohydrates

Carbohydrates perform a variety of vital metabolic and energy-storage roles, and in some organisms they are key structural components. The name, derived from 'hydrate of carbon', was applied originally because the hydrogen and oxygen atoms occur in more-or-less the same ratio as in water. Thus a simple, and approximate, formula for carbohydrates is $(CH_2O)_n$ in which the subscript n denotes the number of repeating units. Carbohydrates can be classified according to the size of the molecules. The term **monosaccharide** is used for carbohydrates that cannot be further broken down by hydrolysis (i.e. by the addition of water); **disaccharides** can be broken down in a hydrolysis reaction to yield two monosaccharides; and **polysaccharides** yield several, in some cases thousands, of monosaccharides on hydrolysis. Monosaccharides and disaccharides are generally known as **sugars**.

The most abundant of the monosaccharides have six carbon atoms (and are therefore called hexose sugars), but a few others contain either five or three carbons (pentose and triose sugars, respectively). Glucose, a hexose, is universally important: it is the most common monosaccharide in plants and it is the blood sugar of mammals. Furthermore, the most abundant polysaccharides are based on linked glucose units. Glucose molecules are usually represented diagrammatically as open chains or closed rings (and all the atoms are not normally shown). Glucose is described as being optically active (see **16.6**); most natural glucose is dextrorotatory. The ring structure of glucose, as found in water, is shown in Figure 32.1. Note the subtle difference between

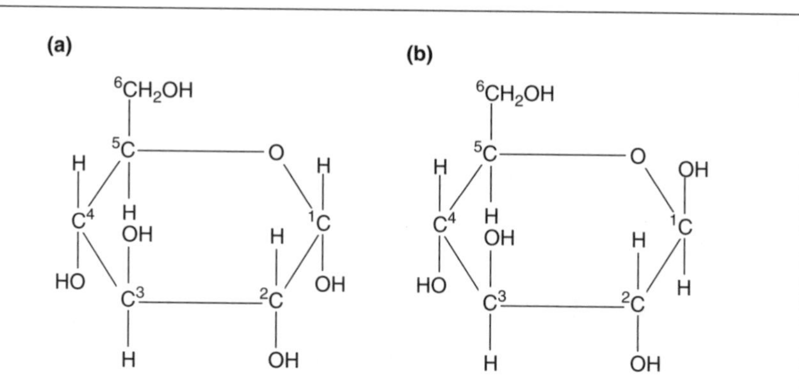

Figure 32.1 The structure of glucose molecules in ring form. (a) Alpha-D-glucose; (b) beta-D-glucose. Note the different arrangement of atoms attached to carbon 1.

parts a and b in the arrangement of the hydroxyl (OH) groups and hydrogen atoms attached to carbon number 1. (The numbering scheme shown is standard.) Carbon 1 is said to be asymmetric, and therefore two **stereoisomers** exist (see **16.6**). If you imagine the ring as a flat plane with the lower side closest to you, then in one form (alpha-D-glucose), the hydroxyl group is below, and the hydrogen is above, the plane. In the alternative form (beta-D-glucose), the position of these groups on carbon 1 is reversed. The reason that this apparently small difference is very important is explained shortly. Fructose, which is found particularly in plants, is another naturally occurring hexose monosaccharide. Fructose is a very sweet substance and is the most abundant sugar in honey. Some non-sugar carbohydrates are made from fructose, not glucose units. Certain five-carbon sugars (pentoses) exist naturally, particularly in association with phosphorus, and are very important in metabolic pathways. One of these, ribose is a key component of the nucleic acids (see **32.3.5**).

Disaccharides consist of two monosaccharide residues, and therefore yield two monosaccharides on hydrolysis. The term 'residue' is used rather than 'molecules' here because two monosaccharide molecules are not simply added to each other to form a disaccharide molecule. Rather, it is a condensation reaction involving the elimination of a molecule of water. (It is thus the reverse of hydrolysis.) A disaccharide formed from two hexose sugars therefore has the molecular formula $C_{12}H_{22}O_{11}$. The numbering of carbon atoms according to position in a monosaccharide (Figure 32.1), together with the alpha/beta scheme for configuration of the hydroxyl groups, allows the nature of the linkage between two monosaccharide units to be specified precisely.

The most abundant disaccharide is **sucrose** (Figure 32.2), which consists of a glucose and a fructose residue. Sucrose is the principal sugar of commerce. It is extracted primarily from sugar cane, which is a tropical grass, and sugar beet, a bulbous-rooted plant whose wild ancestors are quite common in coastal locations in Europe and western Asia. For most plants, however,

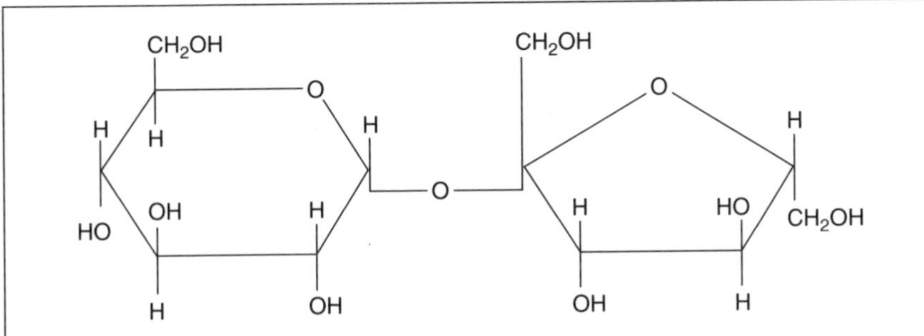

Figure 32.2 The structure of sucrose, a disaccharide. On hydrolysis, sucrose yields glucose and fructose molecules.

sucrose is not the main energy-storage carbohydrate, but it is the principal form in which carbohydrates are transported (**translocated**) from the leaves to other parts of the plant.

Other important disaccharides are lactose and maltose. Lactose, the sugar found in mammalian milk, consists of residues of glucose and another hexose sugar, galactose. Maltose, which is found in germinating seeds and is important in beer production, consists of two glucose residues.

Important related substances are the **sugar alcohols**, which are commonly found in plants, particularly in fruits. One such substance is sorbitol, which is formed from glucose: it is used in the manufacture of vitamin C (ascorbic acid) as well as other pharmaceutical products, foodstuffs and cosmetics.

The term **polysaccharide** is used for carbohydrates which are made up of more than just a few monosaccharide residues. Because each type of polysaccharide is made from the same type of molecule (usually glucose), they can be described as **polymers**. Some polysaccharides are relatively small (typically 20–40 monosaccharide residues), but others consist of hundreds or thousands of sugar residues. The term **oligosaccharide** is sometimes used for the former group. Some of these are quite widely distributed in the plant kingdom (and most are not readily digested by humans). An example is inulin, which is made up of fructose residues.

Two polysaccharides, glycogen and starch, play key energy-storage roles in animals and plants, respectively. Both substances are composed of units of glucose. Glycogen consists of numerous, relatively short chains of glucose units linked by alpha-(1,4) links, while the chains themselves are joined by alpha-(1,6) cross-linkages. The result is a highly branched molecule of very high molecular weight. Glycogen is found principally in the liver and muscle and is readily metabolized to meet energy requirements. Starch consists chiefly of two glucose polymers: amylose and amylopectin. Amylose is made of long, linear chains of up to a thousand glucose residues joined by alpha-(1,4) linkages. In contrast, amylopectin has a highly branched structure. It consists of short chains of glucose units linked in the same way as in amylose, but these linear sections are themselves joined together by alpha-(1,6) links (Figure 32.3). Starch is the principal form in which energy is stored

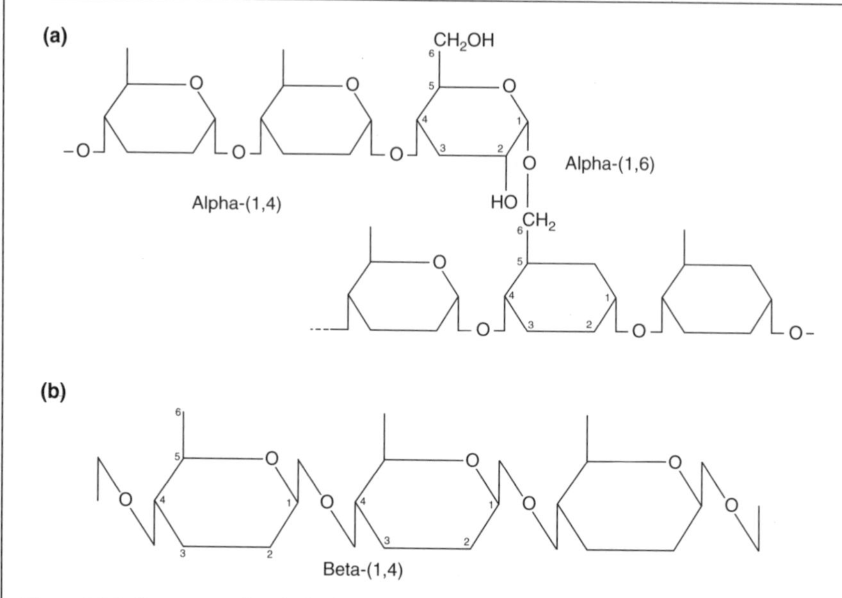

Figure 32.3 Fragments of carbohydrate molecules. (a) Starch (the amylopectin component); and (b) cellulose. Note the alpha-(1,4) glucose units, and linkages, in starch and the beta-(1,4) glucose units, and linkages, in cellulose.

in plants, both in the short term – a few days, say – and for comparatively long periods, typically up to several months, when the plant may be dormant. (Fat plays a similar long-term energy-storage role in mammals.) Starch is also of vital importance as the principal source of food energy for most of the world's human population. It is obtained primarily from cereal grains and the underground parts of certain plants, for example potato and cassava.

Cellulose is another important polysaccharide, and it is also based on glucose residues. It has the distinction of being the most abundant organic substance on Earth, comprising an estimated 60 per cent of the biota's dry weight. Like starch, cellulose is produced by plants (and many algae and some fungi), but, unlike starch, cellulose is a *structural* carbohydrate and does not serve an energy-storage role. Cellulose is produced within cells and is laid down in the surrounding cell wall. As cells increase in age the cell wall becomes thicker, so as plants grow their cellulose content increases both in absolute and relative terms. Cellulose gives structural rigidity to the plant. It is an important raw material for many products, including other organic chemicals and paper. One of the purest forms of cellulose in nature is cotton. The great advantage of cotton over woody plant tissue as a source of cellulose is that it requires relatively little treatment before use, while cellulose must be separated from other substances with which it is intimately associated.

Now, although cellulose, like starch, is a polymer of glucose, there are fundamental differences in the structure of the two substances. These are based on the fact that cellulose is made up of the beta, not the alpha, form of glucose (Figure 32.1) and these are connected by beta-(1,4) links, not alpha-(1,4) links as in the linear parts of starch molecules (Figure 32.3). This apparently small chemical difference gives rise to great differences in the appearance and properties of these two polysaccharides, and it is of enormous biological significance. Cellulose consists of long chains of over 3500 beta-(1,4) linked glucose units, with the chains themselves being held together by hydrogen bonds (see **8.6**) to form microfibrils.

Comparatively few types of organism, notably certain bacteria and fungi, produce the enzyme cellulase which can cleave the beta-(1,4) links between the glucose residues, and thus break down cellulose molecules. In order for the contents of a plant cell to be available to plant-eating (herbivorous) animals, either the cell wall must be ruptured by physical processes or the animals must harbour within their digestive systems the microorganisms that produce the enzyme cellulase. The capacity for physical degradation of the cell wall is limited, so plant-eating organisms (including humans) are in effect restricted to young leafy plants or soft fruits, at least if they want to digest their food. Many herbivores, however, harbour cellulase-producing bacteria in their digestive tracts, which means that they can survive on an exclusively plant diet. The best-developed associations occur between herbivores with a much enlarged stomach (rumen). Examples of animals of this sort, called ruminants, include cattle, buffalo, sheep, goats, antelopes and camels. Horses and their relatives utilize cellulose rather less efficiently because absorption of the digested materials takes place *after* the small intestine, the principal site of food absorption in the gut. The termites, an important group of ant-like insects, particularly in the tropics, also harbour cellulase-producing microorganisms which enable them to digest fibrous material in their diets.

The hemicelluloses are a group of branched polysaccharides. Their structure is rather poorly defined but they are based largely on pentose and hexose sugar units. Hemicelluloses are found, like cellulose, in plant cell walls. Chitin is another structural polysaccharide. It is the principal substance of the exoskeletons of insects, certain crustaceans such as lobsters and crabs, and the cell walls of some fungi. Chitin is based on units of the nitrogen-containing carbohydrate N-acetylglucosamine. When impregnated with calcium it forms a very tough outer protection.

32.3.2 Lignin

Lignin, like cellulose, is a structural component of plants. It is the second most abundant natural organic substance, typically accounting for between 25 and 35 per cent of the dry mass of woody plants. Lignin is largely responsible for making plants 'woody', and therefore is found in only tiny amounts in non-woody plants. (Cellulose, in contrast, is a major component of the cell wall of *all* plants.) Lignin production increases when the cell ceases to expand, with the result that heavily lignified cells have a small interior volume. Lignin is even more resistant than cellulose to biochemical breakdown.

Before wood can be used to produce paper, it is necessary to separate the cellulose from the lignin. Chemically, lignin is described as a polymer of phenyl propene, but because it has a variable molecular weight it is not referred to as a compound.

32.3.3 Lipids

The **lipids**, which perform a wide variety of energy-storage, structural and physiological roles, are another group of biochemical substances of immense importance. Although it is difficult to define this group closely in chemical terms, a characteristic of all lipids is that they are not soluble in water but they do dissolve in organic solvents. The most abundant, and most familiar, of the lipids are the **fats** and the **oils**. The traditional distinction is that fats are more-or-less solid, while oils are more-or-less liquid, at 'room temperature' (around 20 °C): in other words, the fats have higher melting points than the oils. Using this terminology, we tend to associate fats with animals and oils with plants. Exceptions to this generalization are the oils produced by fish. Despite this traditional distinction, it is now common practice among biochemists to refer to both fats and oils as fats, so we use this convention here when dealing with the chemical structure of these substances.

Fat is the major long-term energy-storage substance in animals (compare with glycogen above). It is fat that is primarily responsible for the survival of mammals during periods when food is scarce. Weight for weight, fat contains about twice as much energy as carbohydrate. On oxidation, fat yields nearly 40 MJ kg^{-1}, over twice as much as from the carbohydrate glycogen. Moreover, unlike stored glycogen, stored fat contains little combined water. As a result, fat has over six times the energy density as glycogen in animals. Plants rely much less on fats than on carbohydrates as a store of energy. However, the seeds of some plants are very rich in fat (usually referred to as oil). These include many commercially important plants such as the sunflower, olive, oil-seed rape and soya bean.

Chemically, all fats are composed of substances called **glycerides**. A glyceride is formed from the combination, in condensation reactions, of one molecule of **glycerol** (an alcohol) and either one, two or three molecules of **fatty acids** (Figure 32.4). Glycerides are therefore designated by the prefixes mono-, di- and tri- respectively, although it is the triglycerides which are overwhelmingly predominant. Glycerides are also differentiated by their fatty acid component. In simple triglycerides, all three fatty acids are the same; in mixed triglycerides more than one fatty acid is involved. The fatty acids themselves are differentiated by the number of carbon atoms in the chain (usually between 3 and 19), and also by the extent to which the carbon atoms are saturated (Figure 32.4). If each carbon in the fatty acid is linked to four atoms, the bonds are single and the fatty acid is described as saturated. However, if two of the carbon atoms are linked by a double bond, each must be linked to only two other atoms, in which case the fatty acid is described as unsaturated. By the addition of hydrogen (**hydrogenation**), therefore, saturation can occur. The

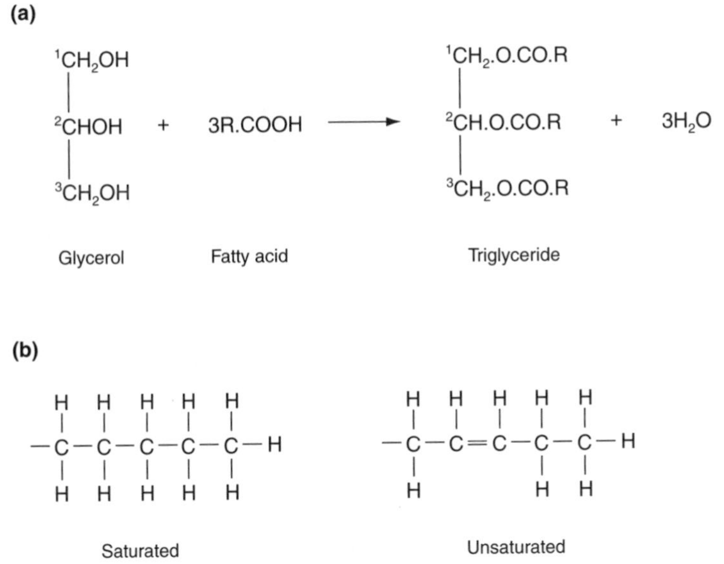

(a)

$$\begin{array}{l} ^1CH_2OH \\ | \\ ^2CHOH \\ | \\ ^3CH_2OH \end{array} + 3R.COOH \longrightarrow \begin{array}{l} ^1CH_2.O.CO.R \\ | \\ ^2CH.O.CO.R \\ | \\ ^3CH_2.O.CO.R \end{array} + 3H_2O$$

Glycerol Fatty acid Triglyceride

(b)

Saturated: –C(H)(H)–C(H)(H)–C(H)(H)–C(H)(H)–C(H)(H)–H

Unsaturated: –C(H)(H)–C(H)=C(H)–C(H)(H)–C(H)(H)–H

Saturated Unsaturated

Figure 32.4 (a) The formation of a triglyceride from glycerol and three fatty acids. R represents the hydrocarbon chain on the fatty acid; (b) a saturated and an unsaturated part of a fatty acid.

fats from land animals are usually much richer in saturated fats than those from plants. The terms 'saturated fats' and 'unsaturated fats' have become familiar in recent years because of the debate concerning the role of the former in heart disease.

The **phospholipids** are another important group of lipids. These substances are major components of the membranes of cells and subcellular structures (organelles). In most phospholipids, a phosphate, which itself is often attached to another group, is linked to glycerol instead of a fatty acid.

Some lipids are not based on glycerol. Familiar and abundant lipids in this category are the **waxes**. These are based on certain 'fatty' alcohols and fatty acids. Waxes are differentiated by the structure of the alcohol unit and by the composition of the fatty acid component. Waxes serve a protective function in nature: they are commonly found on the surfaces of leaves and fruits, and are also produced by some animals. The protective properties of waxes have long been appreciated: thus their use for automobiles and wooden furniture, although these waxes are now produced synthetically.

32.3.4 Proteins

Proteins, which are present in every living cell, perform a variety of important functions, but most importantly they are responsible for orchestrating the metabolic processes that define life itself. They perform this role principally as **enzymes**, often in association with a non-proteinaceous **cofactor**.

Enzymes, which are considered in more detail elsewhere (see **33**), are in essence biochemical catalysts. (A catalyst is a substance which greatly accelerates a chemical reaction, but which is not consumed during the process.) Earlier we used the example of cellulase, the enzyme that is necessary to break the bonds linking the glucose units of cellulose molecules.

In addition to their role as enzymes, proteins perform other vital functions. Antibodies, which protect the body against harmful invading agents, are proteins; many hormones are largely proteins; the substance haemoglobin, which transports oxygen around the body, is an iron-containing protein. All these proteins perform their roles within cells. In shape, proteins are roughly round, and these are aptly referred to as globular proteins. Other proteins, known as fibrous proteins, although produced within the cell, are principally found in the cell wall or between cells. Their function is usually protective or structural. Familiar examples include keratin (in wool, fingernails and toenails), elastin (in arteries and tendons), and collagen (connective tissue). Protein is not normally considered an energy-storage substance, but when carbohydrates and fats are in short supply protein may be metabolized for this purpose.

Chemically, proteins are complex, high (often exceedingly high) molecular weight substances. All proteins, though, are built up from quite small molecules of substances called **amino acids**, which are linked together. Only about twenty amino acids are found in protein molecules, but because proteins can be made up of thousands of amino acid units, and because the amino acids can occur in virtually any sequence, the number of different permutations is in effect infinite.

In an amino acid there is a central carbon atom which is surrounded by an amino (NH_2) group (in one case, NH), a carboxyl (acidic) group (COOH), a side chain (which is conventionally represented by R) and a hydrogen (Figure 32.5a). The R group distinguishes one amino acid from another. In one of the amino acids, R is simply a hydrogen atom but in all other cases the R group forms a ring structure.

The carboxyl group can give up a proton and the amino group can take up a proton. An amino acid can thus be both an acid and a base. Amino acids can carry charges of opposite polarity but be electrically neutral (Figure 32.5b). Such a dipolar substance is called a **zwitter ion**. For an amino acid in solution, it is the pH that determines the nature of the charge: at low pH values (acidic) the molecule exists primarily as a cation, but at high pH values (alkaline) the molecule exists primarily as an anion. For each amino acid there is a pH value, known as the **isoelectric point**, at which it is electrically neutral. These properties of amino acids mean that they are effective buffers (see **12.6**), i.e. they resist changes to pH.

The names of the common amino acids are shown in Table 32.2. Cysteine is worth noting because it contains sulphur, as a sulphydryl group (S–H). The significance of this point will become clear shortly.

Amino acids are linked to each other by **peptide bonds**. This is a covalent bond between the carboxyl group of one molecule and the amino group of another (Figure 32.5c), with the elimination of a water molecule. (It is thus

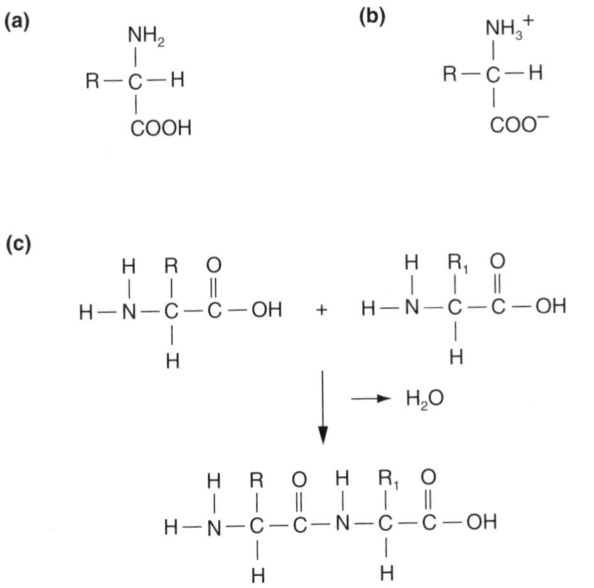

(a)

```
      NH2
      |
  R — C — H
      |
      COOH
```

(b)

```
      NH3+
      |
  R — C — H
      |
      COO−
```

(c)

```
     H   R   O                   H   R1  O
     |   |   ||                  |   |   ||
 H — N — C — C — OH    +    H — N — C — C — OH
         |                           |
         H                           H

                  |
                  |  ——► H2O
                  ▼

         H   R   O   H   R1  O
         |   |   ||  |   |   ||
     H — N — C — C — N — C — C — OH
             |           |
             H           H
```

Figure 32.5 Amino acids. (a) General structure. Amino acids are differentiated by the structure of the R group; (b) an ionized amino acid: the carboxyl group has given up a proton; the amino group has acquired a proton; (c) the formation of a peptide bond between two amino acids in a condensation reaction.

yet another example of a condensation reaction.) What results is called a **peptide**, composed of two amino acid *residues*. The number of linked amino acid residues can be specified, as in *di*peptide and *tri*peptide, while several amino acid units form a *poly*peptide. Proteins consist of one or more polypeptides, although the distinction between polypeptide and protein is not closely defined.

The sequence of amino acids in a protein is known as the **primary level** of organization. However, the strands of amino acids that form a polypeptide adopt characteristic shapes which are due to hydrogen bonds (see **8.6**). Such structures form the **secondary level** of protein organization. One

Table 32.2 The amino acids.

Alanine	Glycine	Proline
Arginine	Histidine	Serine
Asparganine	Isoleucine	Threonine
Aspartic acid	Leucine	Tryptophan
Cysteine	Lysine	Tyrosine
Glutamic acid	Methionine	Valine
Glutamine	Phenylalanine	

common form is called the **alpha helix** (coil-shaped), which is formed by hydrogen bonds between amino acids *within* a single polypeptide. Another is the **beta pleated sheet** (zig-zag shaped), which is formed by hydrogen bonds *between* different, adjacent polypeptides. In addition, proteins have a third, or **tertiary level** of organization, maintained partly by hydrogen bonding but also by ionic and covalent bonding and by hydrophobic interactions. Particularly significant are the covalent disulphide links between the sulphydryl groups on different cysteine residues (recall that this amino acid contains sulphur). The resulting molecule has a folded structure which is specific for each protein.

Globular proteins (essentially the 'metabolic' proteins) have a fourth, or **quaternary level** of organization. This involves the linking of two or more polypeptides to form a unit with a specific and precise biochemical function. Such a unit will maintain its structural integrity, and hence its functional competence, only as long as conditions are suitable. A change in temperature, pH or the chemical environment can easily cause structural alterations and the loss of its functional capacity, in which case the protein is said to **denature**.

What should be clear from this basic outline of protein structure is the complexity of proteins: a complexity which reflects their fundamental role in the organization of life itself.

Returning to the basic structure of amino acids, it is a fact that only certain types of organisms, notably plants, fungi and bacteria, are able to combine hydrogen and nitrogen to form the amino group. Such organisms can thus exist on an inorganic nitrogen supply, provided they have all the other essential elements and sufficient energy. And very importantly, they can therefore synthesize protein. Animals, in contrast, do not produce the enzymes necessary for the manufacture of amino acids from inorganic constituents. Protein must therefore be present in the diet. The great exceptions are those herbivores (most notably the ruminants such as cattle and sheep) which harbour bacteria within their stomachs. These bacteria utilize inorganic nitrogen, just as free-living bacteria do, to manufacture amino acids and proteins. As these bacteria die, their proteins can be broken down to their constituent amino acids, which can then be absorbed across the gut wall and made available to the host animal. Such animals can therefore survive on diets containing no protein.

A final but important point is that differences between types of organism are largely a result of differences in their complement of proteins. In fact, the evolutionary or genetic similarity between organisms can be measured in terms of the proteins they produce.

32.3.5 Nucleic acids

Like the proteins discussed above, **nucleic acids** are found within every cell of every organism, and practically define life on Earth. The substance was first isolated, from the cellular nucleus, in 1869, and accordingly given the name nuclein. When the acidic properties of this substance became apparent, the name nucleic acid was applied. However, nucleic acids are not confined to the nucleus: they are found in other cellular organelles, including the mitochondrion and the chloroplast, and in the cytoplasm. There are two nucleic acids: **deoxyribonucleic acid** (**DNA**) and **ribonucleic acid** (**RNA**).

Each cell of an individual organism carries identical DNA molecules. These are tightly coiled on thread-like structures called **chromosomes**. A species has a characteristic number of chromosomes: for humans the total number is 46, each parent having provided 23. The DNA acts as a repository of information, in chemical coded form, for the types of protein to be produced by the cell. A fragment of DNA with a specific functional outcome, usually in terms of protein production, is referred to as a **gene**. Genes determine the characteristics of organisms and they are the units of heredity. Crucially, DNA is a self-replicating molecule. So when an organism grows, primarily by cell division, a new copy of DNA must be produced. The capacity of DNA for self-replication is also essential for reproduction of organisms.

DNA does not do the actual work involved in protein production: that is the function of the other nucleic acid, RNA, of which there are several variants. RNA molecules 'read' the information stored in chemical form on DNA molecules, convey the information to the sites in the cell where proteins are manufactured, and carry out the assembly process. These processes are discussed further below.

The chemical structure of DNA is represented schematically in Figure 32.6. We can see immediately why the structure can be likened to a ladder. Nucleic acids are made up of units called **nucleotides**. Each nucleotide is itself made up of three subunits: a phosphate group, a five-carbon sugar (deoxyribose), and one of four nitrogenous bases. These bases are called adenine, thymine, cytosine and guanine, but are commonly referred to by A, T, C and G, respectively. Adenine and guanine each have a double-ring structure and belong to a class of chemicals called purines. Thymine and cytosine are each composed of a single ring and belong to a class of chemicals called pyrimidines.

Adjacent nucleotides are linked to each other by covalent bonds between the phosphate group on one nucleotide and the sugar of the next nucleotide. The result is two extremely long strands of sugar–phosphate units which form the backbone of the molecule, or the sides in our ladder analogy. Note in Figure 32.6 that the two strands of nucleotides run in opposite directions. Each nucleotide has a base, and the nucleotide bases face each other. Now, on each side of this structure the bases can occur in any order. However, each base *always* has the same 'partner' base in the other strand: adenine always pairs with thymine (A–T), and cytosine always pairs with guanine (C–G). They are accordingly referred to as complementary base pairs. These complementary base pairs are held together by hydrogen bonds (see **8.6**): there are two such bonds between adenine and thymine and three between cytosine and guanine. The pairing of a purine with a pyrimidine is such that the sides of the DNA molecule are parallel.

The two long strands of sugar–phosphate units are twisted around each other to form the familiar double helix (Figure 32.6b). So it is as if the ladder has rope sides. One of the most significant events in the history of biology was the elucidation of the structure of DNA, in 1953: for this work,

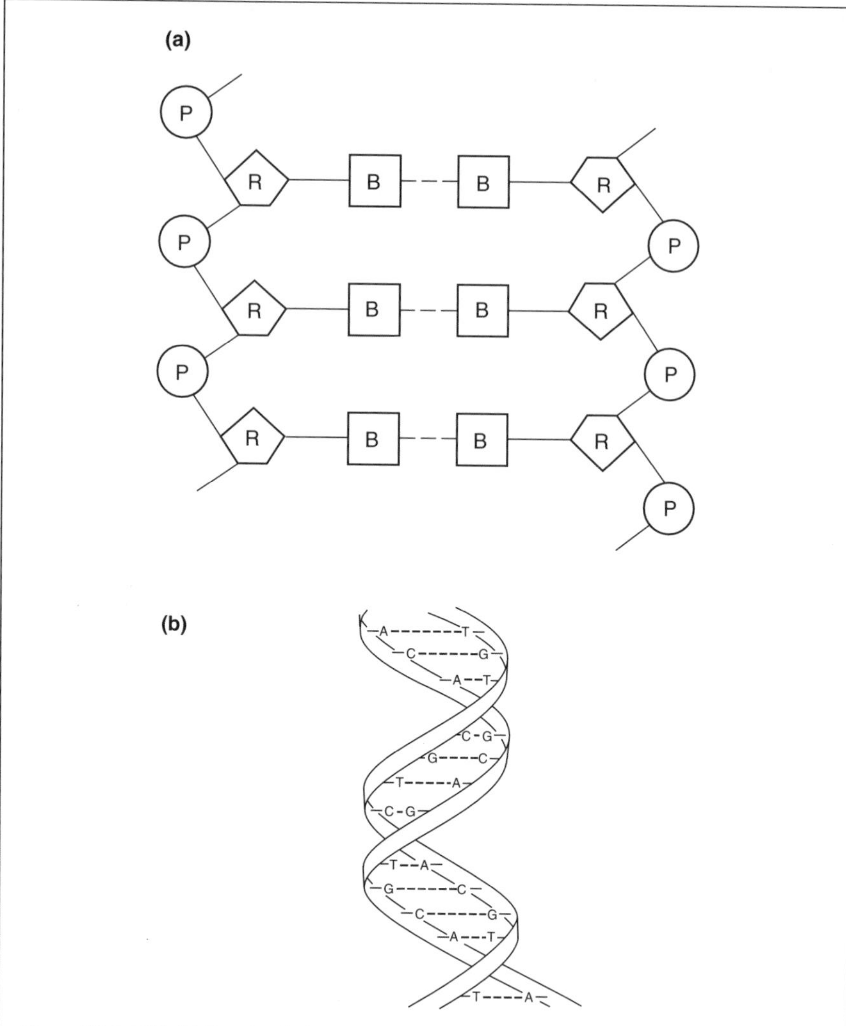

Figure 32.6 DNA. (a) The chemical components of DNA. The 'ladder' has been flattened for clarity. The symbols are: P phosphate, R ribose, B nitrogenous base. The bases are either adenine (A), thymine (T), guanine (G) or cytosine (C); they may be in any order along one strand of the molecule, but A always pairs with T, and C always pairs with G on the opposite strands. The dashed lines denote the hydrogen bonds that hold the sides of the structure together; (b) the characteristic 'double helix' shape of the DNA molecule.

Francis Crick, James Watson and Maurice Wilkins were awarded Nobel prizes.

We mentioned earlier that DNA must be able to replicate itself. This is necessary because each new cell must receive a copy of the genetic material. To appreciate this process we can envisage, first, the uncoiling of the double helical structure of DNA, and secondly, the separation (or 'unzipping') of the molecule so that the nucleotide bases are exposed. (Hydrogen bonds,

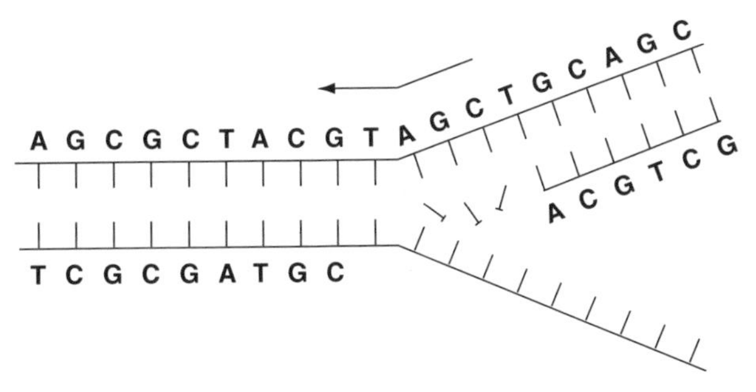

Figure 32.7 Duplication of a DNA molecule. The helical structure first unwinds, and the strands separate. A complementary strand of DNA forms using the existing DNA as a template.

which hold the bases together, are much weaker than covalent bonds.) Then, nucleotides with complementary base pairs are pulled into position on each of the nucleotide strands so that two DNA molecules are produced (Figure 32.7).

The whole complement of DNA present in each cell of an organism is referred to as its **genome**. In June 2000, a 'first draft' of the human genome was announced. The objective here has been to sequence the base pairs on all 46 human chromosomes. The enormity of this task becomes apparent when we consider there are over 3 billion base pairs in the human genome. The job could not have been done in the time – about a decade – without enormous computing power, the use of robotics and rapid technical advances in DNA technology.

The other nucleic acid, RNA, differs from DNA in certain respects. First, RNA exists as a single-stranded molecule, not as a double helix; it is as if the nucleotide bases are permanently exposed rather than being held to other bases by hydrogen bonds. Secondly, the sugar unit in RNA is ribose, not deoxyribose – hence the name. Third, the nitrogenous base uracil (U) substitutes for thymine: they are both pyrimidines and uracil differs only in having one less carbon atom attached to its ring structure. There are a few different sorts of RNA in the cell. They are differentiated by the length of the strands, the shapes they adopt and the functions they perform. This topic is discussed further below.

32.4 The link between nucleic acids and proteins

This topic is absolutely central to understanding life on Earth. Some familiarity of its rudiments will be useful to students of geography and the environment in a variety of contexts. For example, it will be invaluable if you are studying the origin of life and life's early evolution, and it will enable you to better understand the scientific background to the debate about genetically modified organisms. An understanding of the basic structural features of proteins and nucleic acid (see **32.3.4** and **32.3.5**) is necessary to appreciate the vital relationship between them. We begin here with a descriptive account and then look at the mechanisms involved.

In essence, molecules of DNA form a repository of information, in chemically coded form. The code determines which peptides (in effect which proteins) are to be produced. A peptide is defined by the ordering of amino acids. The link between DNA and proteins is that the sequence of the four nucleotide bases (abbreviated as A, T, C and G) on a DNA strand specifies the ordering of amino acids, and hence dictates which peptides are to be produced. However, the DNA is just the information store, it cannot by itself do the job of protein manufacture. It is the RNA that 'reads' the DNA base sequence, carries this information to the sites in the cell where peptides are assembled, and then orchestrates the assembly of peptides. And the different types of RNA have specialized roles in this process. We now elaborate a little on the operation of the process.

A key point is that the 'sides' of DNA molecules are organized functionally into segments, each containing three adjacent nucleotide bases, for example ACG, ATT, GCA, CTG. A particular triplet of bases 'specifies', or codes for, one type of amino acid. In fact, because there are 64 permutations for the order of bases in a triplet, and only 20 acids, more than one triplet can code for the same amino acid. Others carry stopping instructions. The correspondence between base triplets and amino acids forms what is called the 'genetic code', which appears to be the same regardless of the type of organism. By extension, the order of base triplets along a DNA strand serves as a template which determines the order of amino acids in a peptide. The universality of the genetic code makes possible the genetic modification of organisms: a gene that specifies for a particular protein in one type of organism, perhaps a bacterium, will specify exactly the same protein in another, quite unrelated, organism, a plant for example, into whose genome it is inserted. And the protein will imbue the plant with the same characteristic as it did in the bacteria.

So the first task of RNA is to 'read' the sequence of triplets on the DNA molecule, a process referred to as **transcription**. This is the specialized role of messenger RNA (mRNA). For the base sequence along a DNA strand to be read, the double helical structure must first 'unzip' (Figure 32.7), as it does prior to DNA replication. A piece of RNA of complementary base sequence is then produced. So if along a segment of a DNA strand the order of bases is CTG, CGA, TGC, the complementary strand of RNA that is produced will be GAC, GCU, ACG. (Note the U, for uracil, which in RNA substitutes for the thymine of DNA.) The mRNA strand then migrates from the nucleus to the sites in the cell where the peptides are produced (except in the case of prokaryotic organisms as these have just a single chromosome and no nucleus).

The next stage is known as **translation**. It involves the ordering of amino acids, as specified by the order of base triplets (known as **codons**) on the mRNA strand, to produce peptides (Figure 32.8). This occurs at sites known as **ribosomes**, which are made up of protein and another type of RNA, ribosomal RNA (rRNA). Ribosomes and mRNA strands become attached to

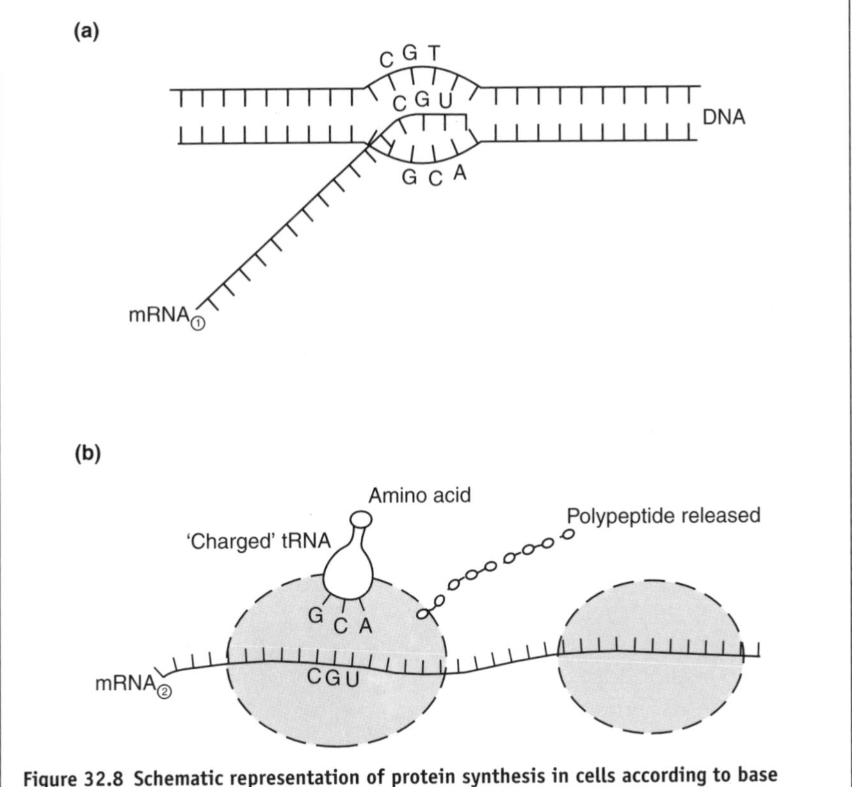

Figure 32.8 Schematic representation of protein synthesis in cells according to base sequence on DNA strand.

each other. Another form of RNA, transfer RNA (tRNA), has the job of 'collecting' the appropriate amino acids in the cell and conveying them to the ribosomes. These tRNA molecules are smaller than other RNA types and have a characteristic bulbous shape (Figure 32.8b). Located at one end of a tRNA molecule is a triplet of unpaired bases, called an **anticodon**. Now, each type of amino acid has its own particular tRNA molecule, distinguished by the three-base sequence of the anticodon. Specific enzymes, which 'recognize' both the anticodon and the amino acid, are involved in connecting an amino acid to a tRNA molecule; the tRNA is then said to be 'charged'. A charged tRNA molecule conveys its amino acid to the mRNA molecule. Recall, the sequence of bases along the mRNA strand is complementary to that of the DNA which was read originally. And since each tRNA molecule has the appropriate amino acid attached to it, the ordering of amino acids is the same as that specified by the DNA.

For the formation of peptides, ribosomes move along the mRNA molecule, and as each triplet of bases is passed, the amino acid brought into position by its tRNA molecule is attached to the lengthening peptide. Relieved of its amino acid, a tRNA molecule is then free to pick up another amino acid of similar type. What we cannot easily convey is the astonishing speed with which these processes occur, with several proteins being synthesized simultaneously within the cell.

Naturally our account is a gross simplification and begs a number of key questions. Of great importance is the regulation of genes. How, and when, are genes turned on and off? Given that each cell contains the same genome, how do brain cells 'know' the cells are brain cells rather than liver cells? Despite the pace of progress in molecular biology, answers to such questions are still proving to be very elusive. Another question, which is of great interest in an evolutionary context, is: which came first, RNA, DNA or protein? Some forms of life – or perhaps we should refer to them as protolife-forms – must have had just one type of molecule. But which one? At present the consensus is that it was RNA, but how RNA originated is also a matter of conjecture and debate.

33 Enzymes

33.1 Overview

Enzymes are biomolecules (i.e. produced by living cells) which accelerate chemical reactions. They are thus often referred to as 'biological catalysts'. Enzymes are mentioned in the introductory section on chemical reactions (**10.2.1**), but their central role in life processes justifies some further comments. Like catalysts generally, enzymes are not consumed during the reactions they catalyse, and hence can be used over and over again. Unlike inorganic catalysts, however, enzymes are **substrate specific**. In effect, each enzyme acts upon only one substance (the substrate) or just a few very similar substances.

Enzymes are necessary for the many chemical reactions that constitute an organism's metabolism. A **metabolic pathway** is a sequence of biochemical reactions such that the product(s) of one reaction serve(s) as the substrate for another reaction, thus:

$$A \xrightarrow{\text{Enzyme 1}} B \xrightarrow{\text{Enzyme 2}} C \xrightarrow{\text{Enzyme 3}} D$$

Here, each of the reactions (for example $A \rightarrow B$) is catalysed by a specific enzyme. Enzymes are also responsible for the digestion (or breakdown) of large molecules, such as starch and protein, that are consumed by animals and other heterotrophic organisms in their food. This process normally occurs outside the cell, either in the digestive tract as in the case of animals, or outside the organism itself as in the case of bacteria and fungi. The products of digestion, for example sugars and amino acids, are then small enough to be absorbed across the wall of the digestive tract of animals or the cell membranes of bacteria and fungi. Certain enzymes occur in nearly all organisms simply because some processes, respiration (**34.3.3**) for example, are common to most organisms. Similarly, all photosynthetic plants have certain enzymes in common. In contrast, other enzymes may be found in a very limited range of organisms, perhaps just a single species.

Chemically, enzymes are wholly or mainly proteins, typically being made from around one hundred amino acid units. Some enzymes consist of a single polypeptide (see **32.3.4**), but others are much more complex and are composed of several polypeptides. Many enzymes contain a non-protein component, termed a **cofactor**. The cofactor may be a metal, for example magnesium or iron, or it may be an organic component, in which case it is often termed a **coenzyme**. Examples of coenzymes are the several substances which comprise the vitamin B complex.

An international code governs the naming of enzymes, which is necessary because of the vast, and increasing, number of enzymes known to science. Six groups of enzymes are recognized, namely oxidoreductases, transferases, hydrolases, lyases, isomerases and ligases. The first part of the name describes the type of reaction which is catalysed: thus hydrolases bring about the addition of water molecules to the products of a reaction. Note that all group names finish with the suffix '-ase'. This is true of the individual enzyme names also. For example, the enzyme maltase catalyses the hydrolysis of maltose, generating two molecules of glucose per molecule of maltose. However, some enzymes are commonly known by their older name, which does not end in '-ase'. An example is pepsin, a digestive enzyme produced in the stomach. Usually, though, the suffix will tell us immediately whether a substance is an enzyme.

33.2 Mode of action

Each enzyme has an area on its surface called the **active site**. This is where enzyme and substrate join together during a reaction. The active site has a unique geometry, and will accept only substrate molecules which fit that configuration. From this developed the 'lock and key' model of enzyme action. Attraction and repulsion is also determined by the electrical charges around the active site of the enzyme and the substrate molecule. The lock and key model of enzyme action suggests a rigidity of both enzyme and substrate. In fact, in many cases a better analogy would be that of a hand entering a glove. This is because parts of the substrate molecule (the hand in this analogy) may induce some changes in the shape of the enzyme (the glove) to accommodate it. Accordingly, the term 'induced-fit model' is applied.

Like all catalysts, enzymes are said to lower the **energy of activation**. This term refers to the amount of energy required to bring about a chemical reaction. Enzymes do not bring about chemical reactions that would not occur in their absence, but without enzymes the reactions would proceed extremely slowly at 'normal' temperatures – so slowly in fact that it would take several decades for you to digest a single meal. The rate at which enzyme activity occurs, i.e. the rate at which the substrate is converted to products, depends on the amount of substrate and also the prevailing environment. A

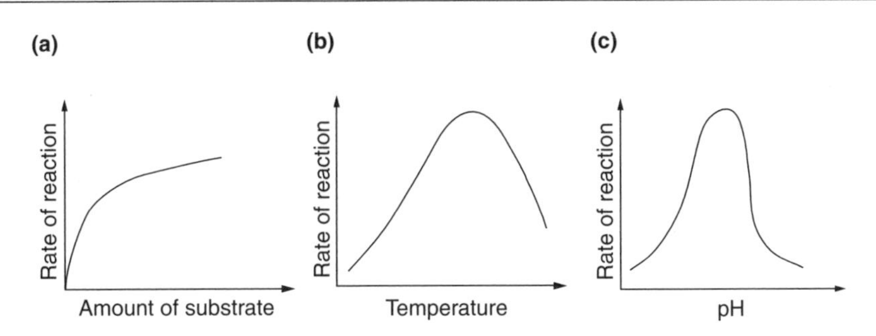

Figure 33.1 Generalized relationships between the rate of enzyme reaction and (a) amount of substrate; (b) temperature; (c) pH. The response curves for temperature and pH are specific for each enzyme.

plot of the relationship between the rate of reaction against the amount of substrate is hyperbolic in form (Figure 33.1a). At low substrate concentrations the active sites of the enzymes are not saturated, and reaction rate and substrate concentration are more-or-less linearly related. As substrate concentration increases, however, the number of unoccupied active sites declines and eventually none is available. The enzyme is then said to be saturated. At greater substrate concentrations the reaction rate is determined by the turnover time, which is the amount of time it takes for an individual reaction to occur.

Individual reactions between enzyme and substrate are determined largely by temperature (Figure 33.1b). Over a particular temperature range the guideline is that, for every 10 °C increase in temperature, there is a doubling in the rate of a chemical reaction. An increase in heat (i.e. an increase in energy in the system) increases the rate of movement of molecules, and thus raises the probability of a successful engagement between enzyme and substrate. The temperature range over which such a relationship holds is limited. Above a certain temperature, the relatively weak bonds maintaining the tertiary structure (see **32.3.4**) of the protein molecules begin to break, and the protein molecule loses its structure: it is said to **denature**. As a result, it also loses its capacity to function as an enzyme. So above this critical temperature the rate of enzyme activity declines very sharply (Figure 33.1b). The specific response of an enzyme or series of enzymes to temperature explains a great deal about the way that whole organisms respond to temperature changes in their environment.

The rate of enzyme–substrate reactions is also affected by pH (Figure 33.1c). For each such system there is a pH at which the rate of the reaction is at its maximum: increasing acidity or increasing alkalinity of the solution serves to reduce the rate at which substrate is converted to product.

34 Energy and life

34.1 Overview

Energy is introduced and dealt with systematically elsewhere (**28**). Here we focus on the energy metabolism of living organisms. Energy, which is informally defined as the ability to do work, is a key topic when considering life processes because all living organisms are working entities. We consider the different modes of energy acquisition and energy use by organisms, the various energy transformations involved and the chemical reactions that are associated with these transformations. This information will contribute to a better understanding of topics as diverse as the origin of life, the function of ecological systems, the productivity of the biosphere, the circulation of carbon and oxygen in the environment, and global warming.

34.2 Life's energy currency: ATP

Throughout the sections dealing with living processes we emphasize the unity of life: for example, the universal nature of the genetic code, and the link between nucleic acids and proteins (**32.4**). In discussing energy and life we have more unifying concepts. Of great importance is the fact that whenever work is being done in living cells, regardless of the type of organism, the substance called **adenosine triphosphate**, or **ATP**, is almost always involved. ATP is the universal short-term energy carrier in living cells: it is sometimes known as 'life's energy currency'. Individual molecules of ATP are not stored in the body, nor are they transported around an animal or a plant, as for example are sugar molecules. Rather, they are continuously produced.

The structure of an ATP molecule is shown in Figure 34.1. It is composed of adenine, ribose (both of which are components of nucleic acids) and phosphate groups. (Note that the individual atoms of the adenine and ribose components are omitted for simplicity.) Our focus here is the three phosphate

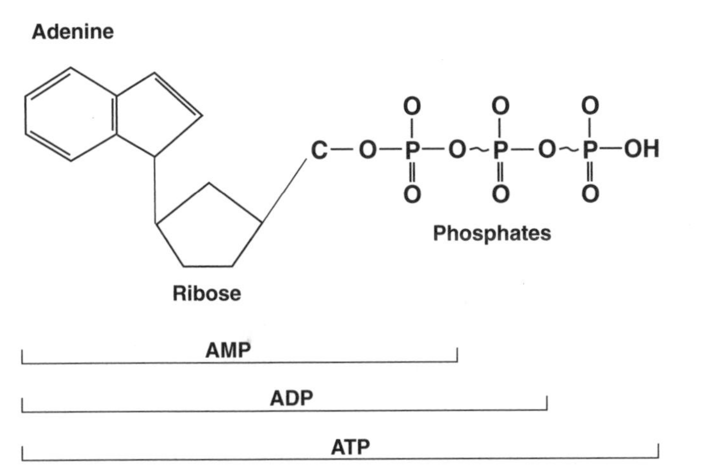

Figure 34.1 Schematic representation of an ATP molecule. The cleavage of the oxygen-to-phosphorus bond (indicated by the 'squiggle') is the immediate source of energy for nearly all endergonic (energy-requiring) reactions in living cells. Hydrolysis of ATP leaves ADP and inorganic phosphate.

groups and the two oxygen-to-phosphorus bonds that link them. These two bonds are important because they are readily broken in hydrolysis reactions to make energy available for work in the cell. On hydrolysis, a molecule of ATP yields a molecule of **adenosine diphosphate** (ADP), inorganic phosphorus, and some free energy. This is a highly exergonic reaction, i.e. it releases energy. The energy made available per mole of ATP is approximately 52 kJ, although this value varies according to the conditions. Most of the energy used in the living cell is provided by the hydrolysis of ATP. (Less commonly, molecules of ADP are also hydrolysed, which results in the formation of **adenosine monophosphate** (AMP).) Crucially, the hydrolysis of ATP is closely coupled to the many and various energy-consuming (endergonic) reactions which are necessary to sustain life. The endergonic reactions in the cell nearly always involve the addition of phosphate to another chemical group, a process known as **phosphorylation**. The addition of phosphate enhances the energy status of the accepting group, which means that the group is more likely to enter into chemical reactions. Endergonic reactions include the synthesis of molecules within cells, the transmission of nerve impulses, the contraction of muscles, and the pumping of blood in animals.

In our discussions so far, we have assumed that ATP is present in the cell. But the formation of ATP, particularly its oxygen-to-phosphorus bonds, requires an input of energy. Now we consider where this energy comes from, and we will also see that it is possible to categorize most living organisms on this basis. These processes are of significance not just for individual organisms – if they were then they would probably not justify consideration in this book. Rather, the energy transformation processes of living organisms are of immense importance for the whole planet, particularly at a time of great concern about changing climate.

34.3 Ways to ATP formation

34.3.1 Photosynthesis

Most of us are aware that Earth's biota (i.e. the totality of life on the planet) is solar powered. What we mean by this statement is that the chemical energy held within organic matter (carbohydrates, fats, proteins, etc.) derives ultimately from the conversion of solar energy. More specifically, we mean that certain types of organism are able to transform some of the Sun's radiant energy to chemical energy. The process involved is called **photosynthesis**. The prefix *photo-* denotes that light (i.e. visible radiation) is necessary for this process (see **22.2.3**). The term **photosynthetically active radiation** (PAR) refers to the wavelengths of electromagnetic radiation (approximately 400–700 nm) that activate photosynthesis.

The organisms responsible for transforming radiant energy to chemical energy by photosynthesis are the green plants, the algae (found mostly in water, and traditionally considered to be plants), and a few types of bacteria. Photosynthetic organisms possess the green pigment **chlorophyll**. Actually there are a few types of chlorophyll (denoted by letter, as in chlorophyll *a*, *b* and *c*) which differ slightly in composition. Different types are associated with different types of photosynthetic organism. In plants and in algae the chlorophyll is contained within special organelles called **chloroplasts**.

As well as being an energy conversion process, photosynthesis also involves chemical reactions. Chemical raw materials are necessary for the manufacture of energy-rich organic molecules during photosynthesis, and by-products result from the reactions. All photosynthetic organisms use carbon dioxide as a source of carbon, and all but a few use water as a source of hydrogen. (Water is said to be the hydrogen donor.) A simple representation of this process is:

$$H_2O + CO_2 \rightarrow (CH_2O) + O_2$$

where (CH_2O) represents the first stable product of photosynthesis. (Textbooks often show glucose here, with the expression written as an equation which means there must be the same number of carbons, hydrogens and oxygens on each side of the equals sign.) During photosynthesis, water molecules are split, the oxygen is released and carbon dioxide is reduced by the hydrogens. This is not a one-step process, however. It involves the formation of ATP and a series of biochemical reactions which lead to the manufacture of energy-rich products. (The details of these processes can be found in any basic general biology or plant physiology text.) Photosynthesis involves both light and dark phases. The generation of ATP, which requires an input of energy, occurs in the light, but the use of that ATP, for splitting water molecules and synthesizing organic molecules, is independent of light. We should note also, then, that the chemical products of photosynthesis, as well as providing a source of energy, supply carbon skeletons for the manufacture of other biomolecules.

A few types of bacteria do not use water as a hydrogen donor: most such organisms use hydrogen sulphide (H_2S) instead. In this case, sulphur, not oxygen, is a by-product of the reaction. Such bacteria are confined to anaerobic environments. It is believed that this type of photosynthesis was the first to evolve, around 3.5 billion years ago. This was a momentous event in the history of life because it meant that, for the first time, ecosystems were powered by a source of energy (sunlight) which was in continuous supply. Another momentous biological 'invention' was that of water-splitting photosynthesis, somewhat later. Henceforth, photosynthesis could occur wherever water was available, providing that the other conditions for photosynthesis (light, heat, nutrients etc.) were met. Water-splitting photosynthesis also transformed the planet because of the free oxygen (O_2) which was released as a by-product. For many hundreds of millions of years, this free oxygen combined with molecules in the atmosphere, and with rocks and minerals at the Earth's surface. But it seems that from around 2.5 billion years ago, oxygen began to accumulate in the atmosphere. Furthermore, from this oxygen, ozone was formed. Because ozone absorbs incoming ultraviolet (UV) radiation (see **22.2.2**) this predisposed the land masses for the colonization by life, which followed much later. Another crucial by-product of the new oxygen-rich environment was the appearance of oxidative energy metabolism, which, as we shall see, is far more efficient in terms of ATP generation than anaerobic metabolism.

Photosynthetic organisms are referred to as **autotrophs**, or **autotrophic** organisms, which literally means they are 'self-feeding'. In terms of their energy requirements, they are independent of other organisms. Ecologists also apply the term **primary producer** to such organisms, because they are responsible for the initial production of organic matter from simple inorganic components.

34.3.2 Chemosynthesis

There is another form of autotrophic energy metabolism, called **chemosynthesis**. This process is confined to certain bacteria (which are sometimes called **chemoautotrophs** to distinguish them from **photoautotrophs**). It is not the energy of sunlight that is harnessed in chemosynthesis but, as the name suggests, the energy contained within simple chemical substances, which are scavenged from the environment. These substances are oxidized and the energy which is liberated is used to generate ATP. As with photosynthesis, this process results in the formation of organic compounds, using carbon dioxide as raw material. A variety of chemical species are used in chemosynthesis, including ammonium ions, nitrite ions, hydrogen sulphide and even hydrogen gas, while the bacteria which meet their energy needs in this way are quite specific in terms of the chemical entities they use.

Chemosynthesis merits our attention for a number of reasons, even though the amount of organic matter produced in this way is vanishingly small compared with that produced by photosynthesis. First, because chemosynthesis involves chemical reactions, it is a key process in the cycling of chemical elements, particularly nitrogen and sulphur, at the Earth's surface. Secondly, in lightless environments, chemosynthesis is the only means available for primary production. If organic matter is not imported from elsewhere, life cannot be sustained in dark environments without chemosynthesis. Ecosystems based on chemosynthetic bacteria have been discovered around submarine hot springs deep on the ocean floor where mineral-rich water is ejected at high temperatures. The bacteria use the hydrogen sulphide which is found in such hot water plumes. Other locations include caves and deep wells hundreds of metres underground. All the other organisms found in such environments depend on the energy-rich organic matter produced by chemosynthetic bacteria. The third reason why chemosynthesis is so important is because of a growing body of support for the view that the very first organisms on our planet used this form of energy metabolism, nearly 4 billion years ago.

34.3.3 Respiration and fermentation

Organisms which are neither photosynthetic nor chemosynthetic (i.e. are not autotrophic) must generate ATP in other ways. As they are unable to manufacture energy-rich organic molecules from raw materials using a non-biotic energy source, such organisms need to procure organic molecules 'ready-made'. They do so by consuming other organisms, their dead remains or their waste products. The organic molecules are then dissassembled in a series of enzymatically controlled reactions, with some released energy being conserved as ATP. Such organisms are thus dependent on other organisms for their energy needs. Because of this they are called **heterotrophs**, or **heterotrophic organisms**, meaning to 'feed on others'. Ecologists call such organisms **secondary producers** to distinguish them from primary producers. Secondary producers include all animals, all fungi, most bacteria and a huge range of single-celled organisms.

Nearly all organisms use free oxygen (O_2) in the generation of ATP. The process is called **respiration**, or **oxidative energy metabolism**. Now, although we introduced respiration in the context of heterotrophic organisms, this process is not confined to heterotrophs. Photosynthetic organisms do not stop using energy at night when there is no light for photosynthesis. And plants do not consist solely of green tissue; they also have non-photosynthetic organs such as roots and stems, and seeds or spores. Respiration occurs in all these organs, and it also occurs within photosynthetic tissues during daylight. So respiration occurs in plants just as it does in animals and all other aerobic organisms. Respiration is both an energy transformation process and a chemical process. Only a proportion of the chemical energy released during respiration is conserved by organisms, the rest is transformed to heat. In chemical terms, respiration can be simply represented as follows:

$$(CH_2O) + O_2 \rightarrow H_2O + CO_2$$

where CH_2O represents an organic molecule, or **energy substrate**, that is broken down in the presence of oxygen. Water and carbon dioxide are the chemical by-products of this process, the water being formed by the addition of oxygen to hydrogens released during some of the chemical reactions. As with photosynthesis, we show in this representation of the process only the chemical raw materials and by-products.

You may have noticed that photosynthesis and respiration appear to differ only in the direction of the reaction, indicated by the arrows. Carbon dioxide is thus a raw material for photosynthesis but a by-product of respiration. We can extend this idea to a single plant, to any defined area of ground or water, or to the entire biosphere. So a plant can grow only if the amount of carbon fixed in photosynthesis exceeds the amount that is lost in respiration. (We would thus expect plants to lose weight at night because they are still respiring but not assimilating carbon in photosynthesis.) For the biosphere as a whole, the balance between photosynthesis on the one hand, and the respiration of *all* organisms on the other, influences the amount of carbon dioxide in the atmosphere. This is shown by intra-annual oscillations that occur in atmospheric carbon dioxide concentrations, which are particularly marked in the northern hemisphere. The concentration dips during the summer because of the large amounts of carbon being sucked out of the atmosphere by photosynthesis, but rises during the winter months because photosynthesis is severely restricted, while respiration continues, albeit at a relatively low rate. The balance between photosynthesis and respiration over longer time periods is crucial in determining the amount of carbon dioxide in the atmosphere, which has a major influence on the planet's surface temperature.

In summary, we have said that respiration involves the breakdown of energy-rich organic molecules in a series of enzymatically controlled reactions, with the conservation of some of the released energy as molecules of ATP. If you want to look in detail at the biochemical pathways involved, they can be found in most college-level biology texts. However, an outline of the scheme is presented in Figure 34.2 because it will assist your understanding of some important environmental phenomena.

It is convenient to think of respiration as a two-stage process. The first stage, called **glycolysis**, involves the conversion of a suitable energy substrate, such as glucose (which we assume here) or fatty acids, through a series of reactions, to two molecules of a three-carbon substance called pyruvate. In the early stages of glycolysis, phosphorus, which is supplied by ATP, is added, and each six-carbon molecule derived from glucose is then split into two three-carbon molecules. From the subsequent series of reactions, ATP is generated. So some ATP is necessary to begin the process, but it is a very worthwhile investment. From one molecule of glucose the ultimate net yield from glycolysis is two molecules of ATP if oxygen is absent and eight molecules of ATP if oxygen is present.

Glycolysis is not a very efficient process for ATP generation. It probably appeared quite early in the history of life, but a more efficient ATP-generating biochemical pathway later evolved, which is outlined below. However, we stay with glycolysis for a moment to consider organisms (they are nearly all

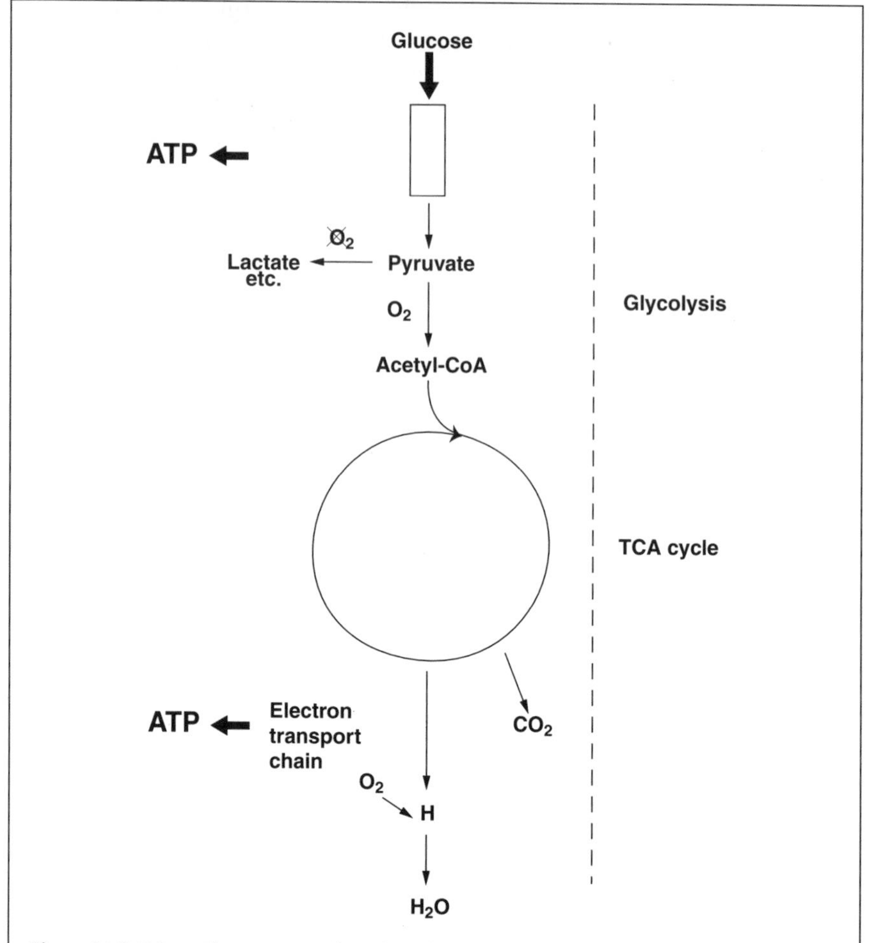

Figure 34.2 Schematic representation of respiration as a two-stage process. Glycolysis, which ends with the generation of pyruvate, does not require oxygen. In most organisms, subsequent reactions lead to the tricarboxylic acid cycle which involves free oxygen.

bacteria) which do not use free oxygen and for whom this gas is actually poisonous. We should point out that such organisms are not poorly adapted in general: indeed they are extremely well adapted to those environments, such as waterlogged peat bogs, where oxygen is absent. Organisms of this type are often called **fermenters**. They convert pyruvate to various by-products (depending on type of organism) including lactate and ethanol (an alcohol). When the cells of aerobic organisms are deprived of oxygen, they too switch to anaerobic energy metabolism: for example, mammalian cells generate lactate as a by-product. But such organisms cannot survive long without oxygen.

Now we return to aerobic energy metabolism, which we left at the point at which pyruvate is generated (Figure 34.2). If oxygen is present, pyruvate is oxidized further, in another series of biochemical reactions. First, pyruvate is used to produce, in a rather complex process, a substance known as acetyl coenzyme A, usually shortened to acetyl-CoA. This two-carbon substance is then fed into the series of reactions known variously as the **tricarboxylic acid (TCA) cycle**, the **citric acid cycle** or the **Krebs cycle** (after Hans Krebs who was largely responsible for first describing it). We use TCA cycle here. The word 'cycle' is used because, after a series of biochemical conversions, a four-carbon molecule results (called oxaloacetate). This reacts with a molecule of acetyl-CoA to repeat the series of reactions, as depicted in Figure 34.2. From the TCA cycle, carbon dioxide is released. In addition, hydrogens are released, and these combine with oxygen to form water. (Recall the by-products of respiration shown in our simple representation of respiration earlier in this section.) More significant from an energetic point of view is the fact that each turn of the TCA cycle yields 30 molecules of ATP per molecule of glucose. This is added to the eight molecules of ATP generated during glycolysis when oxygen is present, thus a total of 38 molecules of ATP per molecule of glucose is produced during aerobic energy metabolism.

Organisms that can tolerate only brief periods of oxygen exclusion (and most fall into this category) are termed **obligate aerobes**, while those that cannot tolerate free oxygen are called **obligate anaerobes**. Just a few micro-organisms are **facultative aerobes** (or **facultative anaerobes**), meaning that they can tolerate both aerobic and anaerobic conditions. The most familiar examples of these latter are yeasts: under aerobic conditions yeasts release carbon dioxide (an effective raising agent, so yeasts are used extensively in baking), but if deprived of oxygen they switch to anaerobic metabolism and produce ethanol (ethyl alcohol), on which the brewing industry is based. It is worth mentioning that many of the terms used here are fairly loosely defined. Fermentation is often applied to anaerobic energy metabolism in general, but sometimes the term is confined to the post-pyruvate stage. The term **anaerobic respiration** is sometimes used to refer to anaerobic energy metabolism in general, but it is also used specifically for a process, confined to certain bacteria, which uses not oxygen but an alternative inorganic species.

The final points in this section refer to the sites in the cell where energy metabolism occurs. Glycolysis occurs in the cellular cytoplasm. In eukaryotic organisms (in effect all organisms except bacteria), the products of glycolysis are then transferred to specialized cellular organelles called **mitochondria** (singular, mitochondrion) where the TCA cycle proceeds.

34.4 The mechanism of ATP formation

We have seen that both photosynthesis and respiration yield ATP. In the former case, visible radiation (sunlight) is the energy source; in the latter case ATP is generated from the enzymatically controlled breakdown and conversion of an energy substrate such as glucose. We have not said much, however, about the *mechanism* of ATP formation. This is quite a complex subject but it is useful to be aware of the nature of this process. ATP can be generated both directly and indirectly. In the former case, a substance containing phosphorus combines with a molecule of ADP under appropriate conditions to form ATP. This process, known as **substrate-level phosphorylation**, is a fairly inefficient way of generating ATP. It occurs during glycolysis, which, as already noted, is a quite ancient process.

An appreciation of the indirect method of ATP formation, or **chemiosmotic phosphorylation**, requires some familiarity with **coenzymes** (see **33.1**) and also oxidation and reduction processes (**14.2**). Coenzymes are non-proteinaceous chemical groups that are necessary for many enzymes (which *are* proteins) to function. Redox reactions are coupled reactions that involve the transfer of electrons from one substance (which is thus oxidized) to another (which is reduced). Moreover, as protons are also often transferred, either on their own or in company with electrons, oxidation can also be defined in terms of the transfer of protons or hydrogens. Hydrogen transfer is actually very important in the present context.

The significance of oxidation and reduction here is that during some of the biochemical reactions of respiration and photosynthesis, electrons are transferred between a number of molecules, each of which is successively reduced and oxidized. Such molecules are called **electron carriers**, while a series of them constitutes an **electron transport chain** (or **system**).

A group of coenzymes serves as one type of electron carrier. They play such a critical role in the processes which lead to ATP formation that they merit some comment. The three most common coenzymes are called NAD (nicotinamide adenine dinucleotide), NADP (nicotinamide adenine dinucleotide phosphate), and FAD (flavin adenine dinucleotide). NAD and NADP play key roles in respiration and photosynthesis, respectively, while FAD is involved in both processes. In the oxidized state, NAD and NADP normally carry positive charges; they are therefore represented by NAD^+ and $NADP^+$, respectively. In the oxidized state they can normally accept two electrons, one of which neutralizes the positive charge while the other attracts a proton. In the reduced state these can be written as NADH and NADPH respectively. As one of the protons remains free, they may also be represented by $NADH + H^+$ and $NADP + H^+$. When NAD^+ or $NADP^+$ oxidizes a substrate (i.e. accepts its electrons), NAD or NADP is formed. FAD behaves a little differently from NAD and NADP. In the oxidized state, FAD accepts two electrons and two protons. As this is equivalent to two hydrogen atoms, the reduced state is represented by $FADH_2$.

Why are coupled oxidation and reduction reactions involving these coenzymes so important to the generation of ATP? The reason is that when a coenzyme is reduced, its free energy status is raised. In this more reactive state it transfers its electrons and hydrogens to a molecule of lower energy status. During some of the biochemical conversions of the TCA cycle, NAD^+ and FAD are reduced, and the electrons they receive are passed along an electron transport chain. Such chains involve other types of carrier molecule, including an exceedingly important group of proteins called **cytochromes**. There are a number of cytochromes, and they are designated by letters, as in cytochrome b. The cytochromes contain a haem (iron-containing) group. Now, iron can occur in different oxidation states. When an electron is picked up by the iron species Fe^{3+} it is converted to Fe^{2+}; in other words it is reduced. This is a reversible reaction: when Fe^{2+} gives up this electron it is converted to Fe^{3+}, i.e. it is oxidized.

During the transfer of electrons along an electron transport chain ATP is generated, using ADP and inorganic phosphate as raw materials. As oxygen is required for the TCA cycle, ATP generation by this route is called **oxidative phosphorylation**. The hydrogens released from the TCA cycle combine with oxygen to form water, which is one of the by-products of respiration. Oxygen is thus said to be the **terminal electron acceptor**. The other chemical by-product of respiration, carbon dioxide, is released quite early in the TCA cycle. Its loss reduces the number of carbons in the intermediary molecules of the TCA cycle, first from six to five and then from five to four. It is the four-carbon substance, oxaloacetate, which combines with the two-carbon acetyl-CoA to begin a new turn of the cycle, as outlined in **34.3**.

35 Chemical equilibria

35.1 An example of chemical equilibrium

Chemistry deals with chemical interactions between materials. These are the reactions that bring about changes in the physical and chemical nature of the materials involved. As discussed elsewhere (see **8**), such interactions involve the outer electrons of the atoms and molecules, and a focus of modern chemistry is the shape and intensity of the electrical fields involved. In the past, however, the primary focus was on the extent to which various substances reacted, and a sophisticated equilibrium theory was developed to quantitatively predict this. In this section we look at the equilibrium concept, and see how far it is useful in explaining chemical reactions in the natural environment.

To illustrate the point, let us consider the reaction between carbonic acid and water. The carbonic acid (H_2CO_3) gives away one of its hydrogen ions to the water, leading to the formation of bicarbonate ions (HCO_3^-) and hydroxonium ions (H_3O^+):

$$H_2CO_3 + H_2O \rightarrow HCO_3^- + H_3O^+$$

The arrow tells us that in this expression it is the H_2CO_3 and H_2O that are reacting (these are termed the **reactants**), and that the products of their reaction are HCO_3^- and H_3O^+ (termed the **reaction products**). The expression does not tell us how far the reaction proceeds. In other words, it does not tell us how much $HCO_3^- + H_3O^+$ we will get, and how much $H_2CO_3 + H_2O$ will remain.

Experimentation with this system would soon show us that the reaction does not go to completion. In other words, there is a considerable amount of H_2CO_3 left in the solution. Indeed, it turns out that we should not consider the reaction as unidirectional at all. Rather, in addition to a continuous reaction of water molecules with carbonic acid molecules to form bicarbonate and hydroxonium ions, there is also a continuous reverse reaction:

$$HCO_3^- + H_3O^+ \rightarrow H_2CO_3 + H_2O$$

What we have here is a **dynamic equilibrium** between the substances, in which formation and destruction are both taking place continuously. Because the competing chemical reactions are proceeding at the same rate, the mixture appears to be static; the proportions of HCO_3^- and H_2CO_3 reach a constant value. This situation is represented in chemical expressions by use of the double arrow:

$$H_2CO_3 + H_2O \rightleftharpoons HCO_3^- + H_3O^+$$

If single arrows are used in chemical expressions it normally indicates that the reaction is very one-sided. Strictly, the reaction will always be a dynamic equilibrium, but it may go so far that the one side is negligible.

What factors will influence the direction of the equilibrium? Let us alter the amounts of the reactants. We cannot, by definition, change the amount of water; in normal aqueous reactions it is always present in such excess that it is assumed to be constant with unit concentration (that is a value of 1). However, we can increase the amount of H_2CO_3, and we find that this leads to an increase in the amount of both reaction products.

This leads to a very simple rule concerning chemical equilibria known as **le Chatelier's principle**. If you impose a change on a system that is in equilibrium, it will shift in such a way as to reduce the impact of the change. In this case, if we put in more H_2CO_3 the reactions move to the right, consuming some of the H_2CO_3, and thereby reducing its impact. Likewise, if we were to add either HCO_3^- or H_3O^+ ions to the system, it would move to the left, turning some of the added ions into H_2CO_3.

This rule is extremely useful, as it tells us the tendency of the reaction. However, we can do better. The relationship between the different species can be formally expressed by:

$$\frac{[HCO_3^-][H_3O^+]}{[H_2CO_3]} = \text{constant} = k$$

where the reaction products are on the top of the expression, and the reactant(s) are on the bottom. What this tells us is that the product of the reaction product concentrations, divided by the product of the reactant concentrations, is constant. Note, the square brackets [ ] tell us that we are considering the concentrations of the substances. Notice that water is left out of this expression because its concentration value is 1 and thus makes no difference. Note also that as $[H_3O^+] = [H^+]$ (see **12.1**), the equilibrium expression could just as well be written with $[H^+]$ rather than $[H_3O^+]$.

The constant k is termed the **equilibrium constant** and its value can be found by experiment. Thereafter, we can use the equilibrium expression and constant to predict the extent of a future reaction.

35.2 Generalizing the equilibrium expression

The example given above is relatively simple because the reactions all involve the formation of single ions. Consider the case of a reaction where two molecules of a given substance are generated. For example, carbonic acid can also react with water to generate carbonate ions:

$$H_2CO_3 + 2H_2O \rightleftharpoons CO_3^{2-} + 2H_3O^+$$

The equilibrium expression treats this reaction as if it were:

$$H_2CO_3 + 2H_2O \rightleftharpoons CO_3^{2-} + H_3O^+ + H_3O^+$$

Thus:

$$\frac{[CO_3^{2-}][H_3O^+][H_3O^+]}{[H_2CO_3]} = k$$

or

$$k = \frac{[CO_3^{2-}][H_3O^+]^2}{[H_2CO_3]}$$

In an equilibrium expression the concentration of a species is raised to a power equal to the number of molecules indicated by the expression (i.e. 2 in the case of $[CO_3^{2-}]$). In general, if we have the reaction

$$a[A] + b[B] \rightleftharpoons c[C] + d[D]$$

then

$$k = \frac{[C]^c[D]^d}{[A]^a[B]^b}$$

Not all chemical systems exist in a state of equilibrium. Indeed, strict equilibrium may be quite the exception in the natural environment. However, there are also many situations where this equilibrium approach can be applied with great success. For an example, and guidance on how to apply the method, see **42**.

35.3 How constant are equilibrium constants?

Equilibrium constants are actually not constant at all, as they vary with temperature. Furthermore, even for a given temperature, they are constant only if the substances are behaving in an ideal manner. The true equilibrium constants can therefore only be used if a correction factor is applied to the concentrations to allow for the non-ideal behaviours. These coefficients are termed **activity coefficients** for solutions or **fugacity coefficients** for gases. More usually, constants are formulated for real concentrations, and are termed **conditional constants**. This means that they are valid for only a narrow range of conditions.

36 Coordination chemistry: coordination numbers, ionic radii and crystals

36.1 Ionic radii

Ionic solids comprise lattices of positively and negatively charged ions (see **8.2**). However, we cannot mix any positively and negatively charged ions and get an ionic solid. Some will simply not combine. One explanation for this is that different ions have different sizes. For a lattice to be stable, the oppositely charged ions must be as close as possible without forcing the like-charged ions into close contact. Some combinations of ion sizes make this impossible.

What do we mean by the size of an ion? We know that an atom consists of a tiny nucleus surrounded by an arrangement of electrons. However, we also know that by an electron we simply mean a wave-like energy form resonating around the nucleus. Does size, in a conventional way, mean anything at all? It turns out that we can avoid these theoretical issues by simply looking at ionic solids. A sodium chloride crystal has a very well-defined size, implying that both Na^+ and Cl^- ions also have well-defined sizes. We know how many atoms there are in a given crystal, and we can thus work out mathematically the average distance between each Na^+ and Cl^- ion by comparing that number with the size of the crystal. In fact, by looking at lots of different types of crystal, we can work out how that distance can be attributed to each of the different ions, and then arrive at a size for each of them. From this it is clear that we can talk sensibly about the size of an ion even though in terms of the atomic model it is not easy to understand what this really means.

So, we can measure the size of any one ion, and this is normally expressed as the radius. However, we have slightly oversimplified things. The actual size varies depending upon how ionic (see **15.3**) the bonding is, and on the valence of the ions. If there is any tendency towards covalent bonding, then the spacing between the ions is different. There is also one further effect. As we will see below, there are a number of different physical arrangements by which a solid lattice can be constructed. This results in different numbers of ions being in contact with each other. In the most closely packed structure each cation could be surrounded by twelve anions. In the least compact structure it could be surrounded by as few as three anions. In the former case the surrounding anions may exert a considerable force of attraction on the electrons of the cation, causing it to swell in size. Thus, the Si^{4+} ion can vary in size between 0.034 nm and 0.048 nm depending upon the kind of lattice in which it is held.

Table 36.1 shows various values for the ionic radii of the common elements, together with estimates of the covalent radii. Some trends are apparent. For atoms which form positively charged ions, the ionic radius is smaller than the covalent radius. This can be explained by the greater attraction of the nucleus for the remaining electrons, which draws them closer to the nucleus. The opposite is true for negatively charged ions. By adding electrons to the atom, the attraction of the nucleus to each one is reduced, and they are

Table 36.1 Atomic/ionic radii for some common elements.

Element	*Van der Waals' radius*	*Radius in metallic or non-ionic solid*	*Ionic radius for specified ionic charge ( )*	*Covalent radius*
H	0.12	–	–	0.037
N	0.15	–	0.171 (3−)	0.070
C	0.185	–	0.016 (4+)	0.077
O	0.14	0.060	0.146 (2−)	0.066
Si	0.224	0.118	0.038 (4+)	0.117
S	0.185	0.095	0.190 (2−)	0.104
F	0.135	–	0.133 (1−)	0.064
Cl	0.180	–	0.181 (1−)	0.099
Li	–	0.152	0.068 (1+)	0.123
Na	–	0.186	0.098 (1+)	0.157
K	–	0.227	0.133 (1+)	0.203
Rb	–	0.248	0.148 (1+)	0.216
Be	–	0.112	0.030 (2+)	0.106
Mg	–	0.160	0.065 (2+)	0.140
Ca	–	0.197	0.094 (2+)	0.174
Sr	–	0.215	0.110 (2+)	0.191
B	0.217	0.079	0.016 (3+)	0.088
Al	0.253	0.143	0.045 (3+)	0.126
Mn	–	0.137	0.080 (2+) 0.066 (3+)	0.117
Fe	–	0.124	0.076 (2+) 0.064 (3+)	0.116
Cu	–	0.128	0.069 (2+)	0.135
Zn	–	0.133	0.074 (2+)	0.131
Pb	–	0.175	0.084 (4+)	0.154

able to reside further from the nucleus. Thus, while covalently bonded sodium is 50 per cent larger than covalently bonded chlorine, a sodium ion (Na^+) is nearly half the size of a chloride ion (Cl^-).

This effect might lead one to expect that ionic radius would increase from left to right across the periodic table. However, this is not particularly accurate. The reason is that the charge of the favoured ion increases towards the middle of the periodic table. This can be illustrated by looking at the ions of row 3 of the periodic table (refer back to Figure 15.1). Thus, while Mg^+ would be larger than Na^+, it is not stable. Instead, magnesium is always divalent, Mg^{2+}, and because of its higher charge its electrons are drawn more strongly towards the nucleus than is the case with Na^+ and its radius is less. Al^{3+} is smaller still owing to its higher electron deficit. and the smallest in the series is Si^{4+}. The next element in row 3 is phosphorus; as the P^{5+} ion it would be smaller still. However, in its favoured ionic form of P^{3-} it is in fact far larger, its radius being more than five times greater than that of Si^{4+}. As we move further to the right in row 3, the ions remain large, but fall slightly in size again as the ionic charge declines from S^{2-} to Cl^-.

Ionic radius also changes as we move down the groups of the periodic table. This is because the number of electron shells increases. Thus, in group 1 the ionic radius is of the order rubidium > potassium > sodium > lithium.

How, in practice, does ionic size affect the formation of ionic solids? For a lattice to be stable the ions must be close enough to minimize the electrostatic potential energy. If the size of one or other of the atoms is such as to physically prevent the other ions from being close enough, the lattice will not form.

36.2 Coordination and packing

If we have lots of spheres of identical size, then we are all familiar with the form that they will take if packed together. If we lay out a single sheet of the spheres there is a single optimal arrangement, with each sphere in direct contact with six other spheres (Figure 36.1a).

If we overlay this sheet with another identical sheet, this will lie comfortably in an offset position (Figure 36.1b). There is only one possible relative position for this second sheet. However, if we add a third sheet, then we have two possible arrangements to choose from. The spheres of the third sheet can be positioned directly above the spheres of the first sheet (case 1), but there is also the possibility of choosing a position that is different from both of the underlying sheets (case 2) In either situation the spheres are packed tightly together, occupying the least possible volume. This arrangement is termed **close packing**, and a body so constructed is said to be **close packed**.

While the two cases of close packing are equally dense, they result in materials which have different physical properties. The arrangement in case 1 leads to a cuboid crystal, and is termed **cubic close packing**. The other arrangement leads to hexagonal crystals, and is termed **hexagonal close packing**. Thus, subtle differences in atomic-scale structures cause great differences at the large scale.

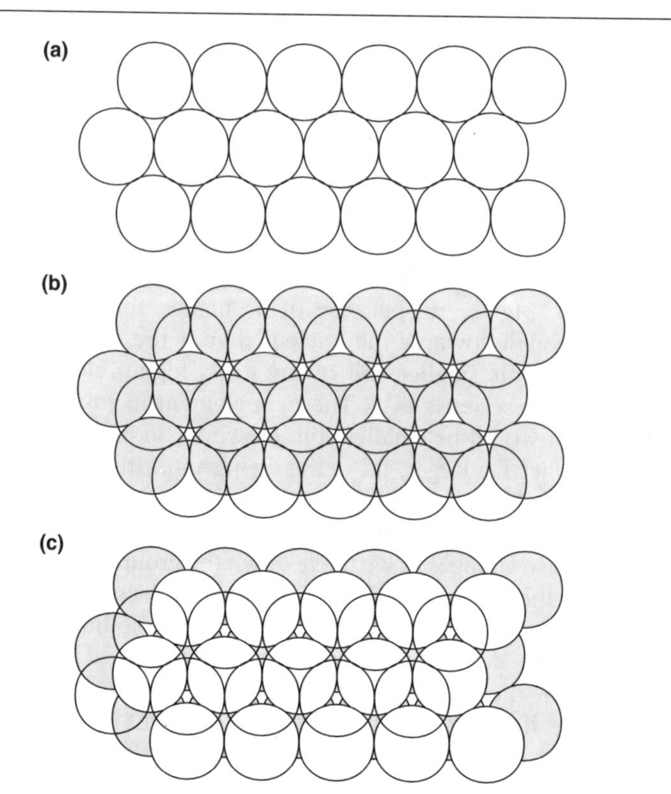

Figure 36.1 Close packing of spheres. (a) A sheet of spheres is packed most closely when each is in contact with six adjacent spheres; (b) a second identical sheet can be placed on the first, each sphere of the second sheet occupying a depression in the first. Each sphere within the sheet is now in contact with nine other spheres; (c) when a third sheet is added there are two choices. Here, the third sheet is placed in such a way that its spheres do not lie straight above any other spheres. It could have been placed such that its spheres lay above those of the first sheet. In either case, each sphere within the middle sheet is in contact with twelve other spheres; it has a coordination number of 12.

In both of these close-packed arrangements each sphere in the structure has an identical configuration in relation to its neighbours, and each is in contact with twelve other spheres: six from its own sheet and six from adjacent sheets (three above and three below). Each sphere is said to have a coordination number of 12, the coordination number referring simply to the number of direct neighbours.

From this we can deduce that close-packed ionic crystals will form if the ionic radii of the cations and anions are sufficiently similar. Thus, we would expect potassium fluoride (KF) to form close-packed crystals with all ions having a coordination number of 12. We can test this by looking at the density of the crystals, their growth habit and their cleavages.

Experimentation shows that if the ionic radii are very close then this kind of form is expected. If, however, one or other of the ions exceeds this percentage,

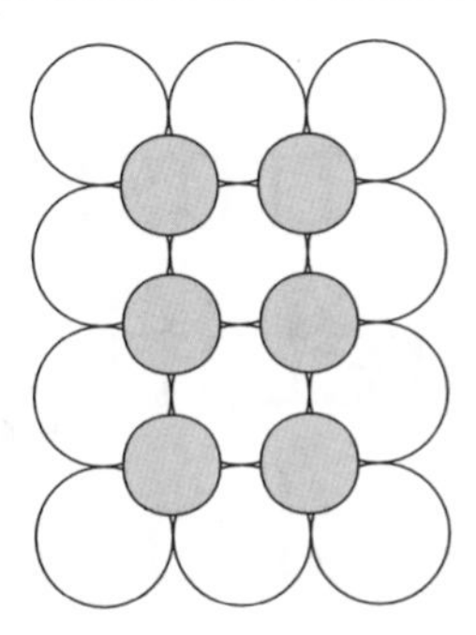

Figure 36.2 When the cations and anions are very different in radius, close packing is not possible. In this example, optimal packing is achieved with alternating 'square' sheets. Each cation (dark) is in contact with eight anions, and has thus a coordination number of 8.

the close-packed form becomes unstable. We can see why if we consider what each layer would look like with uneven ion sizes (Figure 36.2). If the ionic radii of some anions and cations are not compatible with any packing structures, then a stable ionic lattice may simply not form.

37 An electrostatic interpretation for chemical bonding

37.1 Electrons and atoms

At the largest scales in the universe the force between bodies is due to gravity. No other forces operate at great distances. This is because all other forces are both attractive and repulsive, and the effects cancel each other out when the objects are far apart. At the atomic scale, gravity still operates, but only very weakly. Just how weakly becomes apparent when we consider that the repulsive force between two electrons is 10^{42} times greater than the attraction due to gravity. Even stronger forces exist within the nucleus, but these have no bearing on the forces between atoms: just as with electrical forces between planets, the various positive and negative nuclear forces are cancelled out before reaching the edges of the atom. Thus, electrical forces are all we need to consider if we wish to understand the chemical behaviour of atoms.

Electricity may be the only force, but it turns out that there is one more factor which must be considered. The electrons which form the outer shell of an atom have wave-like properties: not only do they revolve around the nucleus, but they must be in **resonance**. A good analogy, suggested by the British physicist Brian Pippard, is a piano string. Vibrations try to bounce back and forth along the length of the string. However, they can only succeed in this if their wavelengths fit the length of the string. This is why a piano string plays only a specific note. Likewise, electrons in orbit around a nucleus are constrained by the geometry of the system, and can only resonate (and thus be stable) at certain specific wavelengths. The energy of an electron is directly linked to its wavelength. Thus, only a limited number of energy levels exist for the electrons of any particular atom. But there is a further constraint. There can only be one electron of a specified type within each orbital, and that means that predictable, well-defined energy levels develop around atoms (see **7**).

Why is it that these electron shells matter? It is all because of the electrical forces between the positively charged central nucleus and the negatively charged electron shells. The electrical potential energy is lowest, and hence the atoms are at their most stable, if their outermost shell or subshell is full. Chemical behaviour at its most basic can be seen simply as the pursuit of complete shells. An element that naturally has complete shells simply does not need to undergo chemical reactions: it is already at its most stable. These elements are called the noble gases (see **8.1**), and they are characterized by their reluctance to react with other elements. All other atoms have outer shells that are incomplete. The rest of chemistry is explained by the various ways that the different atoms find to achieve completion of their outer shells.

The importance of this outer electron shell means that the first stage in understanding the chemistry of an element is to determine the state of this outer shell. This can be done quite simply, as outlined in 7. However, we are spared this exercise by the periodic table (introduced in **15**). In any one group (column) in the periodic table, all elements have identical outer shells. This forms the basis for explaining chemical behaviour. At its simplest level we can use the electronegativity of elements to guide us in understanding the character of bonding.

37.2 Electronegativity and bonding

Electronegativity (see **15**) is a measure of the power of an atom in a molecule to attract electrons to itself. If we react atoms that have very different electronegativities then the more electronegative of the two will take the electron from the less electronegative; thus ionization will occur. The compound formed will be held together by the strong electrostatic forces between the oppositely charged ions. This is ionic bonding, which we introduced in **8.2** and is described in more detail below. But what about the case where the atoms have a similar electronegativity? To start with, let us consider the case of pure elements, where all the atoms have identical electronegativities. The reaction we get depends on the nature of the atoms. If they are electropositive (metals), we can get metallic bonds. If they are electronegative (non-metals), we can get covalent bonds, leading either to solid lattices (as in diamond or quartz) or gases composed of paired atoms (as in dioxygen, O_2), which are covalently bonded pairs of atoms that are only weakly linked to each other. Let us consider the nature of these bonds.

37.3 Metallic bonding

All forces between atoms and molecules derive from electrical attraction. The only difference between the different bond types is the intensity and geometry of these fields. In metallic solids, bonds develop in arrays of metal atoms owing to permanent and strong electrical fields arising from the freeing of the outer electron (or electrons in some cases) from the donor atom. In sodium metal, for example, a packed array of atoms is held together by the cloud of outermost electrons which are shared between the atoms. The

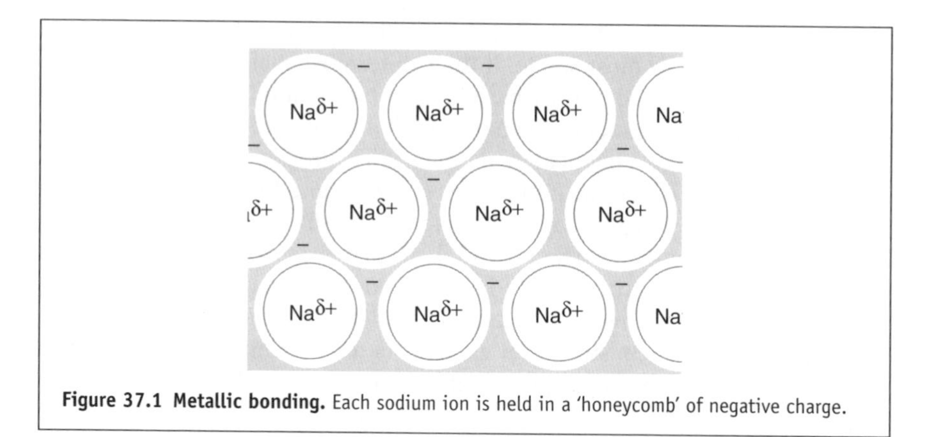

Figure 37.1 Metallic bonding. Each sodium ion is held in a 'honeycomb' of negative charge.

structure resembles an array of Na^+ ions, held in a diffuse, negatively charged 'honeycomb' of mobile electrons (Figure 37.1). The mobility of the electrons explains the special properties of metals: high conductance of both heat and electricity, the ability to reflect light and the ability to deform without losing cohesion. In such a structure the specific bonds between pairs of atoms are negligible: it is the general attraction between each atom and its enveloping cloud of electrons that dominates and explains the capacity of metals to deform without losing their structural integrity.

37.4 Covalent bonding

Metallic bonding can take place only where there are electrons in the valence shell that are sufficiently loosely held that they can free themselves from the parent atom. The more electronegative an atom the less able it is to do this. Atoms of the metal elements have achieved a complete outer shell by losing valence electrons to the electron-shared cloud. Electronegative atoms can achieve shell completion only by gaining electrons. They can partially achieve this by sharing their electrons. In a covalent bond, a pair (sometimes more than one pair) of electrons is shared by the bonded atoms. Thus in H_2 (H–H) one of the outer electrons from each of the hydrogen atoms is shared between them. Just as with metallic and ionic bonding (see below) the forces involved are electrical in origin. However, there are some important differences. In metallic and ionic bonding the forces are relatively unfocused: a Na^+ ion in a NaCl ionic lattice is attracted to all the surrounding negatively charged Cl^- ions in the lattice. In a covalent bond, the electron donated to form the bond is moved away from the atom. This leaves a slight charge deficit for the centre of the atom. At the same time, a region of negative charge has been generated by the shared electrons, which is located between the two atoms (Figure 37.2). Thus, each atom is drawn by electrical forces towards the central point. This gives the covalent bond a direction, which is not seen in the other bond types.

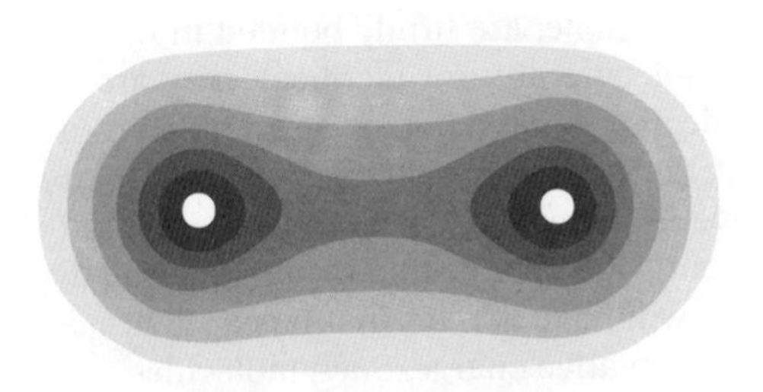

Figure 37.2 Covalent bonding of dihydrogen. The white circles represent the positive charge associated with the H nuclei (bare protons, not to scale). Shading intensity indicates electron density.

Pairs of atoms like these are found with the lighter halogens (Cl_2 etc.) also with molecular hydrogen (H_2), oxygen (O_2) and nitrogen (N_2). These substances are gases because most of the electrical energy is devoted to holding the pairs of atoms together; with little remaining to attach the molecules to each other. The other non-metals show slightly different behaviour. Sulphur, for example, can form more than one covalent bond permitting it to build a three-dimensional covalent lattice. The non-metals capable of doing this form solids rather than gases. Furthermore, those that form the most stable covalent lattices, such as carbon in the form of diamond, are the hardest materials and have exceptionally high melting points.

37.5 London and van der Waals' forces

If a covalently bonded atom-pair such as dinitrogen (N_2) is cooled sufficiently, it will form a liquid. This tells us that there are some forces of attraction between the molecules, even though they are very weak. In non-polar molecules (described below) these forces are commonly termed **London forces** (or London dispersion forces). They arise from chance variations in the distribution of electrons in the outer shell of molecules. These variations lead to short periods when electrons will be concentrated in one part of the molecule, leaving a deficit elsewhere. This gives rise to temporary electrical fields which affect neighbouring molecules. The neighbours tend to harmonize their fluctuations, minimizing repulsion and maximizing attraction. This causes weak temporary electrostatic forces of attraction, and it is these which hold the molecules together. It is such forces that allow even noble gases to form liquids and solids.

In the cases we have considered, the bonding is between atoms or molecules of the same kind. When we start to look at mixtures of different atoms and molecules, then we have a number of different mechanisms leading to bonding. For elements that have similar electronegativities we can get covalent bonding which is very similar to that seen for dinitrogen

(N_2) or dioxygen (O_2). Molecules like carbon monoxide (CO) and hydrogen chloride (HCl), for example, are firmly bonded in this manner. However, as the atoms in the molecule are not identical, they are very likely to have a slightly different degree of attachment to the electrons. Simply, the more electronegative of the pair will hold the electrons more strongly than the less electronegative of them. This leads to the formation of a polar covalent bond, where one part of the molecule is permanently depleted in electrons. Thus, the molecule will have an end that has a slight positive charge and an end that has a slight negative charge. Such molecules, which have permanent *dipoles* (ends with opposite charges), experience much stronger forces of electrostatic attraction than do non-polar molecules (which experience only temporary dipoles). When molecules with permanent dipoles come into contact, they arrange themselves in such a way as to maximize the separation of the like charges and to minimize the separation of the opposite charges. A special case of this **dipole–dipole** attraction is the **hydrogen bond** which is introduced in **8.6** and described in detail in **41.3**. These electrostatic forces and those caused by the temporary dipole effects (London forces) are sometimes lumped together under the title **van der Waals' forces**.

37.6 Ionic bonding

In the sections above we saw that unequal attraction of atoms to shared electrons explains the existence of some forms of bonding. But what happens if the attraction to the electrons is still more different between the atoms? In other words, what do we get if we mix atoms which have very different electronegativities? In such cases the electrons are not shared but exchanged. Instead of getting a covalent bond, we get an ionic bond.

We show earlier (see **8**) how groups of sodium atoms (and those of other group 1 elements) readily share their electrons to form metallic solids: each atom in the metallic solid has an identical attraction to the electrons. However, atoms of other elements possess different affinities for electrons. For example, we may mix sodium (which is weakly attached to its outer electron) with a substance with a strong attraction to electrons (such as chlorine). Instead of sharing its outer electron with its neighbours, as in metallic sodium, this electron is completely transferred to a chlorine atom. As a result we have a mixture of entirely separate positively charged Na^+ ions and negatively charged Cl^- ions. The electrical force between these separate entities is extremely strong. In a hypothetical cloud of Na^+ and Cl^- ions an inverse square law exists for the force between ions. Pairs of sodium ions (Na^+–Na^+) are strongly repelled and thus distance themselves from each other. The same is true of pairs of chlorine ions (Cl^-–Cl^-). But the forces of attraction between Na^+ and Cl^- pairs increase geometrically as they approach each other. It is easy to see why such a cloud is highly unstable and would rapidly condense to form a solid (Figure 37.3). The ionic array, or lattice as it is termed, organizes itself to minimize the average distance between Na^+ and Cl^- ions and to maximize the distance between Na^+–Na^+ pairs and Cl^-–Cl^- pairs. No bond

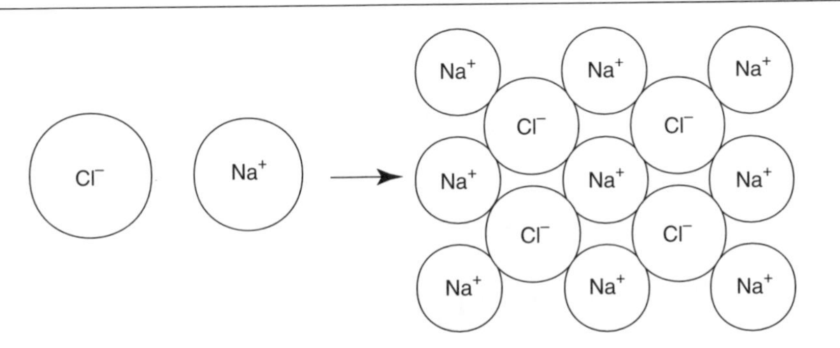

Figure 37.3 Ionic bonding of Na^+ and Cl^- ions. The complete capture or loss of electrons results in ion formation. Ions behave as charged spheres, which are strongly attracted to ions of the opposite charge. In the absence of water, lattices form in which ions alternate to minimize the spacing between ions of opposite charge, and maximize spacing between ions of the same charge.

exists between any two ions; instead a complex pattern of attractive and repulsive forces exists, depending upon the distances between ions.

Superficially the ionic lattice appears similar to that of a metal. However, there are some important differences. In the metal, an array of positive point charges is held in place by a diffuse negatively charged network of electrons. In the ionic solid a regular structure of positive and negative point charges is arranged in such a way as to minimize the overall energy (to maximize its stability). This contrast explains the differing physical properties of such substances. Any slight deformation in the ionic lattice must be accommodated by increasing the separation of the ions, which requires a considerable force. This explains why ionic solids are hard and rigid. On the other hand, if sufficient force is applied to cause separation, the attractive forces fall with the inverse square of their separation, which explains why ionic solids are brittle. Incidentally, by considering the shape of an ionic lattice, it is easy to see why there are planes across which attractive forces are weaker than on average. This explains why cleavages exist in mineral lattices.

38 Silicate minerals: structure and formation

38.1 Silicon and silica tetrahedra

The element silicon and the minerals it forms deserve our special attention. This is because most rocks are composed of minerals based on oxides of silicon (**silicates**). Here our intention is to introduce you to the basic chemistry of this important class of substances which you will repeatedly encounter when studying rocks, minerals and soils.

Silicon (Si) has four electrons in its valence shell, each occupying a separate orbital. The high ionization potential and low electron affinity (see **15**) mean that ionization is difficult, and covalent bonding is favoured. Silicon forms strong covalent bonds with oxygen: and as each silicon atom has four valence electrons it readily forms bonds with four oxygen atoms (Figure 38.1). The oxygens are bonded using one of their two valence electrons; the other electron is readily removed to allow ionic interactions between the SiO_4^{4-} tetrahedron and other ions. This basic structure is an essential building-block in silicate minerals.

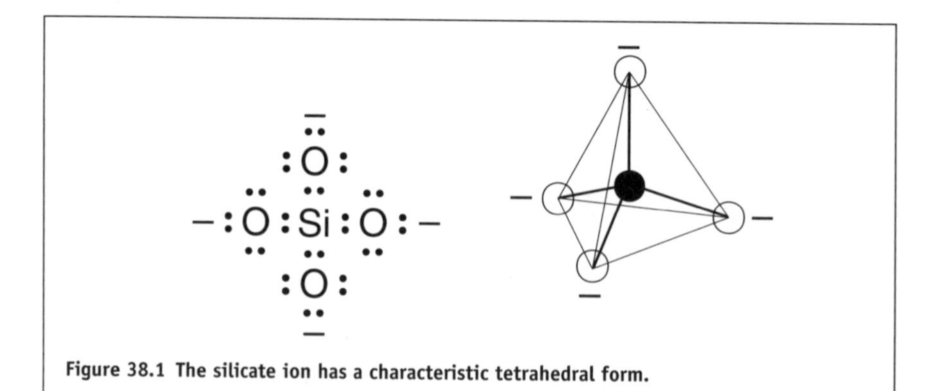

Figure 38.1 The silicate ion has a characteristic tetrahedral form.

While the building-block seems simple, there are a great many ways in which it can be utilized to form minerals. It is this which leads to the great diversity of silicate minerals. The way that they are constructed has implications for their physical and chemical properties, and must be appreciated if the properties of Earth surface materials are to be understood.

Silicate minerals are conventionally classified into four main structural groups according to the nature of the linkages between the silica tetrahedra; their names are: **island**, **chain**, **sheet** and **framework**. Here we merely outline the structure of these groups; for a more detailed yet still accessible account, we recommend the following text: Deer, Howie and Zussman *Introduction to Rock Forming Minerals*, 2nd edition, Longmans, 1992.

38.2 Island silicates

In these minerals, pure silica tetrahedra behave as discrete ions, balancing their excess charge by forming a lattice with positively charged ions (Figure 38.2). The lattices are most stable if every tetrahedron is matched with two divalent cations. These comprise the **olivine** family of minerals. They may comprise mixtures of different divalent ions, but the pure end-members are also worth remembering. *Forsterite* is the magnesium end-member (Mg_2SiO_4), while *fayalite* is the iron end-member (Fe_2SiO_4). Naturally occurring olivines may be better represented as $(Fe,Mg)_2SiO_4$ indicating

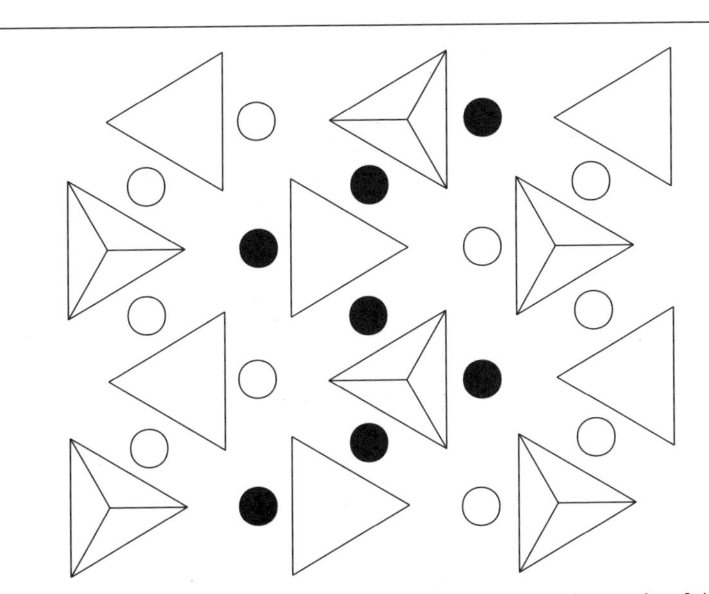

Figure 38.2 Structure of an island silicate: olivine. This can be viewed as a series of sheets. In each sheet half of the tetrahedra point upwards. The closed circles represent Mg^{2+} ions lying above the plane of the diagram. The open circles represent Mg^{2+} ions lying within the plane of the diagram.

that the sum of the iron and magnesium must be 2, but that the proportions may vary. The island structure maximizes the number of oxygen atoms held by each silicon atom, which therefore minimizes the number of silicon atoms. Olivines thus have among the lowest silicon concentrations of the silicate minerals.

Olivines are more ionic in character than most other silicate minerals. As with all ionic lattices, there are many planes of weakness along which the mineral may split, or cleave. More importantly, the ionic character of olivines makes them exceptionally susceptible to weathering: water readily attacks the lattice (see **39**). These minerals fall into the ultrabasic category of geological classification. Although relatively depleted in silicon, they are rich in bases (magnesium and iron), and are the dominant mineral of most ultrabasic rocks.

While these minerals are only minor, if common, constituents of basic volcanic rocks, they are the dominant silicate of the Earth's mantle. Thus, while scarce in crustal rocks, in global terms olivines comprise a high proportion of all silicate minerals.

38.3 Chain silicates

An alternative arrangement for the silica tetrahedra is for them to line up in chains. This way, instead of each silicon atom having four of its own oxygen atoms, it has two of its own, and two that are shared with each of the adjacent members of the chain (Figure 38.3). This way, each silicon atom balances with an average of three oxygen atoms; these minerals thus have a higher silicon content than island silicates. The chains so formed can be single or double (Figure 38.3), where in the double form each silicon atom shares slightly more oxygen atoms than in the single form. Either way, the bonds within the chain are covalent, creating an elongated structure with a net negative charge. This charge is again balanced by surrounding positively charged ions, and the minerals formed are like two-dimensional lattices: stacks of chains bound to each other by intervening charged ions. The intervening ions are predominantly divalent. The minerals which form these structures are known as **pyroxenes** (single chain) and **amphiboles** (double chain). The Si : O ratio of 1 : 3 leaves a net charge of −2 per silicon atom; thus a typical pyroxene formula is $(Fe,Mg,Ca)SiO_3$. In amphiboles the ratio is 4 : 11, and the formula is made more complex by the fact that amphiboles incorporate hydroxyl ions (OH^-) into their structures, as in $(Fe,Mg,Ca)_7Si_4O_{11}(OH)_2$. The best-known pyroxene is **augite**, while the best-known amphibole is **hornblende**.

The physical properties are influenced by the chain structure. Cleavage across the chains is difficult; thus both cleavage and crystal growth favour an elongate form. The simple ionic binding of the chains to each other leaves both amphiboles and pyroxenes susceptible to chemical weathering. They weather more readily than most other silicates, but less so than the olivines.

While olivines dominate the ultrabasic rocks, pyroxenes and amphiboles characterize the basic and intermediate types. Pyroxenes are characteristic

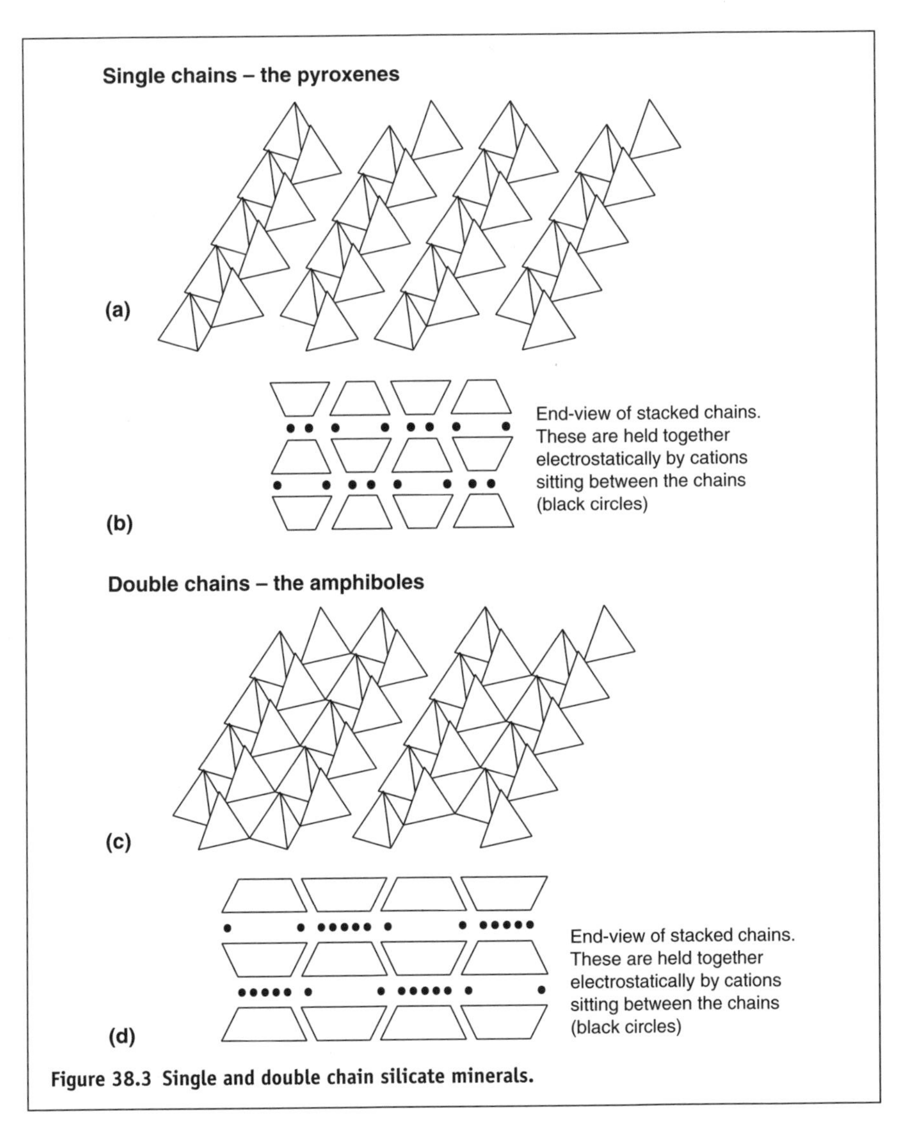

Figure 38.3 Single and double chain silicate minerals.

of the basic igneous rocks, notably basalts and their coarse-grain counterparts (dolerites and gabbros). Amphiboles are more abundant in intermediate igneous rocks, i.e. andesite and its coarser counterparts (diorite, granodiorite). They are also present in granitic rocks.

38.4 Sheet silicates

Sheet silicates have more shared linkages between the silica tetrahedra than do chain silicates. Complete sheets of covalently bonded tetrahedra are found (Figure 38.4a). This can be viewed as a further extension of the double chains of amphiboles. Each silicon atom shares three adjacent

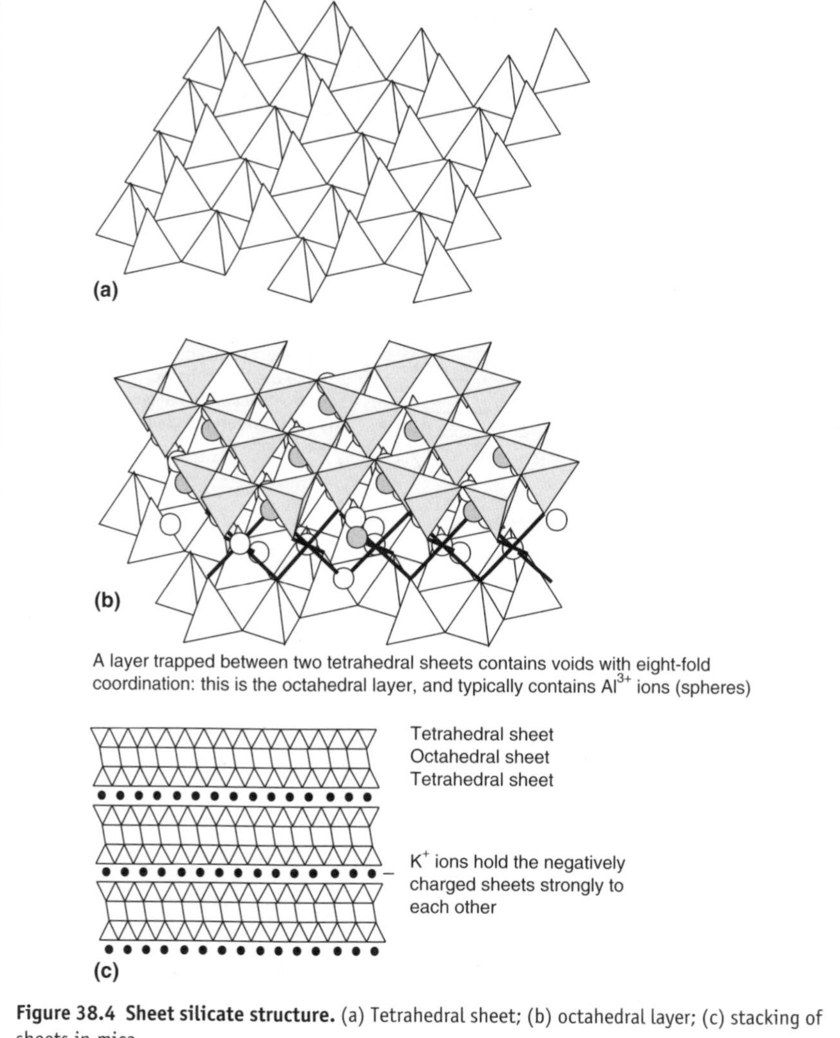

Figure 38.4 Sheet silicate structure. (a) Tetrahedral sheet; (b) octahedral layer; (c) stacking of sheets in mica.

oxygen atoms from the sheet, and has only one of its own. This leaves a Si : O ratio of 2 : 5.

The similarity with double-chain silicates ends there. The sheets of silica tetrahedra do not occur alone. They are always attached to aluminium-rich sheets, which comprise aluminium ions surrounded by six O^{2-} or OH^- ions (Figure 38.4b). Such layers are called octahedral sheets because ions lie at the six corners of a hypothetical octahedron. It is also sometimes referred to as the gibbsite layer, after the mineral gibbsite ($Al_2(OH)_6$) which has an identical structure.

Sheet silicates also differ from the island and chain silicates in the diversity of minerals found. The diversity arises from the variety of ways in which the sheets can be bound to each other, and the effect that this has on the chemical and physical properties. Part of the explanation lies in the weaker charges involved, which allow more types of ion to reside between the layers. The unit charge for the basic sheet shown in Figure 38.4 is only +1. In contrast, the unit charge per silicon atom in olivine is +4, leading to the formation of strong lattices with only the most suitable (divalent) ions.

Sheet silicates can be subdivided into two main classes: **micas** and **clays**. Of the two, micas are generally simpler and more strongly held together. Clays, which are held together relatively loosely, are more variable.

38.4.1 Micas

Micas comprise composite sheets, each made of two tetrahedral silica sheets sandwiching an octahedral sheet (Figure 38.4b). However, the pure tetrahedral sheet has been altered; one out of every four silicon atoms in the sheet has been replaced, or substituted, by Al^{3+}. This leaves the sheet structurally identical, but with an increased negative charge. To compensate for this, monovalent ions (chiefly potassium) sit between each of the composite sheets (Figure 38.4c). This provides powerful forces of electrostatic attraction which bind the composite sheets to each other. The two common micas are **muscovite** and **biotite**. Muscovite closely fits the model described above. However, in biotite some of the aluminium in the octahedral layer has been replaced by Fe^{2+}, which further increases the negative charge on the composite layers. Compensation for the more negative sheet charge means that biotites are richer than muscovite in potassium, in addition to being iron-rich.

38.4.2 Clays

One class of clays is similar to muscovite. **Illite** has the same composite layers, each made of three sublayers. However, there is less aluminium substitution in the tetrahedral layer, and thus fewer potassium ions between the layers. From a chemical perspective, illite can be viewed simply as potassium-poor muscovite. The main effect of the reduced potassium concentration is reduced electrostatic attraction between the composite layers: illite is physically weaker than muscovite.

Another member of this class of clay minerals is **montmorillonite** (a member of the **smectite** family). This is similar to illite, but has still less aluminium substitution in the tetrahedral layer, and may indeed have none at all. There is thus relatively little potassium between the composite layers, which are held together only by weak forces. Montmorillonite is thus even weaker than illite, and is characterized by variable spacing between the composite layers. When dry, montmorillonite has the same interlayer spacing as illite, but when exposed to moisture, water molecules are drawn into the spaces between the composite layers, and the gap expands to permit this. This gives rise to the variable volume of soils rich in montmorillonite, and other smectite members, which is such a problem for buildings and other structures.

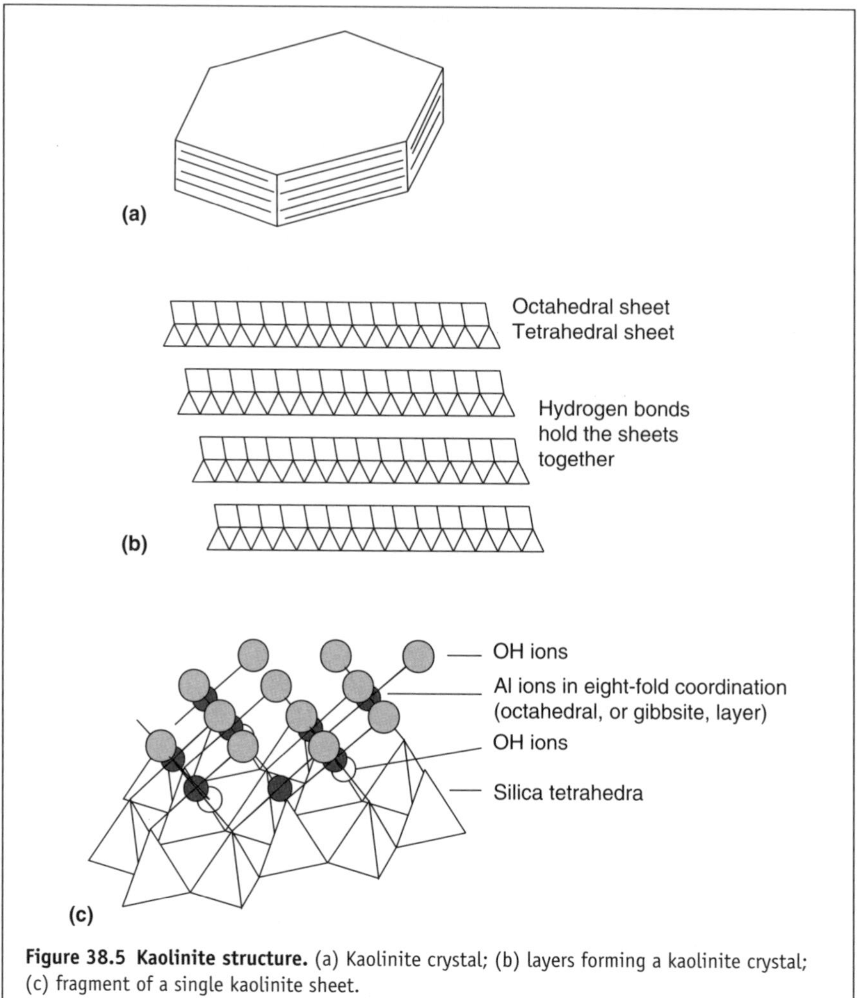

Figure 38.5 Kaolinite structure. (a) Kaolinite crystal; (b) layers forming a kaolinite crystal; (c) fragment of a single kaolinite sheet.

A separate group of clay minerals is represented by **kaolinite** (a member of the **kandite** family). Each composite layer comprises only two sublayers: one tetrahedral and one octahedral (Figure 38.5). Substitution in kaolinite is minimal, but the composite layers are quite firmly attached to each other by hydrogen bonds between the OH^- ions of the octahedral layer and oxygen ions of the adjacent tetrahedral layer.

Many of the properties of sheet silicates have already been described. All share a dominant platy cleavage. All have high specific surface areas (area of surface divided by mass). The micas are rich in potassium and are relatively weatherable. In granitic areas, the potassium released by weathering of micas accounts for a major part of the acid neutralizing capability. The clays are typically weathering end-products, and thus have little scope for further weathering. Illite is an exception to this rule, and weathering of potassium from illite may be significant. The main feature shared by all clays is their high ion exchange potential relative to most silicate minerals, a feature which is of major significance in soils (see **13**).

Micas are abundant in granitic rocks and in sediments derived from them. They are, however, present only in chemically immature sediments, i.e. those that have been little subjected to chemical weathering. Micas tend to produce illite on weathering.

Clay minerals form through the chemical weathering of certain precursor (i.e. parent) minerals, and may be formed directly in soils under suitable circumstances (see **39**). This leads to their high abundance in soils. Their resistance to further weathering and their fine grain sizes (via mechanical sorting) lead to the accumulation of clays in deep water sediment in both terrestrial and marine environments.

38.5 Framework silicates

In framework silicates, a three-dimensional covalent lattice is formed, with each silicon atom sharing all of its oxygen atoms. This leaves a Si : O ratio of 1 : 2, giving these types of mineral the highest silicon concentration of all silicate minerals. There are two common classes of framework silicates: **silica minerals** and **feldspars**.

38.5.1 Silica minerals

Most silica minerals are composed almost entirely of silicon and oxygen, and a very stable and strong three-dimensional framework results (Figure 38.6). Slight differences in the precise arrangement of the atoms lead to the different

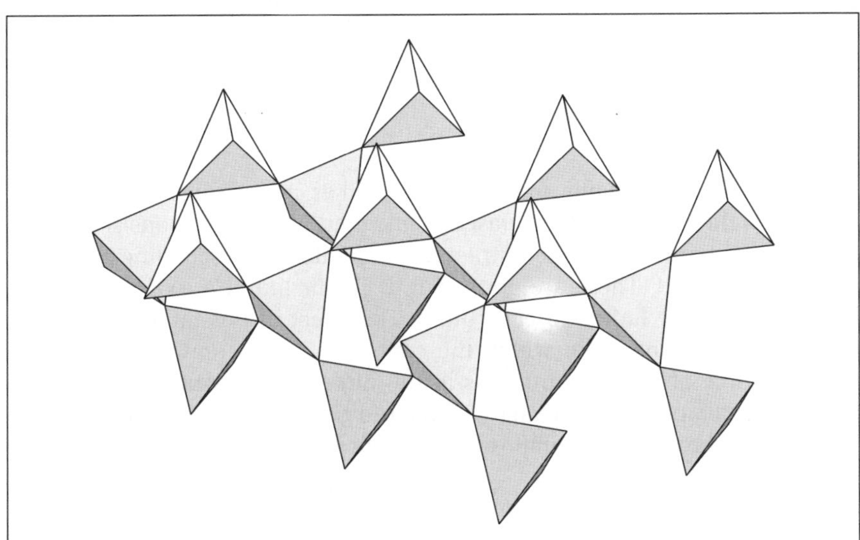

Figure 38.6 Beta quartz structure. An example of a framework silicate where silica tetrahedra are linked in a rigid three-dimensional framework.

minerals. The common minerals are **quartz** and **cristobalite**, both of which have the formula SiO_2. An exception to this pattern is found in amorphous opal, known as **opal-a**, which is very similar to cristobalite, but is crypto-crystalline. This means that it comprises crystals that are so small that they can barely be detected. Opal-a has many small cavities filled with water, which accounts for up to 10 per cent of the mass. The formula is given as $SiO_2.nH_2O$, the n indicating that the amount of water is not fixed.

The covalent framework of silica minerals leads to an absence of cleavage, great physical strength and resistance to chemical weathering. Quartz is therefore the ultimate example of what is termed a resistate mineral (meaning it resists destruction by physical or chemical weathering). In most situations, severe weathering or physical abrasion will lead to a residue that is rich in quartz. Only in the wet tropics is quartz removed in solution more rapidly than other minerals. Opal-a is much less resistant than quartz, and readily dissolves in sediments and in soil.

Quartz is one of the most ubiquitous minerals of the Earth's surface. It is absent only from the most basic of igneous rock. It never occurs in ultrabasic rocks, and is thus absent from the mantle (Figure 43.1).

The properties outlined above explain why quartz is concentrated on beaches. Much of the quartz on a beach has not only survived erosion of the source rock, transport to the beach and continuous grinding by wave action, but has done so through more than one geological cycle, summarized as beach → subaqueous sediment → burial and lithification to sandstone → uplift → erosion → beach.

Opal-a is produced by many forms of life, including diatoms. These are unicellular organisms of marine and freshwater environments. Diatoms can become concentrated in terrestrial or marine sediments, provided they are not swamped by the normally high accumulation rate of non-biological sediment types. Diatom enrichment thus typically happens only where total sediment accumulation rates are low.

38.5.2 Feldspars

The feldspar family is large, and can be viewed as comprising two subgroups: the **alkali feldspars** and **plagioclases**. While they, like quartz, consist of three-dimensional frameworks, their structures are different, resembling a mesh of interlinked chains. These are not separate chains like those of amphiboles, but are covalently linked to each other. The effect of this internal structure is to leave void spaces in which the cations K^+, Na^+ or Ca^{2+} can occur. Further, some of the silicon atoms in the linked tetrahedra are substituted by Al^{3+}, which provides the charge deficit to balance the additional cations. Chemically, the different compositions can be expressed by the formulae:

$$KAlSi_3O_8 \qquad NaAlSi_3O_8 \qquad CaAl2Si_2O_8$$

The sodium form of the mineral is an end-member in both feldspar families. In the alkali feldspars it forms a solid solution series with the potassium end-member. With the plagioclase feldspars it forms a solid solution series

Table 38.1 Variation in sodium and calcium in plagioclase feldspars.

% Na	% Ca	Name
100	0	Albite
80	20	Oligoclase
60	40	Andesine
40	60	Labradorite
20	80	Bytownite
0	100	Anorthite

with the calcium end-member. All possible mixtures of sodium and calcium are found in the plagioclase family (Table 38.1). The alkali feldspars tend to show dominance of either the sodium or the potassium form, though mixed forms do exist.

Further complexity arises from the many different possible structures that fit the compositions above. The K-spars (a common compositional abbreviation for the potassium feldspars) include orthoclase, microcline, sanadine and anorthoclase, each with variable amounts of sodium and with a different structure. The plagioclase feldspars have an ordered series of names depending upon the proportions of sodium and calcium they contain.

The chain-like internal structures of feldspars make them weaker than quartz, and they do display cleavage. They are, however, physically more resistant than most common minerals, and are found abundantly in some sandstones. They are susceptible to chemical weathering to a degree that increases with the calcium content: alkali feldspars are less weatherable than calcic-plagioclases.

Feldspars are the most abundant of the crustal minerals, leading to the dominance of silicon and aluminium in crustal rocks. In acidic igneous rocks, alkali feldspars are important, with many rocks containing both K-spars and albite. Basic rocks also typically contain two types of feldspar: albite and a more calcium-rich plagioclase. The calcium content of the calcic-plagioclase plays a part in the classification of the more basic igneous rocks. Thus intermediate igneous rocks have andesine, while basalts typically have labradorite or bytownite. In sedimentary rocks the presence of substantial amounts of feldspar (which confers the name arkose) is usually taken as evidence for minimal chemical weathering, owing to either dry conditions or proximity to the source area.

39 Mineral weathering: clays and their precursors

39.1 Introduction

Chemical weathering is ultimately driven by the conflict between the acidic nature of the atmosphere and the basic character of minerals and rocks. Thus, an understanding of weathering requires a good grasp of acid/base reactions. In addition, water balance has a controlling influence on the overall rates of weathering: dry conditions inhibit chemical reactions, while high water fluxes enhance the removal of both dissolved and particulate products of weathering. Furthermore, we need to consider oxygen because this gas plays a major role in promoting some chemical reactions. The nature of the chemical reactions varies with the minerals present in the ground. First let us consider some of these chemical reactions.

39.2 Acid/base reactions in the ground

Most soil waters are rich in carbonic acid because of the high partial pressures of CO_2 in soils. (This CO_2 is generated primarily by the microbial decomposition of organic matter.) It is thus useful to represent weathering reactions as reactions with carbonic acid.

Quartz is acidic in character and will undergo hydrolysis in water, thus:

$$SiO_2 + 2H_2O \rightleftharpoons Si(OH)_4$$

However, this reaction proceeds only very slowly at Earth-surface temperatures, and plays only a minor role in either the removal of quartz from soils or the supply of silica to surface waters. The little $Si(OH)_4$ that is released has a negligible impact on the water acidity; the pK_A (see **12.5**) for the first acidity constant for this very weak acid is 9.5 (thus the pH of the water must be as high as 9.5 before even 50 per cent dissociation occurs).

The dissolved silica that is released by weathering comes instead from reactions with other silicate minerals. The simplest reactions can be illustrated using olivine, thus:

$$Mg_2SiO_4 + 4H_2CO_3 \rightarrow 2Mg^{2+} + 4HCO_3^- + Si(OH)_4$$

In this reaction, carbonic acid, being the stronger acid, donates hydrogen ions to the silica tetrahedra to form silicic acid. This frees up the base, bicarbonate. The solution represented by the left-hand side of the expression is slightly acidic, while the solution on the right-hand side has bicarbonate alkalinity (see **12**). The essence of the acid/base reaction here is:

$$SiO_4^{4-} + 4H^+ \rightarrow Si(OH)_4$$

It is thus the silica tetrahedra that exhibit basic behaviour (i.e. they are proton acceptors). This seems to contradict the typical view of magnesium as the base, and silica as the acid. This is not the case, however. It is the presence of Mg^{2+} (or other cations) in the silicate that allows the simple SiO_4^{4-} tetrahedra to be present, and this situation owes itself to the basic properties of metallic magnesium. The amount of Mg^{2+} released during the reaction is a measure of the amount of base released, even though the ion itself is a weak acid (i.e. $Mg^{2+} + H_2O \rightarrow Mg(OH)^+ + H^+$). It is in this way that all the cations (Mg^{2+}, Ca^{2+}, Na^+ and K^+) found in silicate minerals indicate the amount of base present.

Many other minerals react in the same way as olivine. They are said to show **congruent dissolution**, meaning that the sum of dissolved reaction products equals the amount of dissolved reactant. This distinguishes the reaction from others where only a part of the dissolved material emerges in solution and another part of the reaction product is in solid form, commonly a clay mineral.

A typical example of an **incongruent dissolution** reaction is:

$$4KAlSi_3O_8 + 4H_2CO_3 + 18H_2O$$
$$\rightarrow 4K^+ + 4HCO_3^- + Al_4Si_4O_{10}(OH)_8(\text{solid}) + 8Si(OH)_4$$

In this reaction, a potassium feldspar reacts with carbonic acid and water to release potassium ions, bicarbonate, silicic acid, and a hydrated sheet silicate, kaolinite (see **38.4.2**). The essence of the acid/base reaction is exactly as described for the olivine reaction. This kind of reaction is typical for feldspars and micas. In the case of the micas, it is the clay mineral illite rather than kaolinite that is produced.

39.3 Oxidation reactions

Many minerals contain the lower oxidation forms of iron and manganese. Conditions at the surface of the Earth are not suitable for such minerals, and there is a tendency for them to undergo oxidative reactions. The oxidizing reagent in this reaction is usually molecular dioxygen:

$$2Fe_2SiO_4 + O_2 + 6H_2O \rightarrow 4FeO.OH(\text{solid}) + 2Si(OH)_4$$

In this reaction with an Fe-olivine, called fayalite, we see hydrolysis of the mineral components, but also oxidation of the Fe^{2+} to Fe^{3+}. The heart of the oxidation reaction is:

$$4Fe^{2+} + O_2 \rightarrow 4Fe^{3+} + 2O^{2-}$$

The O^{2-} ion rapidly undergoes hydrolysis, generating the OH^- ions of the reaction products. This is an acid/base reaction. Thus the oxidative hydrolysis of Fe^{2+}-bearing silicate minerals is not purely a redox reaction, but a mixture of redox and acid/base effects.

The FeOOH is a mixed oxide and hydroxide, sometimes referred to as an oxyhydroxide. Several minerals have this form, the best known being **goethite**. These minerals are highly insoluble under oxidizing conditions. Therefore, there is a solid product comparable with the incongruent dissolution described above.

40 Chemical properties of the elements: an overview

40.1 Introduction

Before embarking on this topic we would advise you to be familiar with atomic structure (see **4** and **7**) and the periodic table (see **15**). Our purpose here is to describe systematically some essential chemical properties and common chemical reactions of the different elements. Although there are 92 naturally occurring elements, this topic is less complex than it appears owing to the similarity of chemical behaviour within groups of elements; in fact it is these groups which constitute the columns in the periodic table. We start with hydrogen, which has unique properties, and then move through the groups of elements that comprise the main block of the periodic table.

40.2 Hydrogen

Hydrogen does not fit the periodic table well: a fact illustrated by its isolated position. With one electron in its outer (and only) shell, it could be interpreted as a member of group 1. However, it is also one electron short of a noble gas configuration, and thus might be seen as a member of the halogens. In reality it shares few properties with either group and is best treated alone.

One of the most important features of hydrogen is that on losing its valence electron, which it does readily, it becomes simply a proton. This gives it an exceptionally high electric field, which means that it reacts strongly with other substances. In normal situations, H^+ does not exist on its own, but combines with the materials that surround it. Thus H^+ in water forms a hydrated ion (see **12.1**). The high reactivity of hydrogen means that it forms more chemical substances than any other element. Water is the most abundant of these. Hydrogen is also one of life's key elements.

A key factor in the chemistry of hydrogen is its tendency to cause acidic behaviour (see **12**). Its very high field strength means that it must react with its surroundings. H^+ ions can exist only when dissolved in a liquid that binds to protons. Water is such a liquid, thus:

$$H^+ + H_2O \rightarrow H_3O^+ \quad \text{and} \quad H^+ + 2H_2O \rightarrow H_5O_2^+ \text{ etc.}$$

H_3O^+ is called the **hydroxonium ion** and is a reasonable representation of what H^+ is like in water. However, H^+ in water is usually expressed as the simple hydrogen ion for convenience. The equilibria (see **35** and **42**) can all be expressed in such a way as to allow for the fact that this is not strictly true.

Hydrogen forms a very stable diatomic molecule (H_2), which is its dominant form in the atmosphere. Hydrogen gas is sufficiently light that its velocity due to thermal energy allows it gradually to escape the Earth's gravitational field. That there is any dihydrogen at all in the atmosphere is owing to its continuous formation through the photodissociation of water, thus:

$$2H_2O + UV \rightarrow 2H_2 + O_2$$

Dihydrogen is also readily formed by reacting acids with metals. For example:

$$2Na + 2H_2O \rightarrow H_2\uparrow + 2Na^+ + 2OH^-$$

Most hydrogen on Earth is in the form of water (H_2O). (For a more detailed discussion of the chemical properties of water, see **42**.) Hydrogen also forms strong bonds with carbon, to form the hydrocarbons. There are a great many different hydrocarbons. The simplest, methane (CH_4), can be used to illustrate the linkage between hydrocarbons, dioxygen, water and carbon dioxide. A comparable expression could be written for any of the hydrocarbons:

$$CH_4 + 2O_2 \rightarrow 2H_2O + CO_2$$

This reaction represents combustion or respiration, while the reverse represents photosynthesis (see **34.3.1**). In the form of hydrogen ions (H^+), hydrogen plays a central role in acidity, a topic discussed in more detail elsewhere (see **12.1**).

40.3 Group 1 (IA) elements: the alkali elements

The group 1 elements are lithium (Li), sodium (Na), potassium (K), rubidium (Rb) and caesium (Cs). Each alkali element can be seen to be very similar to one of the noble gases; sodium differs from neon only in having an additional unit of charge in the nucleus, and a corresponding additional electron, which sits alone in the outer orbital. This outer electron is relatively well shielded from the nucleus by the electron shell beneath, and is thus quite easily removed (it is said to have a low first ionization potential). Losing an outer electron means acquiring a net positive charge. This permits very strong forces to exist with respect to other charged species. The strong possibility of bonding with other species leads to two very important properties of the alkalis, which they share in part with other metallic elements. First, as pure elements they form metallic bonds (see **8.5**); and secondly, on ionization they readily form ionic solids (see **8.2**).

In metallic form, the alkalis tend to react with water to liberate hydrogen, thus:

$$M + H_2O \rightarrow M^+ + OH^- + H_2\uparrow$$

where M refers to any alkali element.

The metals will also react directly with non-metals to form simple ionic solids, thus:

$$2M + Cl_2 \rightarrow 2MCl$$

In solution, the ions are hydrated. The ions Li^+, Na^+ and K^+ are quite small relative to the water molecules, and tend to adopt the form $[M(H_2O)_4]^+$. The metal holds strongly to the mantle of water molecules and does not strongly interact with other dissolved ions. Consequently, the alkalis tend to remain dissolved in water and do not readily form solid precipitates.

40.4 Group 2 (IIA) elements: the alkaline earth elements

The group 2 elements are beryllium (Be), magnesium (Mg), calcium (Ca), strontium (Sr), barium (Ba) and radium (Ra). In elemental form the members of group 2 are metallic solids, and they react readily with electron-attracting elements such as chlorine to form ionic solids. In this respect they are similar to the alkalis. They differ, however, in having two electrons in the outer shell. These outer electrons are shielded less well from the positive charge of the nucleus than is the case with the alkalis, and in consequence are harder to remove (i.e. the ionization potentials are higher than for the alkali elements). Nonetheless, while it takes quite a lot of energy to strip away both the outer electrons, the electrical forces experienced by the resulting 2+ ions are such that this does always happen. In other words, the 2+ ions form very stable ionic lattices and hydrated ions, which pays back the high cost of double ionization.

As with all of the groups of the periodic table, there are differences between alkali earth elements according to their masses. The lighter elements in any one group generally have higher forces of attraction holding their outer electrons in place. Thus it is much harder to ionize beryllium than it is to ionize magnesium. This results in very different chemical properties among elements. In this case, beryllium has a much greater tendency than the other alkaline earth elements to form covalent bonds (see **8.3**). For example, beryllium hydrolyses water to a significant extent, and is thus a relatively strong Lewis acid (see **12**).

While beryllium shows a number of covalent characteristics, the other elements of group 2 are ionic in character and more similar to each other. Their higher ionic charge and slightly smaller ionic radius lead to greater stability of ionic lattices when compared with the alkali elements. They form particularly stable crystalline carbonates and sulphates. For this reason, the group 2 elements are more easily removed from solution than are the alkalis: in other words they are less soluble.

In solution these elements form stable hydrated ions (i.e. $Mg[H_2O]_4^{2+}$), like the alkalis. Following the convention by which such ions are represented without their mantle of water molecules, most of the group 2 ions can be treated as entirely free in water (i.e. Mg^{2+}). However, they are less electronegative than the alkalis, and have some tendency to act as Lewis acids, 'accepting' electrons from water by partially binding to the OH^- ions:

$$M^{2+} + 2H_2O \rightleftharpoons M[OH]^+ + H_3O^+$$

where M is any group 2 element. This effect is greatest for beryllium and decreases with atomic number. Magnesium is typically only 5 per cent as $Mg[OH]^+$, and the heavier group 2 elements are still less so.

The smaller ionic radius of the group 2 elements compared with the alkalis, and the greater unit charge of their ions, lead to a greater tendency to interact with particle surfaces. Alkaline earth elements thus make up a large proportion of the cations adsorbed on mineral and organic particles in the environment, particularly in soils and sediments.

Barium deserves a special mention on account of its abundance. The general tendency is for the abundance of elements to decrease with an increase in atomic number. But barium (atomic number 56), although a relatively heavy element, is very much more abundant than other elements of similar atomic number.

40.5 Group 13 (IIIA) elements

The group 13 elements are boron (B), aluminium (Al), gallium (Ga), indium (In) and thallium (Tl).

Group 13 elements have three outer shell electrons. The relatively high nuclear charge and the reduced shielding by inner shells are such that it is difficult to remove any of these electrons. Once again, this effect decreases as atomic number rises. Except for boron, the elements in this group are metallic in character. For boron the energy 'cost' of stripping electrons is considerable; the energy gained by forming a lattice or a hydrated ion is not enough to allow it to take place. Thus boron does not react by simply gaining or losing electrons to form metallic or ionic bonds; rather, it shares electrons with neighbouring atoms or molecules to form covalent bonds (see **8.3**). In atoms of the four metallic members of group 13, the outer electrons are less strongly bound to the nucleus than is the case with boron. Nonetheless, they show more covalent behaviour than the metals in groups 1 and 2. All group 13 elements occur predominantly as trivalent ions. All will form an oxide of the form M_2O_3, and all except thallium form a hydroxide of the form $M(OH)_3$, where M represents any of the metals being considered. In addition, all of these elements form hydrated ions of octahedral form when in solution: $[M(H_2O)_6]^{3+}$. In solution, these ions act as acids, thus:

$$[M(H_2O)_6]^{3+} + H_2O \rightleftharpoons [M(H_2O)_5(OH)]^{2+} + H_3O^+$$

An important consequence of the high charge associated with trivalent ions is that the group 13 elements bind strongly with particle surfaces. They also, in solution, form strong complexes with natural organic acids (for example, humic substances), and with fluoride and sulphate. Thus the solution chemistry of the group 13 elements is far more complex than it is for group 1 and 2 elements which tend to occur in the environment as free (though hydrated) ions.

40.6 Group 14 (IVA) elements

The group 14 elements are carbon (C), silicon (Si), germanium (Ge), tin (Sn) and lead (Pb).

All group 14 elements have valence shells with four electrons. These electrons are relatively little shielded from the nucleus; accordingly they have very high ionization potentials and it is difficult to remove them. As with the other groups, the least metallic of the elements are those which are higher up the table. Thus carbon and silicon form only covalent bonds, while tin and lead are metallic.

Compared with the groups discussed so far, there is a greater role for variable oxidation states in the elements of group 14. Recall that the ionic forms of elements in groups 1, 2 and 13 are dominated by a single valence state (+1, +2 and +3, respectively). In group 14 elements, in contrast, the ions can have more than one common state, and consequently more diverse chemical properties. Carbon and silicon show some similarities in their reactions with hydrogen and oxygen. However, carbon chemistry is one of the major, and to a large extent distinct, branches of chemistry (see **16**), while the chemistry of life is dominated by the chemistry of carbon-based compounds (see **32**).

One class of carbon compounds must be treated differently from the rest. Carbon forms strong covalent bonds with oxygen to form the carbonate ion. This can be viewed as a continued trend of increasing Lewis acidity as the ionic charge increases. This is illustrated in Figure 40.1. The carbonate ion is very stable, and acts as a single unit in many ionic lattices. More information about the inorganic chemical behaviour of the various carbonate ions is provided in **42.4**.

Silicon, in oxide form, is the key element in most rock-forming minerals. Like carbon, this element also behaves as a Lewis acid. Silicon forms a wide range of oxides, and these form the basis of the silicate minerals from which most rocks on Earth are constructed (see **38**). The oxides also exhibit a mechanism for the conduction of electrons, which is similar to, if weaker than, that of metals. It is this semiconducting behaviour of silicon which has brought this element into everyday use by the general public.

Germanium has some metallic properties, but on the whole behaves in a manner similar to silicon. Tin and lead are familiar elements with metallic properties. This is partly because they readily exist in a divalent state. If either of these elements were to lose all of their valence electrons they

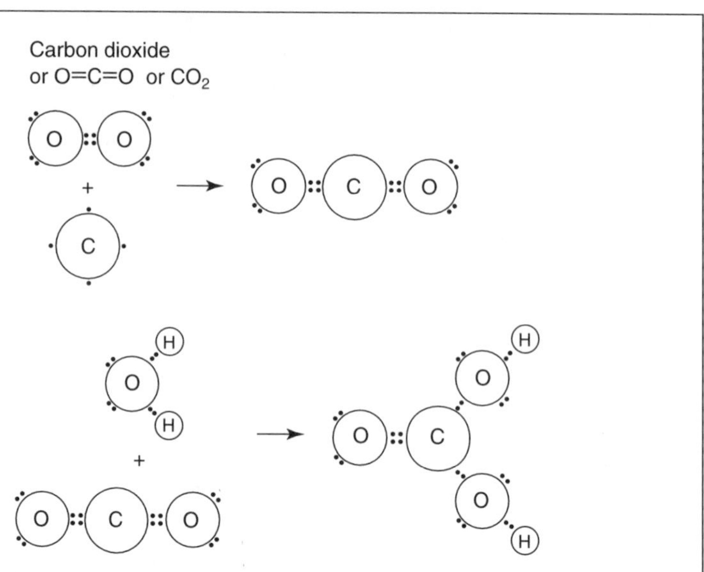

Figure 40.1 Carbon reacts with dioxygen to form the Lewis acid, carbon dioxide. The acidic behaviour of carbon dioxide is illustrated by its reaction with water. The water molecule is broken up, the electrons being accepted by the carbon–oxygen molecule. This leads to the formation of the carbonate ion, shown here bonded with hydrogen atoms to form carbonic acid.

would be unable to form stable ions. Both tin and lead form divalent ionic solids and occur as hydrated cations in solution. Both partially act as Lewis acids, though the effect is far weaker than for silicon and germanium, thus:

$$Pb^{2+} + 2H_2O \rightleftharpoons Pb(OH)^+ + H_3O^+$$

Lead, rather like barium which we mentioned earlier, is far more abundant on Earth than the other elements of similar atomic number. In both cases this is thought to be owing to their exceptional stability, a characteristic conferred by their electron shell configurations. In the case of each element the outer shell orbitals are half filled (contain a single electron) and this is an energetically favoured state.

40.7 Group 15 (VA) elements

The group 15 elements are nitrogen (N), phosphorus (P), arsenic (As), antimony (Sb) and bismuth (Bi).

Each group 15 element has five electrons in the valence shell, and in consequence ionization potentials are very high. This leads to the dominance of covalent bonding, and to a great number of possible oxidation states. As with group 14 elements, the chemical properties of the elements in this group vary considerably.

The first element in the group, nitrogen, is exceedingly important in the environment, particularly as an atmospheric constituent (**43.4**) and because of its key role in biological systems (**32**). Again, as the first element in the group, nitrogen differs from the other group members in various ways. Apart from hydrogen, which is a special case, nitrogen is the first element we have met so far which occurs in elemental form as a gas. In elemental form the covalent diatomic molecule, N_2, is greatly favoured over the atomic form, N. In addition, nitrogen forms a vast number of compounds, many of which are organic. Other important nitrogen-containing species are oxides of nitrogen and ammonia. Ammonia (NH_3) is a base, thus:

$$NH_3 + H_2O \rightleftharpoons NH_4^+ + OH^-$$

In contrast, the oxides of nitrogen are all acidic and covalent. The best known is nitric acid (HNO_3), but the various gaseous oxides (nitrous oxide, N_2O; nitric oxide, NO; nitrogen dioxide, NO_2) are also acidic.

Phosphorus is another very important element in the environment. It is an essential nutrient for all living organisms and its availability is a key factor determining the fertility of soil and water. Like nitrogen, phosphorus is entirely covalent in its chemical behaviour, although many of its compounds (such as the phosphates) are ionic. In the environment, phosphorus occurs principally either as one of a number of oxides or as a component of organic compounds. The oxides, like those of nitrogen, are acidic. Phosphoric acid (H_3PO_4) dissolves in water to release hydrogen ions and orthophosphate (PO_4^{3-}). This partially hydrolyses, leading to a number of additional phosphates, thus:

$$PO_4^{3-} + H_2O \rightleftharpoons HPO_4^{2-} + OH^-$$

$$HPO_4^{2-} + H_2O \rightleftharpoons H_2PO_4^- + OH^-$$

Phosphate also polymerizes (i.e. forms chains made up of the same sub-units) to form heavier polymer molecules.

Arsenic (As) is familiar to most people as a metabolic poison. Like phosphorus, with which it shares many aspects of its behaviour, the chemistry of arsenic is dominated by covalent bonding, and the element shows no metallic behaviour. The common inclusion of arsenic in lists of 'heavy metals' is thus erroneous.

Antimony (Sb), on the other hand, does not form many of the oxides seen for phosphorus and arsenic, but instead shows a number of cationic reactions. However, these occur only in the trivalent state: dissolved ionic SbO^+ is known for example. Bismuth (Bi) shows a wider range of such cationic reactions.

40.8 Group 16 (VIA) elements

The group 16 elements are oxygen (O), sulphur (S), selenium (Se), tellurium (Te) and polonium (Po).

These elements have six electrons in their outer shells. This is only two electrons short of the noble gas outer shell configuration of eight electrons, which is the configuration of maximum stability. The attraction of electrons to the nucleus in group 16 elements is such that they tend to gain electrons rather than lose them. Thus, for oxygen we are starting to see a return to ionic behaviour, not through loss of electrons, but through their capture. The other elements of the group, from sulphur downwards, are more covalent in character, although polonium (Po) shows some distinctly metallic properties.

Oxygen is the most abundant element on Earth, and the third most abundant in the universe (after hydrogen and helium). This key element forms a vast number of compounds, most of which are covalent. We are most aware of the gaseous diatomic form of oxygen, O_2, but in fact this is a minor component compared with the oxygen tied up in the water of the oceans and in the silicate minerals of the Earth's crust and mantle (see **38**).

The covalent oxides which form with the essentially non-metallic elements silicon, nitrogen, sulphur and carbon are all acidic in character. For example:

$$CO_2 + 2H_2O \rightleftharpoons HCO_3^- + H_3O^+$$

In contrast, the ionic, metallic oxides are basic:

$$CaO + H_2O \rightleftharpoons Ca^{2+} + 2OH^-$$

The sulphur-containing oxides and organic compounds are of very great significance in the environment. Plant and animal tissues contain sulphur, mostly within protein, and this commonly comprises the most abundant sulphur fraction in sediments and soils. Microbial decomposition of plant tissues leads to the production and release to the atmosphere of various volatile organic sulphur compounds, including dimethyl sulphide $(CH_4)_2S$.

The oxides of sulphur are all covalent and acidic. Sulphur can also form S^{2-} ions, but does so only in combination with certain elements, for example hydrogen, as in hydrogen sulphide (H_2S) and iron sulphide (FeS). However, in the environment many sulphides do not involve the S^{2-} ion, but rather the S_2^- ion, which consists of two covalently bonded sulphur atoms. An example is the mineral pyrite (FeS_2).

Both selenium and tellurium share many properties with sulphur. For example, as oxides they have similar acid/base behaviour. Polonium is a comparatively rare, short-lived, radioactive element. In contrast with the other group 16 elements, polonium has some distinctly cationic properties: it forms an ionic lattice with oxygen for example.

40.9 Group 17 (VIIA) elements: the halogens

The group 17 elements are fluorine (F), chlorine (Cl), bromine (Br), iodine (I) and astatine (At). Their ionic forms are fluoride (F^-), chloride (Cl^-), bromide (Br^-), iodide (I^-) and astatide (At^-).

Group 17 elements display predominantly ionic properties. As all group members have outer shells containing seven electrons, they readily gain an electron to complete the valence shell. Thus we have returned to the situation in groups 1 and 2 where common valence states are fixed and where there is great similarity between the elements. Not even astatine, one of the rarest of the Earth's naturally occurring elements, shows much evidence of cationic behaviour.

One feature that is shared by all the halogens is that they form simple ions in water (for example Cl^-). In this respect they are quite unlike any of the other non-metal elements, which are polyatomic in common ionic form (for example SO_4^{2-}, NO_3^- etc.). Notice that the ions of the halogens all have names which differ from that of the element (see above). This is typical for the common anions, whereas common cations are simply referred to by the element name followed by 'ion', as in 'sodium ion' and 'calcium ion'.

All the halogens can occur also as diatomic molecules. These molecules are highly reactive (powerful oxidizing agents), and they form compounds (called halides) with most other elements. Their reactivity ensures that un-ionized halogens exist only at very low concentrations. Halides generally form stable hydrated ions, with little tendency to form insoluble salts. As a consequence, halides have accumulated in the oceans.

Once again, the first element of the group stands out from the others. Fluorine is the most reactive of all the elements. Also, as the fluoride ion it has the ability to attack covalent silicate lattices. These properties arise from the exceptionally small ionic radius of fluoride, which strengthens the electrical interaction with other chemical species.

40.10 Group 18 (VIIIA) elements: the noble gases

The group 18 elements are helium (He), neon (Ne), argon (Ar), krypton (Kr), xenon (Xe) and radon (Rn).

The complete outer shells of elements in group 18 mean that it is very difficult to ionize them, and it is consequently difficult to generate strong forces of attraction between noble gas atoms. Thus, unlike the common gases oxygen and nitrogen, the noble gases do not form diatomic molecules. Neither do they readily form liquids or solids: in fact they have to be cooled to within a few tens of degrees of absolute zero to achieve either of these states, and in order to produce a solid state they must be subjected to very high pressures.

The great stability of noble gas atoms means that they do not interact readily with other atoms. Indeed, it was not until 1962 that anyone succeeded in forming any chemical compounds with them. Fluorine, with its very high ionization potential, forms a number of compounds with xenon and some oxides also exist. Many of the species formed with the noble gases are not compounds in the normal sense and tend to be short-lived.

40.11 Transition elements

40.11.1 Overview

When dealing with the transition elements we move away from the approach we have used with the other elements. Rather than describe groups, it is simpler to consider the transition elements by rows, because the transition elements tend to be more similar *within rows* than within columns. The reason for this lies in their electronic structure (see **7**). After the first and second subshells (*s* and *p*) of shell 3 are completed (neon), a new electron orbital type (*d*) comes into play. However, the *d* subshell of shell 3 has a slightly higher energy than the *s* subshell of shell 4. Thus the first two elements after neon (potassium and calcium) gain shell 4 electrons before their shell 3 is complete. The next element after calcium (scandium) has a complete outer shell (4*s*), even though its 3*d* subshell is incomplete (it has only one electron out of a possible ten in its 3*d* subshell). This is approximately the case for all of the ten elements in the first row of the transition elements: they have two electrons which are relatively easily removed (actually, some have only one, as some exchange between the 4*s* and 3*d* subshells takes place). This is in marked contrast to the other groups of the periodic table where there is a simple increase in the number of valence electrons between columns. Consequently, there are great similarities in the chemistry of the transition elements, and accordingly they are treated together.

40.11.2 The first row of the transition elements

The first row comprises scandium (Sc), titanium (Ti), vanadium (V), chromium (Cr), manganese (Mn), iron (Fe), cobalt (Co), nickel (Ni), copper (Cu) and zinc (Zn). Most of these elements are extremely important, not least because they are essential for life processes, albeit in tiny concentrations. These elements have similar atomic weights, and similar valences, and therefore share some chemical and physical properties. However, there are also differences, which arise from the way the *d* orbital electrons can be brought into play. These can be predicted by considering the interplay of the *d* and *s* subshells (Table 40.1).

The transition elements have a wider range of stable oxidation states than most other elements. Chromium, manganese, iron and cobalt all show more

Table 40.1 Valence shell electron configuration of the first row, first transition elements.

Element	*Electronic structure*	*Valences*
Sc	d^1, s^2	2+, 3+
Ti	d^2, s^2	2+, 4+
V	d^3, s^2	2+, 5+ (covalent only)
Cr	d^5, s^1	2+, 6+ (covalent only)
Mn	d^5, s^2	2+, 4+, 7+ (covalent only)
Fe	d^6, s^2	2+, 3+ (covalent only)
Co	d^7, s^2	2+, 3+??
Ni	d^8, s^2	2+
Cu	d^{10}, s^1	1+, 2+
Zn	d^{10}, s^2	2+

than one common oxidation state in the environment. In their higher oxidation states they are less ionic in chemical behaviour and more likely to form insoluble oxides or hydroxides. Thus their oxidation state has a powerful influence over their mobility in the environment. This in turn is controlled by the redox status of the environment. The impact of redox status on iron is illustrated in **14.5**.

The transition elements often form coloured ions. This means that they can absorb radiation in the visible spectrum. This property arises from the electronic state of the ions owing to interaction between the *s* and *d* subshells.

40.11.3 Rows 2 and 3 of the first transition series

The second and third rows of the first transition contain elements which are extremely dilute in natural materials. Most are strongly bound within silicate minerals (they are said to be **lithophile**), and are only very sparingly soluble. Their behaviour in the environment can be seen as passive; they simply imitate the more abundant elements. The best known of these are niobium (Nb), yttrium (Y) and zirconium (Zr), which are relatively common in silicate rocks and are widely used as rock tracers in geochemical studies.

An exception to this passive behaviour is seen in the last column of the first transition. Zinc (Zn), cadmium (Cd) and mercury (Hg) do not really belong to the transition elements. Their complete 3*d* subshell means that they behave as simple divalent ions. Indeed, they show considerable similarity in chemical behaviour to the group 2 elements. All the elements in this group are of biological significance. Zinc is an important nutrient but is toxic at high concentrations. Cadmium is confused by the body for zinc, and can be taken up, with serious consequences. The toxicity of mercury is well known.

40.11.4 The 2nd and 3rd transition series

The elements of the second transition are also known as the **lanthanides**, after their first member, lanthanum (La). The first row of these are also known as the **rare earth elements**. They are generally lithophile in character, and are thus concentrated in silicate minerals and rocks. Like the elements of the first transition they share aspects of chemical behaviour with each other. Geochemists have taken advantage of this phenomenon by searching for anomalous relative concentrations of these elements. This can be done by comparing what is measured in a particular sample (for example rock, sediment or water) with the known values in average crust. For example, seawater is strongly depleted in cerium (Ce) relative to crustal rocks because it is efficiently scavenged by deep-sea manganese nodules. As a consequence it is possible to distinguish between sediments derived from seawater (showing cerium depletion) and those derived from the continents (without cerium depletion).

The elements of the third transition (the **actinides**) are rarer still. Most do not occur naturally. Of the natural ones, thorium (Th) and uranium (U) are relatively abundant, and have been widely used as natural tracers in the environment. Of the non-natural actinides, some have been released into the environment by nuclear bomb testing and nuclear power generation. Americium (Am) and plutonium (Pu) are widely dispersed in soils and sediments, and provide a valuable dating tool owing to a peak in their release to the environment during the early 1960s.

41 The physical properties of water

41.1 Introduction

Water is an extraordinary substance. It is found in all environments on Earth, and its concentration always has a major influence on physical, chemical and biological properties. Thus, whether we are studying desert environments, climatic feedbacks in the atmosphere, or formation of basaltic magma in the crust, water concentration is a master variable. If we want to understand the environment, we must understand water. The approach we have chosen to achieve this is to present a model for water. But first, what do we mean by 'model'? In science, this term is shorthand for a package of ideas and rules which allows us to explain and predict behaviour under a variety of circumstances. The value of a model is in simplification. Instead of learning all the complex patterns of behaviour, we can learn a few simple rules that allow us to work out what will happen. Before looking in detail at the properties of water, let us first look at its distribution on Earth.

The atmosphere contains approximately 14.5×10^{15} kg of water. Although this represents only 0.001 per cent of the water found on Earth, it has a profound influence on the properties of the atmosphere. Water vapour is the most important 'greenhouse' gas; it is not a better radiation absorber than carbon dioxide, but it is an order of magnitude more abundant (water vapour makes up on average 0.28 per cent of the atmosphere, while carbon dioxide is less than 0.04 per cent). Yet, it has only to condense into cloud droplets to reverse its effect on the Earth's heat budget. Further, the water content of the atmosphere influences water fluxes to and from the land surface, and modifies heat fluxes between the atmosphere and the Earth's surface.

The biota contains 2×10^{15} kg of water, or rather more than one-tenth of that in the atmosphere. In fact, water accounts for a high, often very high, proportion of the mass of practically all organisms. Water plays a key role in all physiological reactions, and water abundance is a key ecological variable at many scales.

The hydrosphere contains most of the water on Earth: 1.5×10^{21} kg. The greater part of this is in the oceans (96 per cent), and most of the remainder is in the ice caps (3 per cent) and groundwater (1 per cent). Only 0.025 per cent of the hydrosphere is as surface freshwater. Water in the oceans and ice caps plays a central role in the regulation of Earth surface temperatures.

The lithosphere is not commonly thought of as a major reservoir of water, yet, on average, crustal rock contains 1.2 per cent water bound within its minerals. This concentration may be low, but the great size of the crust compared with the hydrosphere means that this reservoir is significant. Given a mean crustal thickness of 16 km and a density of $2.9\,g\,cm^{-3}$ the amount of mineral-bound water in the crust can be estimated as 2.8×10^{20} kg, making it the second-largest reservoir of water after the oceans. This crustal water plays only a very minor part in the hydrological cycle; its residence time is similar to that of rock cycling owing to plate subduction and ocean crust formation. Nevertheless, this water has a major impact on a number of processes within the crust. For example, it modifies the melting temperatures of crustal rocks and governs the concentration of available oxygen in molten rock. Both of these effects influence the mineral assemblages present in the crystallizing magmas, and hence the composition of volcanic rock released from the crust.

From these examples, we see that water is both abundant and important. Some aspects of water are described when we look at states of matter (**18** and **19**). Here we look in more detail at the fundamental properties of water, and explore a physical model that helps to explain them. We then go on to look at some of the consequences of these properties for the natural environment.

41.2 The unusual properties of water

Before formulating the model, let us look at the strange properties of water and attempt to explain them.

41.2.1 High melting and boiling points

We are so familiar with the liquid and solid forms of water that they do not seem strange. Yet, when compared with other similar chemical compounds – the other covalent non-metal oxides – it is apparent that water is unusual. Consider melting and boiling points (Table 41.1). The other oxides are all gases at Earth-surface temperatures. Why is water different? Melting and boiling points provide a measure of intermolecular attraction. The higher the temperature the more strongly attracted the molecules are to each other. From Table 41.1 we can see that carbon monoxide has relatively weak intermolecular forces. In contrast, water has higher intermolecular forces than all the other oxides. Any physical model for water must explain this.

Table 41.1 Melting and boiling points of some common covalent oxides.

Substance	*Melting point (K)*	*Boiling point (K)*
Water, H_2O	273	373
Carbon monoxide, CO	68	82
Carbon dioxide, CO_2	195	217
Sulphur dioxide, SO_2	198	263
Nitrous oxide, N_2O	182	185

41.2.2 High heat capacity

When heat is given to a body, its temperature rises. This is because the kinetic energy of molecules in that body increases. However, the rise in temperature caused by a given addition of heat varies between substances. To raise the temperature of one gram of the mineral quartz by one degree Celsius takes only 0.9 joules of heat. For one gram of liquid water, however, it takes 4.2 joules to achieve the same rise. This is due to the difference in the property of a substance termed **heat capacity**. Of these two substances it is water that is unusual; it has an exceptionally high capacity to absorb heat without changing its own temperature very much. This is an important characteristic of water and has numerous environmental ramifications. But why would heat capacity vary between substances? Well, if the addition of heat to a body causes little rise in temperature, then the heat must be doing work on some other part of the system. The physical model for water must be able to explain where this energy is going.

41.2.3 High latent heats

If we have a mixture of ice and liquid water, any added heat does not alter the temperature at all. Instead, it causes melting of the ice. Because the added heat does not change the temperature, it is referred to as **latent heat**. For water, the latent heat of fusion is $320\,J\,g^{-1}$, which is substantially higher than for the other non-metal oxides. The latent heat of vaporization for water is higher still at $2260\,J\,g^{-1}$, which is also far higher than for most other substances. Thus, water has exceptionally high lattice energy (the energy gained by holding the molecules in a rigid, ordered solid), and exceptionally high intermolecular attraction. Once again, we see that intermolecular forces are vital to understanding water.

41.2.4 High dielectric constant

This is a very important property of water. The **dielectric constant** is a measure of the ability of a substance to alter electrical capacitance. Capacitance is a measure of the amount of energy stored in an electric field. For example, if we hold two flat metal plates close to each other, then we can store energy by giving one of the plates an electrical charge. It turns out that the amount of charge which can be stored is dependent upon what it is that occupies the space between the plates. Compared with air, some waxes and plastics greatly increase the amount of charge which can be stored, an effect which is known as the **dielectric effect** (and is used in the manufacture of electrical components). The effect is caused by the capacity of the intervening material to store electrical energy itself. The dielectric constant for a given material is the ratio of the charge which it allows to be stored compared with that for a vacuum. Water has an exceptionally high dielectric constant, which means that it has the ability to store charge. This property of water is very important as it plays a crucial role in the ability of water to dissolve ionic solids. Our physical model must therefore also explain why water is so good at influencing electric fields.

41.2.5 Thermal expansion

Most substances expand as they get warmer. We are so familiar with this phenomenon that we do not appreciate how odd it is. Water, however, does not exhibit this property at all temperatures. Ice behaves as expected, and so does water which is warmer than a few degrees Celsius. However, between 0 °C and 4 °C water behaves very unusually. Normally, when a solid melts, the liquid molecules take up more space, and therefore pieces of the solid will sink in the newly formed liquid. But, when water melts, the molecules take up *less* space: consequently, the solid floats on the liquid. Furthermore, as you warm water from 0 °C up to 4 °C, it continues to contract, with the result that water colder than 4 °C floats on water at 4 °C. Above these temperatures the normal pattern reappears, and warm water floats on cold water. Any physical model for water must allow for this peculiar thermal expansion.

41.3 A physical model of water

Elemental analysis reveals that water molecules comprise two hydrogen atoms and one oxygen atom. The very low electrical conductivity of pure water shows that the amount of dissociation (splitting into component ions) is very low, indicating that the bonding is dominantly covalent (see **8.3**). Further study shows that the atomic arrangement is H–O–H. This formula does not, however, tell us anything about the *shape* of the molecule. In fact, using the Lewis dot notation (see **8.1**) for representing covalent bonds, we can see why water is not a linear molecule (Figure 41.1).

The Lewis representation of the water molecule seems to allow the molecule to be flat. However, when considered in three dimensions, we see that the possible electron pairs in oxygen lie at the corners of a tetrahedron (Figure 41.2). When hydrogen atoms are added to form a water molecule the tetrahedron is distorted: the two hydrogen atoms repel each other slightly such that the angle between them is greater than would be predicted from a simple tetrahedron.

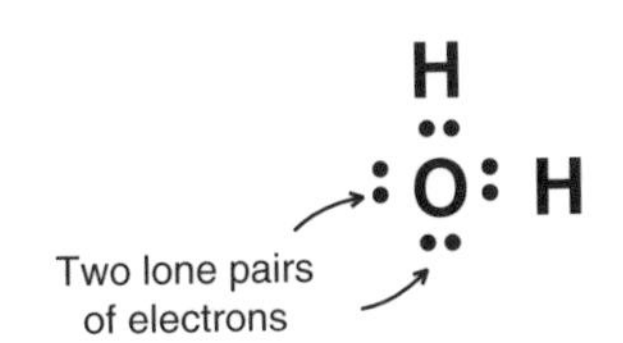

Figure 41.1 The Lewis notation shows why water is not a linear molecule; this is not obvious from the standard formula H_2O.

The covalent bonds in water are not symmetrical. The oxygen atom attracts the electrons more strongly than does the hydrogen atom (oxygen has greater **electron affinity**). Thus the shared electrons are, on average, located closer to the oxygen end of the bond. This asymmetry gives the oxygen atom a slight negative charge and the hydrogen atom a slight positive charge. If the H–O–H molecule were linear this would not lead to a polarized molecule (a molecule is electrically polarized if its ends have opposite charges). However, because the water molecule is bent, we do have an electrical polarization (Figure 41.3).

This polarization is very important. In substances such as carbon monoxide (C=O), the carbon and oxygen have similar electron affinities, and thus there is very little electrical polarization. This causes only very weak forces between the molecules; consequently, carbon monoxide is a gas at normal Earth temperatures. However, liquid water comprises a compact mass of electrically polarized H_2O molecules. The oppositely charged ends of the molecules

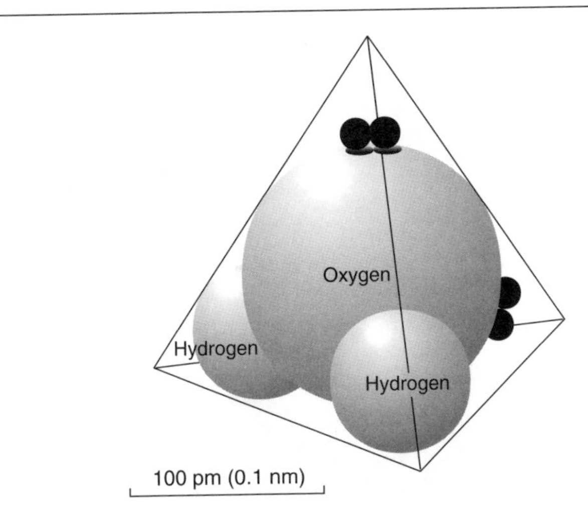

Figure 41.2 The approximately tetrahedral shape of the water molecule. The black spheres indicate the position of the electron lone pairs. This physical form for the electrons is not realistic. The hydrogen atoms and electron pairs are sited at the corners of a distorted tetrahedron. A slight positive charge on the hydrogen atoms, and consequent electrostatic repulsion, means that they are slightly further apart than predicted if the molecule shape were truly tetrahedral.

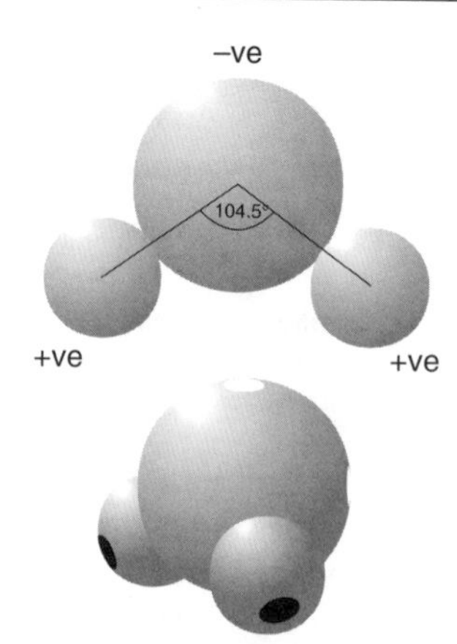

Figure 41.3 The polarization of charge across a water molecule derives from its asymmetry. The high electron affinity of oxygen draws the shared electrons away from the hydrogen atoms, leaving them with a net positive charge. The electrons spend slightly more time than average at the opposite side of the oxygen atom, focused at the corners of the tetrahedron. This leaves a slight negative charge. Centres of charge are indicated by the circles: black for positive and white for negative.

attract, while the similarly charged ends repel. Thus, the liquid water rapidly settles into an ordered, if fluctuating, arrangement of molecules in which the like charges are as far from each other as possible, and the opposite charges as close as possible (Figure 41.4). In this ordered arrangement the

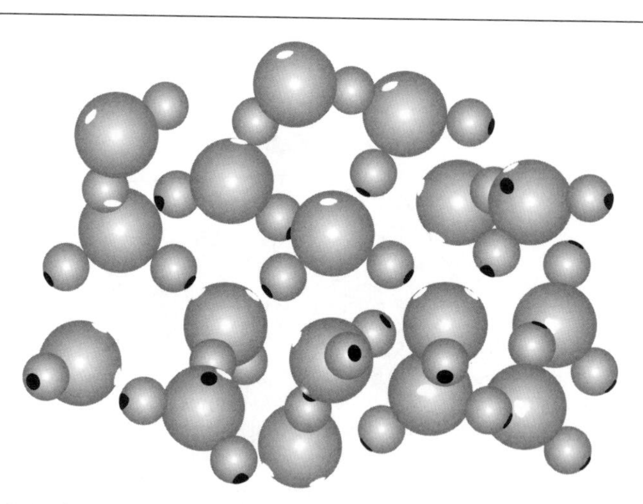

Figure 41.4 A schematic representation of liquid water. In liquid water the molecules are free to move around each other. The mass looks disordered. However, the electrical attraction between the regions of positive (black circles) and negative (white circles) charge prevents a random arrangement. Instead a complex, constantly changing structure develops in which opposite charges are as close as possible, and like charges as distant as possible.

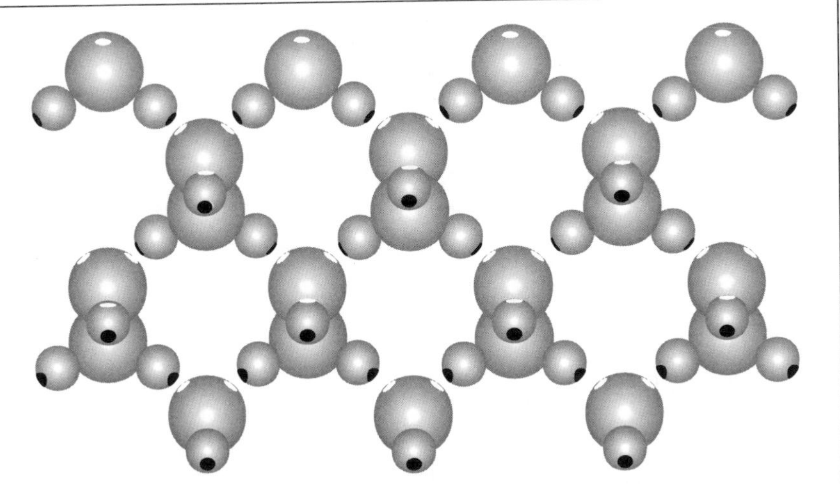

Figure 41.5 A schematic representation of crystalline water. Crystalline water results from electrostatic (largely) attraction between the positively (black circles) and negatively (white circles) charged regions of the water molecules. This holds the molecules in an open lattice with regular 'holes'.

forces of attraction far outweigh the forces of repulsion. This effect, therefore, explains the unusually high intermolecular attraction of water.

The situation illustrated in Figure 41.4, where all the molecules are allowed to wander freely through the liquid (subject to charge constraints), is not entirely accurate. Studies of the properties of water have shown that short-lived, small, three-dimensional structures continually form within the water, held together by the electrostatic attractions. Figure 41.5 illustrates the kind of structure envisaged. The small, short-lived lattices contain voids. These structures are consequently less compact than freely moving molecules at the same temperature. The lattice thus has a low density, and would float on the liquid if it existed long enough. In liquid water, however, this does not happen because the lattice structures are transient, continually forming and breaking up.

Thus, in liquid water we have a mixture of freely moving molecules, and temporary, low-density, rigid clusters of molecules. At high temperature, these clusters make up a small part of the water. However, at temperatures close to freezing point, they become abundant enough to lead to lower than expected densities. In cold water, therefore, we have two competing processes. As the water gets colder, the individual water molecules take up less space, increasing the density. At the same time, low-density clusters are becoming more abundant, which has the effect of reducing the density. At 4 °C the effects balance, but below this temperature it is cluster formation that dominates. Figure 41.6 shows the relationship between density and temperature in water.

In its solid form (ice), all the water molecules take on the low-density structure described above (Figure 41.5). Thus, ice has a lower density than 0 °C liquid water in which a substantial proportion of the molecules are free of the lattice structures.

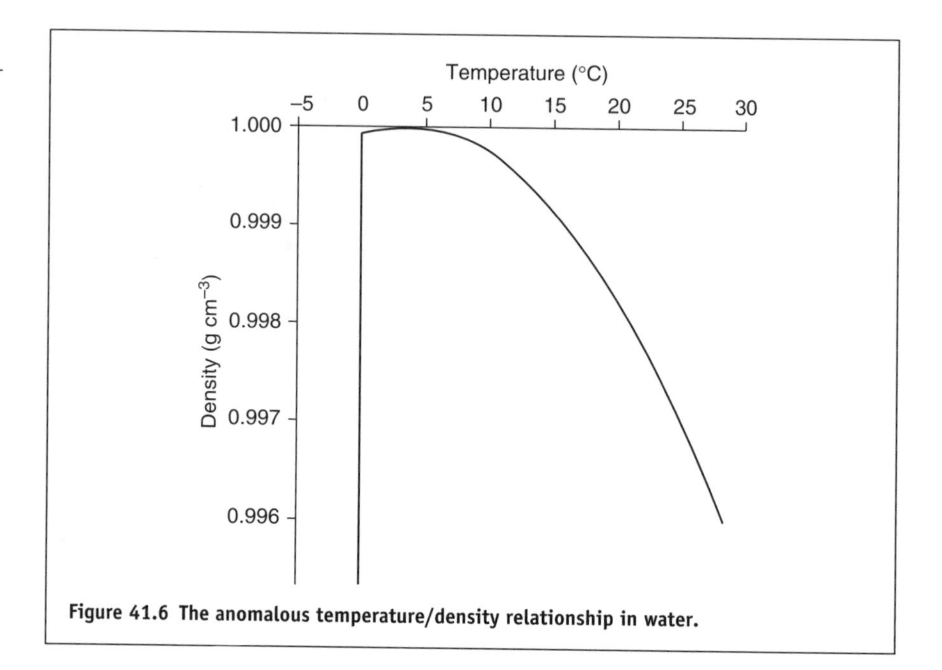

Figure 41.6 The anomalous temperature/density relationship in water.

The linkages which cause the effects described above are known as **hydrogen bonds** (see **8.6**). They result from electrical attraction between polarized water molecules. Such bonds have only about one-tenth the strength of the O–H covalent bonds. Until recently, they were thought to be entirely electrostatic in nature. However, it is now known that they are partially covalent, and can exchange electrons with the bonds within each water molecule.

A key point is that hydrogen bonding lies behind all of the anomalous physical properties of water. The high melting and boiling points and the high latent heats can all be explained by the increased intermolecular forces generated by hydrogen bonds. The high heat capacity of water is owing to a part of the added thermal energy being used to break these bonds. The anomalous thermal expansion can also be attributed to the formation of these low-density structures in the water. What about the high dielectric constant? This, too, is explained by the polarization of the water molecule. If we imagine a two-dimensional representation of our capacitor plates with water in between, we see that the field distorts the pattern formed by the polarized water molecules. The molecules are partially rotated into alignment with the electric field of the capacitors, which increases their potential energy, as they are no longer in their favoured, low-energy orientation. This stores additional energy in the electrical field. All of these effects can thus be explained by the polarity of the water molecules, and consequent hydrogen bonding.

In conclusion, a simple physical model of water based on the Lewis valence bond theory predicts the existence of hydrogen bonding, and the structures which result from hydrogen bonding explain all of the anomalous physical properties of water. The physical model described can account for all the unusual features of water.

As a caveat we might note that the model holds well for water at normal temperatures and pressures, but not for all conditions. For example, below −50 °C we can cause the water to switch from its preferred low-density state to the high-density state (lacking hydrogen bonds) by raising the pressure above 1000 atmospheres or so. However, this effect is not relevant to our concerns with Earth-surface processes.

41.4 Special properties of water: the role of water in the environment

41.4.1 Radiation absorber

The structure of the water molecule is such that it preferentially absorbs long-wave radiation. For a substance to absorb electromagnetic radiation it must resonate with an energy comparable to the energy of the radiation. The lengths of the covalent bonds in many molecules are just right to interact with electromagnetic radiation in the UV, visible and infra-red spectra (see **20.2**). The structure of water causes it to be a good absorber of red and infra-red radiation, but a poor absorber of blue and UV radiation. This leads, for example, to the sky being blue. More importantly, it has the effect of raising the temperature of the Earth's surface. The temperature increase is that necessary to bring the system back into equilibrium. If there were no water vapour in the atmosphere, the temperature at the surface of the Earth would be much lower than it is.

41.4.2 Weathering

The ability of water to dissolve a wide range of inorganic substances means that it is an ideal medium to attack mineral matter. However, the success of water as an agent in weathering (see **39**) owes much to factors other than simple chemical reactions. That water is a liquid is crucial. Only a liquid has both the condensed nature to allow high concentrations of reactants, and the mobility to bring the reactants close to each other and to carry away the reaction products.

Perhaps still more important is the link between water and life. Chemical and physical weathering are strongly enhanced by the presence of living organisms. The physical disruption caused by root growth and animal burrowing, combined with the organic acids that are so effective at attacking silicate minerals, powerfully enhance the ability of water to promote weathering.

41.4.3 Anomalous thermal expansion of water

This is very important for many reasons; without it, aquatic life could not survive in the cooler regions of the Earth. Consider the example of a deep

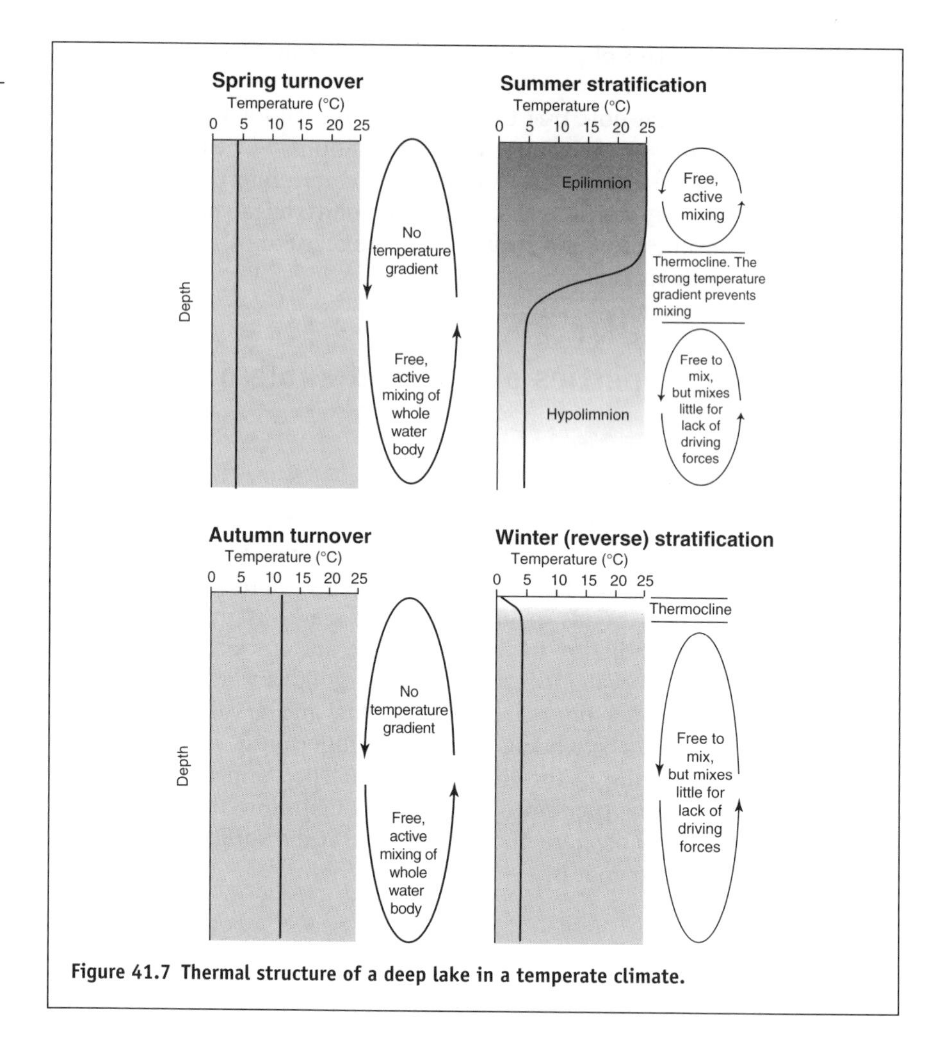

Figure 41.7 Thermal structure of a deep lake in a temperate climate.

lake in a cool, temperate climate (Figure 41.7). Let us assume that in early spring all of the water is at 4 °C. Under these conditions, all of the water has the same density and can mix freely. Over the spring and summer, solar radiation will warm the surface of the lake. The surface water will become warmer than 4 °C, and, being less dense than the colder underlying water, it will form a warm upper layer. Once this has happened, we have a temperature (and therefore density) gradient (**thermocline**) between the upper and lower layers of the lake (**epilimnion** and **hypolimnion**, respectively). Now, the layers cannot mix freely because of the high energy needed to mix the **stratified** water. Once summer stratification has occurred, it becomes easier to maintain: further added heat has only to warm the upper layers. Thus, the upper layer warms more rapidly than before, and the thermocline becomes stronger, severely limiting the transfer of both heat and nutrients between the epilimnion and hypolimnion. (Lake surface water of 16 °C has been observed just 3 m above water of 4 °C.)

In the autumn, the situation is reversed. Cooling of the epilimnion continues until it has the same temperature as the hypolimnion. In this condition, the whole lake can mix freely, an event termed the autumn turnover. After this, the whole lake cools until the temperature reaches 4 °C. Then, any further cooling produces water that is less dense than the underlying water. Thus a reverse, or winter, stratification develops. The deep water remains at 4 °C protected by the stratification, but the surface rapidly cools. If the lake surface freezes, protection is further enhanced. Any ice and snow insulate the lake from rapid cooling, and the ice prevents wind stress from mixing the water. Thus, aquatic life is protected in a 4 °C environment throughout the winter. If water behaved like other substances, it would freeze from the bottom up (and more rapidly because it would have lower heat capacities and latent heats). This would make it much harder for life to survive. On melting of the ice and warming of the water in the spring, the reverse stratification breaks down, and eventually a spring turnover occurs.

Lakes that conform to the pattern described above are termed **dimictic**: they mix twice each year. This is the norm for deep lakes at low elevation in the mid-latitudes. However, many lakes do not behave in this way. Shallow lakes may simply not be deep enough to stratify. At high latitudes, or high elevations, a lake may never become warm enough to develop summer stratification. Instead, a late spring turnover runs through the summer into an early autumn turnover. The absence of summer stratification means that the lake experiences a single prolonged period of mixing, and is thus termed **monomictic**. As this condition arises owing to low temperature, the lake is described as **cold monomictic**. If conditions are cold enough, the lake may never mix (owing to permanent reverse stratification), in which case it is termed **amictic**.

In warmer climates, the opposite situation may be found. Here, lakes stratify well in the summer but are never cold enough to cause reverse stratification. Such lakes mix throughout the colder part of the year. As winter stratification is prevented by the high temperature, such lakes are termed **warm monomictic**. A different situation is encountered in tropical regions. These warm lakes can stratify (if deep enough), but the stratification is weak. There is no cold season to bring about a regular mixing period, but occasional mixing is caused by wind stress. Such erratically mixed lakes are termed **polymictic** or **oligomictic** (rarely mixed) depending upon mixing frequency.

41.4.4 Global heat exchange and buffering

The equatorial regions receive far more solar radiation than the poles. They are warmer as a consequence, but not in proportion to the imbalance in radiative fluxes. A number of different processes are involved, but all take advantage of water's special properties. Its exceptional heat capacity means that advection of water (mass movement via flow) is an effective mechanism for transferring heat. Global oceanic circulation, with bottom water formation at the poles and surface backflow from the equatorial regions, drives heat towards the poles from the equator. It is this process which makes the high latitudes habitable.

On a smaller scale, water vaporizes when the temperature is high, and condenses when the temperature is low. The exceptional latent heats of water mean that this is a very effective mechanism for the transfer of heat from warm to cold areas, particularly in transferring heat from low to high altitude.

However, it is not only spatial inequalities that water helps to moderate. Temporal variations in heat are also reduced by water. Again, this is evident at a number of scales and involves a variety of mechanisms. At the longest timescales, the coexistence of three phases of water provides the greatest heat buffering. The latent heat stored in solid and liquid water is such as to reduce temperature fluctuations by orders of magnitude. At shorter timescales, it is the heat capacity of water that provides a buffering effect. The ameliorating effect of the ocean on climate is well known, although this has as much to do with the magnitude of the oceans as with the high heat capacity of water. A better example is the effect of water on seasonal variation in soil temperature. The high heat capacity of water means that wet soils warm far more slowly than dry ones. For this reason, effective drainage of soil in spring is of great significance to farmers in colder areas because it permits early planting of crops and promotes their early growth.

41.4.5 A driving force in geomorphology

The role of water as an agent in Earth-surface processes is well appreciated. However, this is mostly viewed simply in terms of water being a liquid capable of driving erosion and transporting particulate and dissolved materials. But, how did water get its energy in the first place? This takes us back to the coexistence of multiple states of water. Evaporation of water through the agency of solar radiation gets water off the ground and into the air. More solar radiation, driving turbulent mixing of the lower atmosphere, allows that vapour to gain altitude. Condensation back to liquid water in the cooler upper parts of the troposphere completes the process. Liquid water has been given an enormous amount of gravitational potential energy. Some water will evaporate again, but much will fall to the ground and work its way downhill, acting on rock, soil and sediment on the way. Once again, we owe all this to the hydrogen bond. Without these bonds water would be a gas, and geomorphology would be as sluggish as it is on Mars.

42 The chemistry of water

42.1 Introduction

The physical properties of water are described, and a physical model to explain these properties is provided in **41**. Here, we consider the behaviour of water from a chemical point of view. As with physical properties, water is chemically unusual in a number of respects. Water is one of the few inorganic substances which is liquid at room temperature. However, there are three other aspects of the chemical properties which are striking:

- Water is extremely good at attacking metallic and ionic solids and bringing them into solution.
- Because of its ability to carry ionic substances in solution it is, in its impure state, very good at conducting electrical currents.
- For reasons related to both of these points, water is very good at enhancing chemical reactions.

Why can such a wide range of substances be dissolved in water? A solution results from the uniform dispersal of one substance – the **solute** – within another – the **solvent**. By dispersal, we mean that a substance is broken into molecules and held indefinitely in solution. The solute can be solid, gas or liquid, whereas the solvent is usually a liquid. In order for dissolution to occur, there must be stronger forces of attraction between solute and solvent than between solvent alone or solute alone. Clearly, water can attach very strongly to ions, and any chemical model for water must be able to explain this.

Because of its ability to dissolve ions, water is very effective at conducting electrical charges. Thus, natural water is a good electrical conductor. (In its pure state, water is a poor conductor, but it is rarely pure.) The conductivity of water has a great impact on its ability to promote oxidation and reduction reactions. The conduction of charge by dissolved salts is also central to the physiology of all living organisms.

We are familiar with the ability of water to increase reaction rates, as for example in rusting, the decay of organic matter and the decomposition of mortar. These are not related to a single phenomenon, but to a combination of effects. These effects include the ability of water to dissolve substances directly; to bring dissolved reactants into close proximity, thus allowing them to react with each other; to conduct charges, promoting some kinds of chemical reaction; and to remove reaction products, which would otherwise slow down the reactions.

42.2 Dissociation of water

In the physical model of water presented in **41**, water is assumed to be made up of a large number of identical water molecules. Some of these move freely, while others are temporarily bound in ice-like structure. We now look at this model in more detail. (If you are not familiar with the basic handling of chemical equilibria, then refer to **35** before reading on.)

The first thing which emerges from closer examination is that chemical species other than water molecules are present even in pure water. This is because water undergoes self-ionization (also termed autohydrolysis), thus:

$$\mathrm{H{-}O{-}H} \rightleftharpoons \mathrm{H^+ + OH^-}$$

In other words, some of the water **dissociates** (splits up) into hydrogen ions (H^+) and hydroxide ions (OH^-), by the breaking of one of the covalent H–O bonds. Dissociation occurs continuously. However, the reaction also goes in reverse, re-creating water molecules, so we do not have a net build-up of the ions. Instead, a dynamic equilibrium exists between the water molecules and tiny, equal proportions of H^+ and OH^-, with continual formation and destruction of water molecules. For every billion (10^9) water molecules there are only about 200 ions in total.

How realistic is this model? In a physical sense, it is poor; hydrogen and hydroxide ions do not exist as lone entities. Instead, because they are charged, they have a mantle of appropriately orientated water molecules (Figure 42.1). This 'mantling' by water is termed **hydration**. However, it turns out that predictions of chemical behaviour can still be made if this hydration is ignored. We can therefore treat the ions as if they are not hydrated, which simplifies the chemical model.

42.2.1 The dissociation constant

At the heart of any model for water lies the **dissociation constant**. Consider the reaction below, which describes the dissociation (break-up) of a water molecule into smaller fragments (dissociated ions):

$$\mathrm{H{-}O{-}H} \rightleftharpoons \mathrm{H^+ + OH^-}$$

Notice that expressions such as $H{-}O{-}H \rightleftharpoons H^+ + OH^-$ *show the relationship between the reactants and their products. They tell you which compounds*

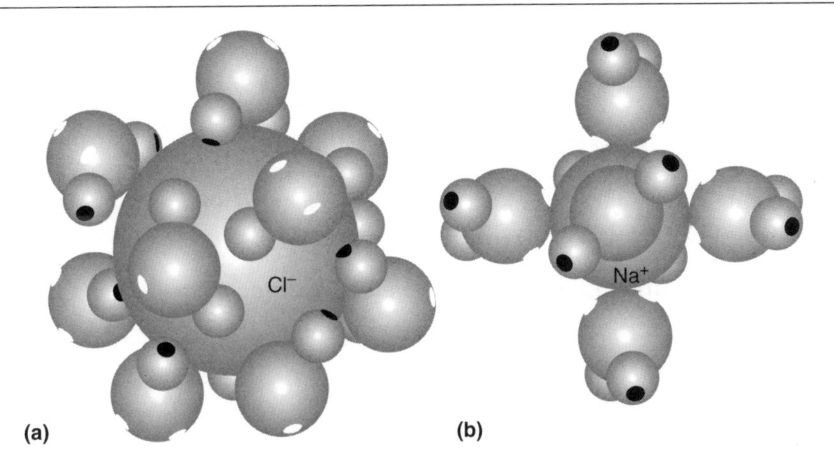

Figure 42.1 Ions dissolved in water have a mantle of water molecules each orientated to minimize the electrical potential energy. (a) The large chloride ions are surrounded by many water molecules; (b) the smaller sodium ion has space for six water molecules, which take up an octahedral arrangement.

react, and which compounds are produced. In this example it is reversible; the expression thus tells us that:

- *One H^+ ion reacts with one OH^- ion to produce one H–O–H molecule.*
- *One H–O–H molecule reacts to produce one H^+ ion and one OH^- ion.*

This information is termed **stoichiometry**. *In contrast, the expression $[H^+]$ refers to the concentration (strictly,* **activity**, *which is a corrected concentration allowing for non-ideal behaviour; in normal dilute freshwater* activity $\approx$ concentration*) of hydrogen ions. When an expression is given in concentration notation, i.e. by use of square brackets, there will normally be a known coefficient which can allow us to constrain the concentrations.*

At any given temperature, the product (in the mathematical sense of the result of multiplication) of the dissociated ions from the reaction above is constant. This can be expressed as:

$$[H^+] \times [OH^-] = k_W$$

where k_W is known as the dissociation constant of water. The value of k_W varies with temperature but is relatively constant at normal temperatures and is usually quoted as 10^{-14}.

This expression is more useful than it looks. It tells us three very important things:

- It tells us that the $[H^+]$ and $[OH^-]$ concentrations are not independent of each other.
- If we know the value of one, we can calculate the value of the other, i.e. $[H^+] = k_W/[OH^-]$ and $[OH^-] = k_W/[H^+]$.

- In pure water it allows us to calculate the concentrations of both $[H^+]$ and $[OH^-]$.

This last calculation is quite straightforward. In pure water:

$$[H^+] = [OH^-] = \sqrt{k_W}$$

This is then $\sqrt{10^{-14}}$ which is equal to 10^{-7}. Therefore:

$$[H^+] = [OH^-] = 10^{-7}$$

From this we can find the pH of pure water, because:

$$pH = -\log_{10}[H^+]$$

Thus, if $[H^+] = 10^{-7}$, then:

$$pH = -\log_{10}[10^{-7}]$$

which evaluates to:

$$pH = -(-7) = 7$$

(The arithmetic manipulations carried out here are introduced in **26**.)

This makes a complete chemical model for pure water if allowance is made for the effect of temperature on k_W. However, as earlier remarked, water is almost never pure. To make the model useful we need to introduce gases, salts and minerals. First, we need to consider the reactions that take place between water and these other substances.

42.3 Water and ionic solids

42.3.1 Solubility

If we take a crystal of an ionic solid such as sodium chloride (NaCl) and immerse it in water, we get very rapid dissolution. On the surface of the crystal, particularly on the corners, there are Na^+ and Cl^- ions which are very exposed. Polarized water molecules are able to partially mantle these ions. The high dielectric constant of the water (see **41.2.4**) reduces the strength of the electric field linking the edge ions to the rest of the ionic lattice. Now we have two systems competing for these ions: the lattice in which they started, and the water surrounding the crystal. The lattice is stable because the arrangement of the ions minimizes the electrical potential energy. However, hydrated ions are also of relatively low energy (and therefore favoured) because of the high dielectric constant of the water. In the case of NaCl, the hydration energy is greater than the lattice energy, so the water 'wins'. As Na^+ and Cl^- ions are broken from the edges of the crystal and surrounded by a mantle of water molecules, fresh ions are exposed on the surface of the crystal. This process continues until the crystal has entirely dissolved, or until a **saturated** solution of the salt is attained.

As **saturation** is approached, the concentration of dissolved hydrated ions is such that a significant proportion are reuniting to form a crystalline lattice again. Why would they do this if they have been soluble before? Well, one can view dissolution as competition between two reactions.

$$Na^+Cl^-_s \rightarrow Na^+_{aq} + Cl^-_{aq} \quad (1)$$

$$Na^+_{aq} + Cl^-_{aq} \rightarrow Na^+Cl^-_s \quad (2)$$

Notice that in these chemical equations the water molecules of the hydration mantle have been left out. (The aqueous state is indicated by the subscript aq; the solid state of the NaCl is indicated by the subscript s.) Reaction 1 is favoured by there being very few dissolved ions. Reaction 2 is favoured by there being very many ions.

This follows qualitatively from a key rule in chemistry known as **le Chatelier's principle**: a chemical system at equilibrium will oppose any changes forced upon it. In other words, if we add more ions to a solution, the system will oppose this by causing some of the ions to **precipitate** and settle out of the solution.

The concentration of dissolved ions in a saturated solution varies with temperature. At any given temperature, the concentration can be found using the **solubility product**, K_{SP}. Saturation is reached when the product of the dissolved concentrations equals the value of solubility product for the ambient temperature. The solubility product for NaCl at 20 °C is 38. The solution is said to be **supersaturated** with respect to NaCl when:

$$[Na^+][Cl^-] > K_{SP}$$

A supersaturated solution (dissolved ion concentrations exceeding the solubility product) can exist without **precipitation** of the salt occurring, but for the purposes of this model, it can be supposed that precipitation will occur such as to maintain concentrations at the solubility product. Strictly, the solubility model outlined here works only when the salts are in isolation. However, when there is a mixture of ions in solution, we have to allow for the phenomenon called the **common ion effect**. This effect is important in studying evaporates but has little significance for those substances that normally precipitate from freshwater (that is calcium carbonate, and hydroxide/oxides of iron and manganese).

42.3.2 Salts in solution

The dissolved cations and anions found in freshwaters are commonly referred to as dissolved salts. It is important to realize, however, that the ions in solution cannot generally be attributed to any particular salt. Na^+ in seawater is no more 'attached' to the Cl^- than it is to SO_4^{2-} (also abundant in seawater). This is very important to bear in mind as there is widespread confusion about this point. While the idea of dissolved salts makes sense when considering chemical leaching at the landscape scale, it is not a helpful concept in relation to water chemistry. Instead, it is better to think of the water as holding a mixture of unrelated dissolved ions. Another reason why 'salts', as opposed to dissolved ions, are relatively unimportant in natural waters is that very few are near saturation (unless the formation of evaporites under hypersaline conditions is being considered). The most important substances which do precipitate from at least some normal natural waters are:

- calcium carbonate, $CaCO_3$
- iron III hydroxide, $Fe(OH)_3$
- manganese oxides.

42.3.3 Hydrolysis and complex ions

So far in our discussion of the interaction of ionic solids with water, the ions released by dissolution have not reacted strongly with the water. When $NaCl_s$ is dissolved in water, the water molecules actively tear ions from the lattice, and then surround those ions. The water molecules and the ions are unchanged; they have simply been rearranged. The ions are still bound by attractive electrostatic forces, but now they are bound to a mantle of water molecules rather than to an ionic lattice. The dissolved ions are said to be **aquo** ions, reflecting this mantling by water. This is an example of **hydration**. But consider what happens when a metal oxide such as CaO_s is dissolved in water:

$$CaO_s + H_2O \rightleftharpoons Ca^{2+}_{aq} + 2OH^-_{aq}$$

This is very different. CaO, like NaCl, is an ionic solid. However, O^{2-}, unlike Cl^-, is not stable in aqueous solution. Instead, a very great force of attraction exists between it and the weak positive charge on the water's hydrogen atoms. Such is the force involved that it tears a hydrogen atom away from the water molecule, breaking one of the covalent O–H bonds. The reaction can be expressed as

$$O^{2-}_{aq} + H_2O \rightarrow 2OH^-$$

This breaking up of the water molecule by chemical attack is termed **hydrolysis**. Notice that hydrolysis can also be viewed from the perspective of Lewis acidity (see **12.1**). Instead of simply getting a rearrangement of water molecules and ions, as in hydration, we get disintegration of a water molecule and bonding with one of the ions (O^{2-}). The Ca^{2+} ion becomes hydrated, but otherwise remains unchanged.

In the case of ionic oxides (i.e. those with O^{2-} as one of the ions) it is the H^+ which is taken from the water. This means that the water becomes less acidic. Consequently, this kind of hydrolysis is sometimes referred to as **basic hydrolysis**. However, some cations are also capable of causing hydrolysis. They do this not by attaching to the hydrogen but by binding to the hydroxide ion. For example, aluminium ions react with water:

$$Al^{3+} + H_2O \rightleftharpoons Al(OH)^{2+} + H^+$$

This has the opposite effect on the acidity of the water, tending to acidify. The term **acid hydrolysis** is sometimes applied for this reason. Notice that this differs slightly from hydrolysis of O^{2-}. The reaction here has led to the formation of a complex ion. The term 'complex ion' is applied where an ion comprises a combination of smaller ions or molecules. For example, NaCl dissolves to form **simple ions** (Na^+ and Cl^-); while $Al(OH)_3$ dissolves to form a mixture of simple ions (Al^{3+} and OH^-) and **complex ions** ($Al(OH)_2^+$, $Al(OH)_3^{2+}$ etc.). A simple ion is not strongly combined with other ions or molecules, but will be mantled by water molecules.

The tendency of cations to cause hydrolysis, and thus to behave as acids, increases with ionic charge. For monovalent ions, the effect is negligible. For most divalent ions the effect is also minor, although Be^{2+} and Mg^{2+} ions, because of their small size and hence high electrical field strength, do bind with hydroxide ions to a significant extent. However, all of the trivalent ions form a series of hydroxide complexes. From the point of view of natural waters, it is Fe^{3+} and Al^{3+} that are most important in terms of acid hydrolysis.

If we dissolve Al^{3+} ions in water, a series of hydrolysis reactions takes place (a similar series could be drawn for Fe^{3+}):

1. $Al^{3+} + H_2O \rightleftharpoons Al(OH)^{2+} + H^+$
2. $Al^{3+} + 2H_2O \rightleftharpoons Al(OH)_2^+ + 2H^+$
3. $Al^{3+} + 3H_2O \rightleftharpoons Al(OH)_3^0 + 3H^+$
4. $Al^{3+} + 4H_2O \rightleftharpoons Al(OH)_4^- + 4H^+$

Notice that each of these reactions releases hydrogen ions, making the solution more acidic. Following le Chatelier's principle (see **42.3.1**) we can deduce two properties of aluminium ions in water:

- The more acidic the solution, the more the above equations are driven to the left (favouring free Al^{3+} ions over the hydroxide complexes).
- Because the reaction in equation 1 generates only one H^+ ion, it is more likely to happen in an acidic solution than the reaction in equation 4 which generates four H^+ ions. Thus, mildly acidic solutions have Al^{3+} and $Al(OH)^{2+}$ ions, while basic solutions have $Al(OH)_3^0$ and $Al(OH)_4^-$.

The effect of pH on the relative proportions of the different aluminium species is shown in Figure 42.2.

In highly acidic water, with pH values below 4, aluminium is predominantly in the Al^{3+} form. This is intuitively reasonable if we think about the dissociation constant for water. At pH 4 the concentration of OH^- ions is:

$$[OH^-] = \frac{k_W}{[H^+]}$$

where

$$[H^+] = 10^{-pH} = \frac{10^{-14}}{10^{-4}} = 10^{-10}$$

The hydroxide concentration is thus only 0.1 parts per billion, and it is hardly surprising that hydroxide complexes (ions formed by joining hydroxide ions

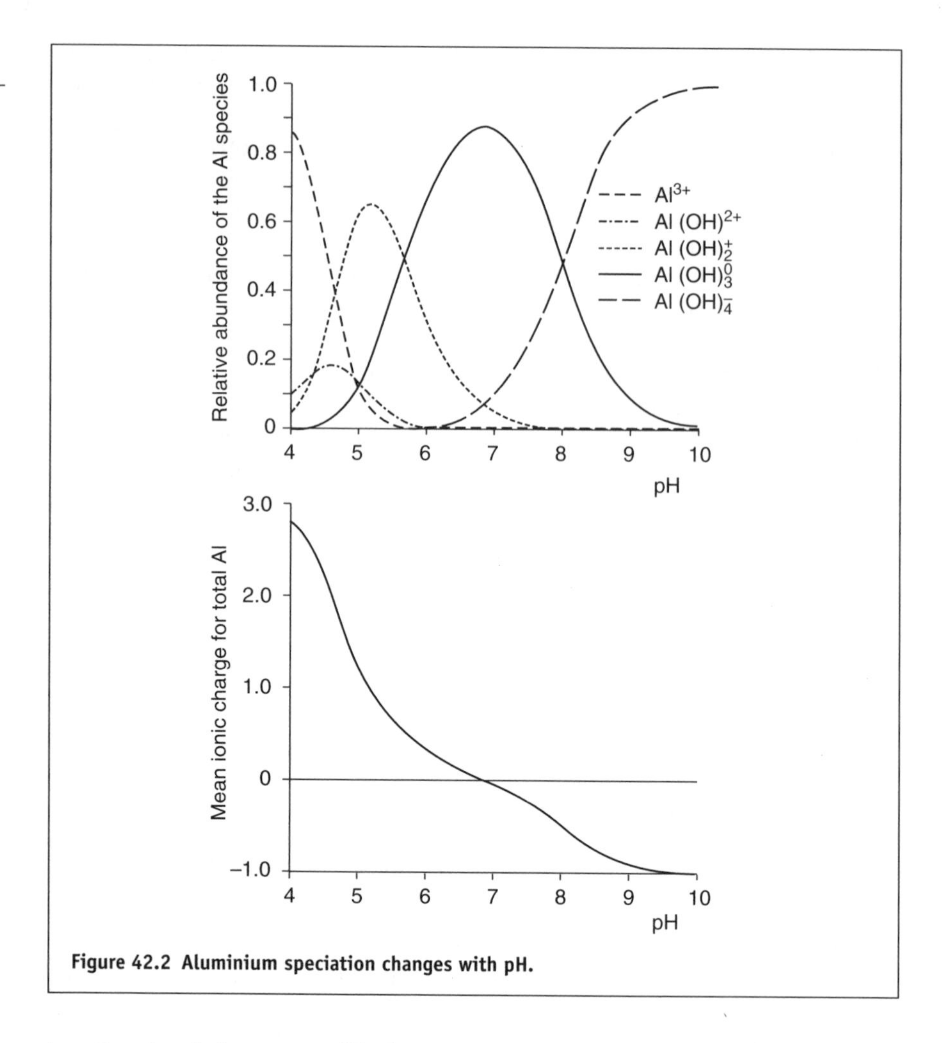

Figure 42.2 Aluminium speciation changes with pH.

to other ions) do not readily form. For every whole pH unit increase, the concentration of hydroxide ions increases ten-fold, making hydroxide formation easier and easier.

The binding of hydroxide ions to aluminium (or any other element) can be calculated easily from equilibrium equations if the appropriate equilibrium constants are known. Take, for example, the first hydrolysis reaction for aluminium:

$$Al^{3+} + H_2O \rightleftharpoons Al(OH)^{2+} + H^+$$

The concentration ratios for the different aluminium species are given by the equation:

$$\frac{[Al(OH)^{2+}][H^+]}{Al^{3+}} = k_1$$

where the first hydrolysis constant, k_1, is $10^{-4.97}$.

Table 42.1 Hydrolysis constants of iron and aluminium.

Reaction	*log K*
$Al^{3+} + H_2O \rightleftharpoons Al(OH)^{2+} + H^+$	−4.97
$Al^{3+} + 2H_2O \rightleftharpoons Al(OH)_2^+ + 2H^+$	−9.3
$Al^{3+} + 3H_2O \rightleftharpoons Al(OH)_3^0 + 3H^+$	−15.0
$Al^{3+} + 4H_2O \rightleftharpoons Al(OH)_4^- + 4H^+$	−23.0
$Fe^{3+} + H_2O \rightleftharpoons Fe(OH)^{2+} + H^+$	−2.19
$Fe^{3+} + 2H_2O \rightleftharpoons Fe(OH)_2^+ + 2H^+$	−5.67
$Fe^{3+} + 3H_2O \rightleftharpoons Fe(OH)_3^0 + 3H^+$	<−12
$Fe^{3+} + 4H_2O \rightleftharpoons Fe(OH)_4^- + 4H^+$	−21.6

For any given pH value, the concentration of H^+ can be calculated ($[H^+] = 10^{-pH}$). The ratio of $Al(OH)^{2+}$ to Al^{3+} is then $k_1/[H^+]$.

Some values for hydrolysis constants are given in Table 42.1. They are all for a temperature of 25 °C. When applied in the way shown here they are accurate only if activities and concentrations are equal. Fortunately, this is a fairly reasonable assumption in dilute freshwaters. However, more accurate results are obtained if activity corrections are applied.

42.3.4 Chemical speciation of ions in water

We have already met some complex ions, i.e. ions which comprise two or more individual ions bound together (for example $Al(OH)^{2+}$). And you should now be familiar with the term 'species' as it is used in a chemical context: in essence, 'species' refers to the various chemical forms in which atoms or molecules occur. The process of identifying the nature and concentration of the different chemical species in solution is referred to as **speciation**. (Note that this term is used very differently in biology.) So why does it matter what form the ions take? Why, for example, would we want to know how much OH^- is bound to aluminium?

There are two reasons why quantifying the aluminium species is important:

- Equilibrium calculations are all expressed in terms of the simple ions (i.e. Al^{3+} or Fe^{3+} in the examples above). Yet when we measure Al in solution we get only the total amount, i.e.:

  $$Al_{total} = \sum\{Al^{3+} + Al(OH)^{2+} + Al(OH)_2^+ + Al(OH)_3^0 + Al(OH)_4^-\}$$

  If we want to calculate the solubility of aluminium then we need to be able to calculate the proportions of the different species.

- $Al(OH)_2^+$ is more toxic to fish than $Al(OH)^{2+}$. Similar relationships between speciation and toxicity are found for other elements. Speciation has, therefore, great ecological importance.

Table 42.2 Speciation of some common metals in water.

Element	*Main species in freshwater*
Fe	$Fe(OH)^{2+}$, Fe-organic complexes Fe^{2+} under reducing conditions
Al	$Al(OH)_4^-$, Al-organic complexes, AlF^{2+}
Mn	Mn^{2+}, Mn-organic complexes
As	$HAsO_4^{2-}$, $H_2AsO_4^-$
Cd	Cd^{2+}, $CdOH^+$
Cr	$Cr(OH)_3^0$, CrO_4^{2-}
Cu	Cu^{2+}, $CuOH^+$, Cu-organic complexes
Hg	$Hg(OH)_2^0$, HgOHCl, CH_3Hg family
Ni	Ni^{2+}, $NiCO_3^0$
Pb	$PbCO_3^0$, $Pb(CO_3)_2^{2-}$, Pb-organic complexes
Zn	Zn^{2+}, $ZnOH^+$, $ZnCO_3^0$

Using the method outlined above we can now estimate the importance of the different hydroxide complexes for aluminium. But, what about other elements? Generally, the difficulty of identifying species increases with ionic charge. For example, Na^+ and K^+ do not significantly associate with any other ions at surface water concentrations (complexes with HCO_3^-, CO_3^{2-} and SO_4^{2-} – anion complexes – amount to less than 0.1 per cent of the total dissolved potassium or sodium). In contrast, as stated above, the trivalent ions readily form complexes. In this section, we run through the main ion complexes which are common in natural waters. The special case of speciating dissolved carbonates is left to a later section.

Of the major cations, only Mg^{2+} and Ca^{2+} form anion complexes to a significant extent, and even these can be ignored under most circumstances because they usually amount to less than 5 per cent of the total dissolved species. The major inorganic anions do not form complexes with the major cations to any significant extent. However, many of them do bind strongly with trace metals. See the data in Table 42.2.

In addition to the inorganic anions, there is a large family of organic anions, known as the **humic substances**. Humic substances, which are formed from dead organic matter, form the bulk of the dissolved organic carbon in freshwaters. They are mostly highly coloured, leading to the brown water commonly encountered in peat lands and coniferous forests. They are sometimes subdivided into **humic acids** and **fulvic acids**, the latter being soluble under acidic conditions, and generally of lower molecular weight than the former. Both types show a wide spectrum of acid/base properties and molecular weights. Indeed, the molecular weight range is such that it spans the gap between dissolved and particulate materials. Humic substances have a very great ability to bind to dissolved trace metals, and understanding their chemistry is one of the key objectives of current geochemical research.

42.4 Atmospheric carbon dioxide

So far, the chemical model for water can allow for hydrolysis, complex formation and dissolution of some ions. However, it does not yet include dissolution of gases from the atmosphere, which is a serious omission in the case of carbon dioxide (CO_2). Free exchange of gases normally takes place across the surface of the water, and this has a profound effect upon the chemistry of the water body.

Carbon dioxide dissolves in water to form a hydrated ion (i.e. CO_2 with lightly attached water molecules):

$$CO_2 + H_2O \rightarrow CO_2.H_2O$$

Some $CO_2.H_2O$ undergoes hydrolysis, forming carbonic acid:

$$CO_2.H_2O \rightarrow H_2CO_3$$

For the purposes of calculation, this step is not important and all the dissolved CO_2 can be treated together, thus:

$$^*H_2CO_3 = CO_2.H_2O + H_2CO_3$$

It is what happens next that is important, and can cause difficulty for the beginner. The reactions are simple though. First:

$$^*H_2CO_3 \rightleftharpoons HCO_3^- + H^+$$

then:

$$HCO_3^- \rightleftharpoons CO_3^{2-} + H^+$$

It is the prediction of how far these reactions proceed which causes problems. By applying le Chatelier's principle we can see that acidic conditions will favour the left-hand side (i.e. adding H^+ will cause the reactions to move to the left in order to consume some of the H^+). Thus, in very acidic conditions we should expect most $^*H_2CO_3$ and least CO_3^{2-}. The opposite should be true of basic conditions, and HCO_3^- will be most important in more neutral water.

However, we can do better than this. Studies of chemical equilibria have shown that reactions of the kind just shown are quite predictable. If we multiply the reaction product concentrations (right-hand side), and divide this value by the reactant concentration (left-hand side, in this case only one species), the result is equal to an **equilibrium constant**. (Strictly, this is true only for activities, but it holds reasonably well for concentrations in normal dilute freshwater. Better results can be obtained if the concentrations are expressed as activities. Refer to Stumm and Morgan, 1995, for discussion and procedures.) Thus:

$$\frac{[HCO_3^-] \times [H^+]}{[^*H_2CO_3]} = k_1 = 10^{-6.4}$$

$$\frac{[CO_3^{2-}] \times [H^+]}{[HCO_3^-]} = k_2 = 10^{-10.3}$$

For convenience these can be rearranged as follows:

$$[HCO_3^-] = \frac{[^*H_2CO_3] \times 10^{-6.4}}{[H^+]}$$

$$[CO_3^{2-}] = \frac{[HCO_3^-] \times 10^{-10.3}}{[H^+]}$$

For example, if the pH is 8 (i.e. $[H^+] = 10^{-8}$) and $[^*H_2CO_3]$ is 10 μM (or 10^{-5} M) then:

$$[HCO_3^-] = \frac{10^{-5} \times 10^{-6.4}}{10^{-8}}$$
$$= 10^{-3.4} \text{ or } 4.0 \times 10^{-4}\text{ M}$$

and

$$[CO_3^-] = \frac{4.0 \times 10^{-4} \times 10^{-10.3}}{10^{-8}}$$
$$= 10^{-1.3} \text{ or } 2.0 \times 10^{-6}\text{ M}$$

So, water with 10 μM of dissolved carbon dioxide is expected to contain 400 μM of bicarbonate, but only 2 μM of carbonate. The sum of carbonate species, or the total inorganic carbon, is the sum of all three species, i.e. 412 μM. This tells us that in water at pH 8, the dissolved carbonate is dominated by the bicarbonate ion. We can repeat this calculation for any water pH value. Figure 42.3 shows how the abundance of the three different carbonate species varies with pH.

The concentration value of 10 μM used to illustrate the dissolved CO_2 was chosen deliberately: it is approximately the concentration of CO_2 found in water at equilibrium with the atmosphere. This concentration does not vary

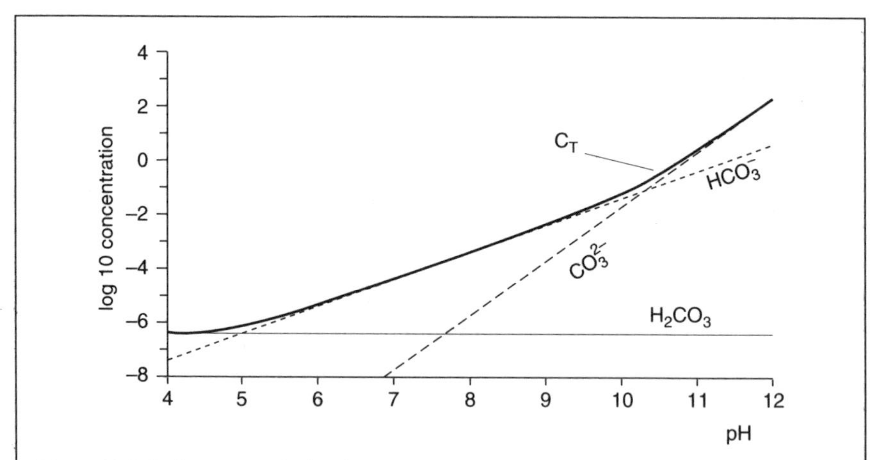

Figure 42.3 Carbonate speciation (here open to atmospheric CO_2) varies with pH. If the concentration of H_2CO_3 (carbonic acid) is held constant, then the concentrations of the CO_3^{2-} (carbonate) and HCO_3^- (bicarbonate) ions vary with pH as shown here. C_T (total carbonate species) is dominated by H_2CO_3 at pH values below 5, by HCO_3^- at pH values between 5 and 10.5, and by CO_3^{2-} at pH values above 10.5.

with the pH of the water. Thus, the graph in Figure 42.3 is a good model for waters that are in equilibrium with atmospheric CO_2.

According to Henry's law, the dissolved concentration of a gas is proportional to its partial pressure in the surrounding atmosphere. Thus, if the partial pressure of CO_2 in the atmosphere is known, then Henry's law can be used to calculate the concentration dissolved in water:

$$[^{*}H_2CO_3] = k_H P_{CO_2}$$

where k_H, the Henry's law constant, is about $10^{-1.4}$.

Is this realistic for all natural waters? Surface waters and atmospheric precipitation are close to equilibrium with the atmosphere, but soil water, groundwater, and deep lake water are often highly enriched in CO_2 relative to equilibrium with the atmosphere. This is because they get most of their CO_2 by the oxidative decomposition of organic matter (i.e. by respiration, see **34.3.3**). For such waters it is necessary to treat the problem differently. The total dissolved carbonate species varies with time, but at any one moment can be treated as fixed.

When water is open to the atmosphere, as in the case shown on the graph in Figure 42.3, the dissolved CO_2 is fixed (in the long run it can change temporarily). This makes calculation simple. However, when water is closed to the exchange of CO_2 the situation is different. In such a case the concentration of dissolved CO_2 can vary, whereas the total dissolved inorganic carbon is constant. This is because an increase in pH will convert some of the CO_2 into HCO_3^- (for example), which is not replaced as it is when it is open to the atmosphere.

The way this is calculated requires knowledge of the total dissolved carbonate species, C_T:

$$C_T = [H_2CO_3] + [HCO_3^-] + [CO_3^{2-}]$$

If we know the pH and the value of C_T then all the other species can be calculated. This is done by expressing all the carbonate species in terms of $[^{*}H_2CO_3]$, i.e.:

$$[HCO_3^-] = \frac{k_1 \times [^{*}H_2CO_3]}{[H^+]}$$

$$[CO_3^{2-}] = \frac{k_2 \times (k_1 \times [^{*}H_2CO_3]/[H^+])}{[H^+]}$$

Therefore

$$C_T = [^{*}H_2CO_3] \times k$$

where

$$k = 1 + \frac{k_1}{[H^+]} + \frac{k_2 k_1}{[H^+]^2}$$

From this:

$$[H_2CO_3] = \frac{C_T}{k}$$

Once we know $[^{*}H_2CO_3]$, the other species can be calculated as above.

It is important to realize that we have been simplifying things. We have assumed that the pH of the water is fixed. The truth is that if we change the CO_2 concentration of water, then under most circumstances the pH will also change. Can we calculate the expected pH values? The answer turns out to be yes, but only if we know a bit more about the water. We thus need to add one more part to the model.

42.5 Mass balance and charge balance: the heart of the chemical model

Our water model now has quite a large number of species: H_2O, H^+, OH^-, H_2CO_3, HCO_3, CO_3^{2-}, plus the many dissolved cations and anions. We also know how to calculate the abundance of a number of these species:

- We can calculate OH^- given H^+ and k_W.
- We can calculate H_2CO_3, HCO_3^- and CO_3^{2-} given H^+ and either P_{CO_2} or C_T.

However, what are we to do with the other ions in natural waters? It turns out that we can simplify the problem by lumping the other ions into four categories, as follows:

- *Strong cations*. These are positively charged ions that are fully dissociated (hence, 'strong') at normal pH values. The important common strong cations are Na^+, K^+, Ca^{2+} and Mg^{2+}.
- *Strong anions*. These are negatively charged ions that are fully dissociated (hence, 'strong') at normal pH values. The important common strong anions are Cl^- (chloride), F^- (fluoride), SO_4^{2-} (sulphate) and NO_3^- (nitrate). Usually less abundant is NO_2^- (nitrite).
- *Organic matter*. This forms an anion, the charge of which varies with pH. In colourless water it can be ignored (from an acid/base perspective), but it may be the most abundant anion in acidic brown-water systems.
- *Weak cations*. Aluminium, iron and ammonia all form hydroxide complexes, to a degree that varies with pH (hence they are weak cations). These weak ions usually have little impact on the ion balance of freshwater. However, in very acidic waters, aluminium and iron can be significant, and ammonia can be important in waters polluted with nutrients.

Now, in a simple model, the weak ions can be ignored. However, the strong ions must be considered. These can be handled easily using the new lumped variables – strong cations (SC) and strong anions (SA) – where, if concentrations are measured in equivalent units (see below):

$$SC = Na^+ + K^+ + 2Ca^{2+} + 2Mg^{2+}$$

$$SA = Cl^- + 2SO_4^{2-} + NO_3^- + F^-$$

Ionic charge is important because water always maintains electrical neutrality. Any temporary charge imbalance leads to very great forces, and an immediate restructuring of the water to reinstate electrical neutrality. This fact is very useful as it tells us that the sum of all positively charged ions must be equal to the sum of all negatively charged ions. Charge balance equations can be most simply solved if all the concentrations are expressed in the unit of equivalents per litre (usually $\mu eq\,l^{-1}$). However, all the carbonate equilibria are expressed in moles. It is easy to muddle units under these circumstances. It may be better to keep all units expressed in moles and to multiply the concentration for each species in the charge balance by its unit charge (usually its ionic charge). Thus the molar concentration of $[CO_3^{2-}]$ must be doubled when determining its contribution to the charge balance. Special care must be taken to allow for ion charge when summing the cations and anions to determine the SC and SA values.

A complete charge balance equation can be written as:

$$[SC] + [H^+] = [SA] + [OH^-] + [HCO_3^-] + [CO_3^{2-}]$$

if measured in equivalent units. This equation is more useful than it looks. It tells us the expected pH of a water body, given that we know strong ion concentrations. This can be done for both open systems (CO_2 open to the atmosphere) and closed systems. One thing that the equation tells us is that the acidity of a solution is ultimately governed by the balance of strong ions, i.e. by the relative importance of strong cations and strong anions. This balance can be represented by a single parameter, formed by subtracting the strong anions from the strong cations. This parameter is commonly termed the **acid neutralizing capacity** (ANC) where:

$$ANC = SC - SA$$

ANC is important because it is a measure of the external forces on the charge balance of the water. ANC is commonly referred to as the **alkalinity**. Alkalinity can be defined in a number of different ways: as used here, it is sometimes referred to as the titration alkalinity.

42.5.1 ANC and acid/base balance

The ANC increases with the excess of strong cations. In other words, a water body with a higher excess concentration of strong cations is less acidic. Conversely, water with an excess of strong anions is acidic. At first appearance, this seems paradoxical; in **12.4** you will see that cations tend to behave as weak acids, and anions as weak bases. The explanation is that the acidic and basic character of these ions is so weak as to be insignificant. We must look instead at what their presence in the solution signifies. It turns out that we can treat all strong cations (M) as if they were added to the solution in the form $MHCO_3$, i.e. $NaHCO_3$, $Ca(HCO_3)_2$, etc. (or MOH in the absence of carbon dioxide). Conversely, all strong anions (A) can be treated as if they were added in the form HA, i.e. HCl, H_2SO_4, etc. The final acidity of the combined solution depends only on the balance of strong ions, i.e. $ANC = SC - SA$. Now we present methods for applying this concept.

42.5.2 Using charge/mass balance qualitatively

If our measurements are in milliequivalents per litre then the sum of negatively charged ions must equal the sum of positively charged ions. This situation lends itself to a simple graphical presentation of the results, which allows a qualitative interpretation. The approach involves drawing pairs of columns where the height of the column represents concentration. One column is for the positively charged ions and the other is for the negatively charged ions. A series of steps to do this is:

1. Select an appropriate scale, such that the total column fits the graph page.
2. Taking care to represent their concentrations accurately (i.e. keeping height proportional to concentration), add the strong cations and anions to the columns.
3. If the anion column is taller than the cation column, then the solution is acidic, and the difference in the column heights is made up by H^+ ions. The pH can be found by reading off the H^+ concentration, scaling it according to the units used (i.e. if the H^+ concentration recorded in $\mu eq\,l^{-1}$, then the value must be divided by 1 000 000 to convert it to $eq\,l^{-1}$), and applying the equation $pH = -\log_{10}[H^+]$.
4. If the cation column is the taller, then the solution is basic, and the difference in the column heights is made up using bicarbonate (HCO_3^-). The approximate solution pH can be found from the expression $pH = 11.4 + \log_{10}[HCO_3^-]$. (Assuming that the solution is in equilibrium with atmospheric carbon dioxide, and that the pH is not greater than about 8). Remember to correct for scaling, as in step 3.

The above method is good for estimating likely pH values from mixtures of anions and cations in solutions open to the atmosphere. It is also excellent for predicting the impact of adding strong acids or bases to water (lakes, streams, etc.) of known composition (Figure 42.4). However, it produces poor results at alkalinities close to zero, when the columns have similar heights. Then, the calculation approach described below is necessary.

42.5.3 Using charge/mass balance quantitatively

To solve the charge balance expressions it is necessary to be able to write the individual terms as functions of $[H^+]$ and P_{CO_2}. This can be done as follows:

$$[OH^-] = \frac{k_W}{[H^+]}$$

$$[HCO_3^-] = \frac{k_1 k_H P_{CO_2}}{[H^+]}$$

$$[CO_3^{2-}] = \frac{k_1 k_2 k_H P_{CO_2}}{[H^+]^2}$$

Values for the constants, which vary with temperature, are given in Table 42.3.

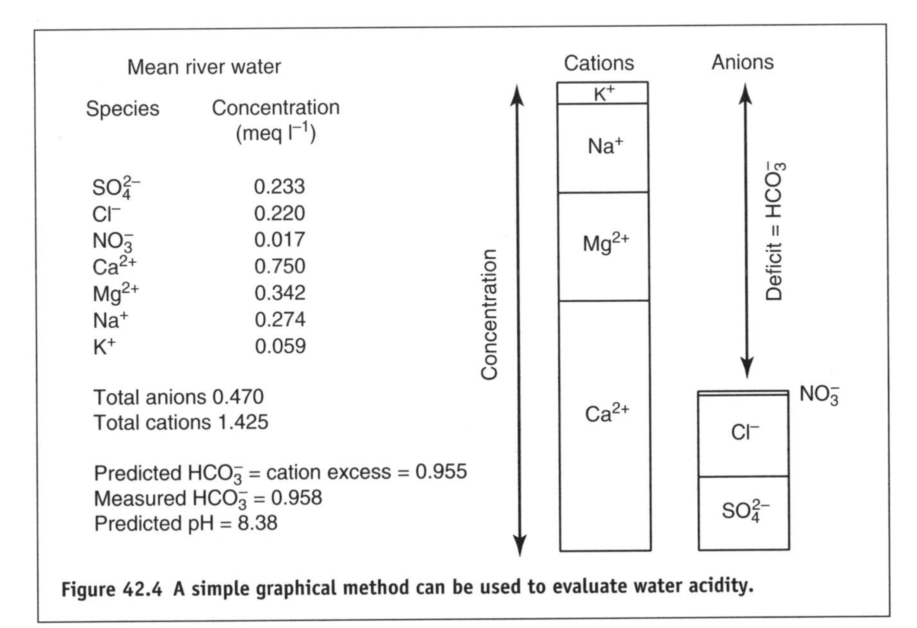

Figure 42.4 A simple graphical method can be used to evaluate water acidity.

To illustrate calculations using the charge balance approach, consider the case of pure water exposed to normal atmospheric carbon dioxide:

- SA = SC = 0; clearly there are no strong ions at all in this case.
- $P_{CO_2} = 10^{-3.5}$
- $[H_2CO_3] = k_H \times P_{CO_2}$ (from Henry's law, see **42.4**).
- The charge balance is: $[H^+] = [OH^-] + [HCO_3^-] + 2.[CO_3^{2-}]$.
- Each term in the charge balance equation can be expressed in terms of $[H^+]$ and $[H_2CO_3]$ from expressions and constants given above:

$$[H^+] = \frac{k_W}{[H^+]} + \frac{k_1 k_H P_{CO_2}}{[H^+]} + \frac{2k_2 k_1 k_H \times P_{CO_2}}{[H^+]^2}$$

- This can be rearranged for simplicity, and solved analytically as a quadratic equation.

Alternatively, the charge balance equation can be solved numerically in a spreadsheet using the following steps:

- In column A of a spreadsheet enter the range of possible pH values (0 to 14). The number of rows depends on the precision required. To obtain a precision of 0.1 pH units, put 0 in row 1, and increment each cell down the column by 0.1.
- In cell B1 calculate the $[H^+]$; make cell equal to $10^{-\text{value in cell A1}}$, i.e. = 10^A1.
- In cell C1 put the right-hand side of the charge balance expression above, taking the $[H^+]$ values from cell B1.

Table 42.3 Values for the constants at different temperature.

	5 °C	*10 °C*	*20 °C*	*30 °C*
k_w	$10^{-14.73}$	$10^{-14.53}$	$10^{-14.17}$	$10^{-13.83}$
k_H	$10^{-1.19}$	$10^{-1.27}$	$10^{-1.41}$	$10^{-1.53}$
k_1	$10^{-6.52}$	$10^{-6.46}$	$10^{-6.38}$	$10^{-6.33}$
k_2	$10^{-10.56}$	$10^{-10.49}$	$10^{-10.38}$	$10^{-10.29}$

Source: Stumm and Morgan, 1995.

- Make cell D1 equal to cell B1 – cell C1.
- Copy the cells B1 : D1 down the columns as far as necessary to allow for all the pH estimates.
- The values in column D will be positive for pH values below the correct answer, and negative for values above. Your estimate of the correct pH is the value corresponding to the column D value closest to zero.

The pH value you will find this way is 5.6. This is indeed the pH you would find if you measured (using appropriate equipment) the pH in distilled water that was open to the atmosphere. It is also the pH you will find in any water open to the atmosphere in which the ANC is zero. For water in which the ANC is not zero, the same procedure can be used but the ANC must be incorporated in the charge balance equation. If Al, Fe or organic matter are to be included in the model, then it is necessary to express their contributions to the charge balance equation in terms of $[H^+]$.

42.6 Summary

The model is now as complete as it can be within the scope of this book. The greatest significant weakness of the model presented here is that it treats concentrations as if they were activities. For dilute surface waters, and for mono- and divalent species this is not unreasonable. However, for trivalent ions, especially in more concentrated solutions, it is necessary to correct the concentrations using activity coefficients. In addition, there has been no treatment of redox conditions. The model as presented is, therefore, applicable only to oxygenated water. If anoxic water is being considered then factors influencing redox must be added.

Nonetheless, although much simplified, the model serves to show that by including a small number of chemical species, and applying a small number of rules governing their behaviour, the basic chemical properties of water can be predicted. To reinforce some of the key principles we show three simple examples of how the model can be applied in realistic situations to provide a quantitative theoretical framework to understanding water.

42.7 Example applications of the water chemical model

42.7.1 Example 1: acidification

What seasonal variation in lake water pH would you expect if:

(a) Summer inflow Ca concentration = 12 μM.
(b) Winter inflow Ca concentration = 8 μM.
(c) Summer pH = 6.5.
(d) Winter nitrate = 10 μM; summer nitrate = 0.

Assume that:

1. The lake adjusts fully to the stream water concentration.
2. That aqueous CO_2 is in equilibrium with the atmosphere, and has a partial pressure of $10^{-3.5}$ atmospheres.
3. That other strong acids and bases do not vary with time.

To solve the charge balance expressions it is necessary to be able to write the individual terms as functions of $[H^+]$ and P_{CO_2}. These are given above in **42.5.3**. Assume a temperature of 20 °C for all the calculations

To solve this problem, we must first find out what the alkalinity of the system is. Then we can find out how much of the alkalinity is due to $[Ca^{2+}]$ and how much to other strong ions. This allows us to answer all of the questions and can be achieved using the following steps.

1. Find the alkalinity for the initial summer conditions. The alkalinity is given by:

   $$\text{Total alkalinity} = [OH^-] + [HCO_3^-] + 2[CO_3^{2-}] - [H^+]$$

   Each term can be evaluated given $[H^+]$ (i.e. 10^{-pH}) and P_{CO_2}, both of which are known for the summer.
   You should get a value of 20.2×10^{-6} (i.e. 20.2 μeq l^{-1}).
2. Find the unknown, fixed, part of the alkalinity. This is simply:

   $$\text{Fixed alkalinity} = \text{Total alkalinity (summer value)} - 2[Ca^{2+}] \text{ (summer value)}$$

   You should get $20.2 - 24 = -3.8$ μeq l^{-1}.
3. Find the total alkalinity for the winter:

   $$\text{Total alkalinity (winter)} = \text{Fixed alkalinity} + 2[Ca^{2+}] \text{ (winter)} - [NO_3^-] \text{ (winter)}$$

   You should get $-3.8 + 16 - 10 = 2.2$ μeq l^{-1}.
4. Find the pH for the winter. This is linked directly with the winter value for the total alkalinity. The charge balance expression is:

   $$\text{Total alkalinity (winter)} = [OH^-] + [HCO_3^-] + 2[CO_3^{2-}] - [H^+]$$

To solve this, insert the P_{CO_2} for the soil water ($10^{-3.5}$) into the charge balance expression in a spreadsheet (see **42.5.3**). By trial and error find the $[H^+]$ value that gives a charge balance of zero. This is $[H^+]$ in equilibrium with the given alkalinity.
You should get 1.41×10^{-6} M (i.e. 1.41 μM).

5. Find the winter pH.

$$pH = \log_{10}[H^+]$$

You should get pH = 5.85.

Therefore seasonal pH variation is 5.85–6.5 for the conditions specified.

42.7.2 Example 2: carbon dioxide and acidity

A lake with a normal average pH value of 8.4 experiences an algal bloom. During the peak of the bloom the P_{CO_2} falls to 5 per cent of the normal atmospheric value owing to active photosynthesis. What is the pH value of the water during this event?

1. Assuming that the pH of 8.4 is in equilibrium with atmospheric CO_2, then the alkalinity of the lake can be found from the charge balance equation:

$$\text{Alkalinity} = [OH^-] + [HCO_3^-] + 2[CO_3^{2-}] - [H^+]$$

This can be solved exactly as in step 1 of the previous example.
You should get a value of 1.317×10^{-3} (i.e. 1.317 μeq l^{-1}).

2. During the algal bloom the P_{CO_2} is very much lower. However, the alkalinity remains unchanged. The new value for the P_{CO_2} is $10^{-3.5} \times 0.05 = 10^{-4.8}$. Given the alkalinity from step 1, and the new value for P_{CO_2}, we can find the new $[H^+]$ using the method from step 4 of the previous example (substituting alkalinity for the total alkalinity).
You should get pH = 9.6.

42.7.3 Example 3: karst chemistry

If soil water draining into a limestone had a P_{CO_2} of 10 times the normal atmospheric value, and a cave atmosphere had normal atmospheric P_{CO_2}, how much $CaCO_3$ would be deposited per litre of drip water in the cave system?

The amount of Ca in equilibrium with a given P_{CO_2} and calcite can be found from the charge balance equation, provided that an expression linking calcite solubility to pH (or $[H^+]$) is included. This is provided by the solubility product for calcium carbonate. The calcium concentration in equilibrium with calcite (a calcium carbonate mineral) can be found from:

$$[Ca^{2+}][CO_3^{2-}] = 10^{-8.3}$$

This can be put into a charge balance equation. If the concentrations are expressed in molar units we get:

$$2[Ca^{2+}] + [H^+] = [OH^-] + [HCO_3^-] + 2[CO_3^{2-}]$$

Expressions for most of the terms, and values for the constants are given in Example 1. The concentration of $[Ca^{2+}]$ in equilibrium with calcite can also be expressed in terms of $[H^+]$ and/or P_{CO_2}:

$$[Ca^{2+}] = \frac{2 \times 10^{-8.3}}{[CO_3^{2-}]}$$

$$= \frac{10^{-8.3}}{k_1 k_2 k_H P_{CO_2}/[H^+]^2}$$

The value for P_{CO_2} in the average atmosphere is approximately $10^{-3.5}$. At ten times this partial pressure the value will thus be $10^{-2.5}$.

1. Insert the P_{CO_2} for the soil water ($10^{-2.5}$) into the charge balance expression in a spreadsheet (see **42.5.3**). By trial and error find the $[H^+]$ value that gives a charge balance of zero. This gives the equilibrium $[H^+]$ value for the defined conditions.
   You should get $[H^+] = 2.218 \times 10^{-8}$ M, thus pH = 7.65.
2. Insert the value you found for $[H^+]$, and the soil water P_{CO_2}, into the expression giving $[Ca^{2+}]$ from calcite solubility. This gives you the concentration of $[Ca^{2+}]$ in the solution before it gets to the cave.
   You should get $[Ca^{2+}] = 1.153 \times 10^{-3}$ (i.e. 1.153 mM).
3. Repeat step 1, but use the P_{CO_2} of the cave ($10^{-3.5}$). This gives the new equilibrium $[H^+]$ after degassing of the water in the cave.
   You should get $[H^+] = 4.81 \times 10^{-9}$ M, thus pH = 8.32.
4. Repeat step 2, but use the new values for the cave water $[H^+]$ and P_{CO_2}. This gives the concentration of $[Ca^{2+}]$ in the drip water.
   You should get $[Ca^{2+}] = 0.542 \times 10^{-3}$ (i.e. 0.542 mM).
5. Subtract the drip water $[Ca^{2+}]$ from the soil water $[Ca^{2+}]$. This gives the amount of calcium, per litre of water, deposited in the cave.
   You should get $[Ca^{2+}]$ deposited = 0.611 mmoles per litre of supplied water.
6. Multiply the amount in moles by the molecular weight to get the answer in mass units.
   You should get $[Ca^{2+}]$ deposited = 24.4 mg per litre. This is equivalent to 61.1 mg of calcium carbonate.

43 Profiles of the Earth

43.1 Introduction

During this book we frequently refer to features of the Earth, particularly the various zones of the solid earth, the oceans and the atmosphere. It is the purpose of this section to provide a context for the phenomena introduced elsewhere. This section also reinforces our intention that this book is not meant to be an end in itself, but has been written to enhance your appreciation of those processes which are occurring in the world around us.

43.2 The solid Earth

A cross-section of the Earth's interior reveals a layered structure. The three principal layers – **core**, **mantle**, **crust** – shown in Figure 43.1 are distinguished by their chemical structure. The inner core is believed to be composed principally of iron, with some nickel. Owing to the great pressures experienced, the inner core is solid. Pressure is not so great in the outer core although the temperature is very high: it is thus believed to be molten. The surrounding mantle consists mainly of silicon, oxygen, iron and magnesium. The crust forms a skin which varies in thickness from 5 to 60 kilometres. In addition to silicon and oxygen the crust is rich in aluminium, sodium, calcium and potassium (refer back to Table 3.3).

The physical characteristics of materials, which are due to temperature- and pressure-dependent processes, also give rise to a layered structure. In this scheme, three layers are recognized. From the surface inwards these are **lithosphere**, **asthenosphere** and **mesosphere**. The relationship between the two schemes near the surface is shown in Figure 43.1b. The lithosphere is composed of crust and upper mantle and forms a rigid, but brittle layer. Its thickness varies considerably, from around 45 to over 120 kilometres. Owing to its brittle nature the lithosphere is fragmented into numerous

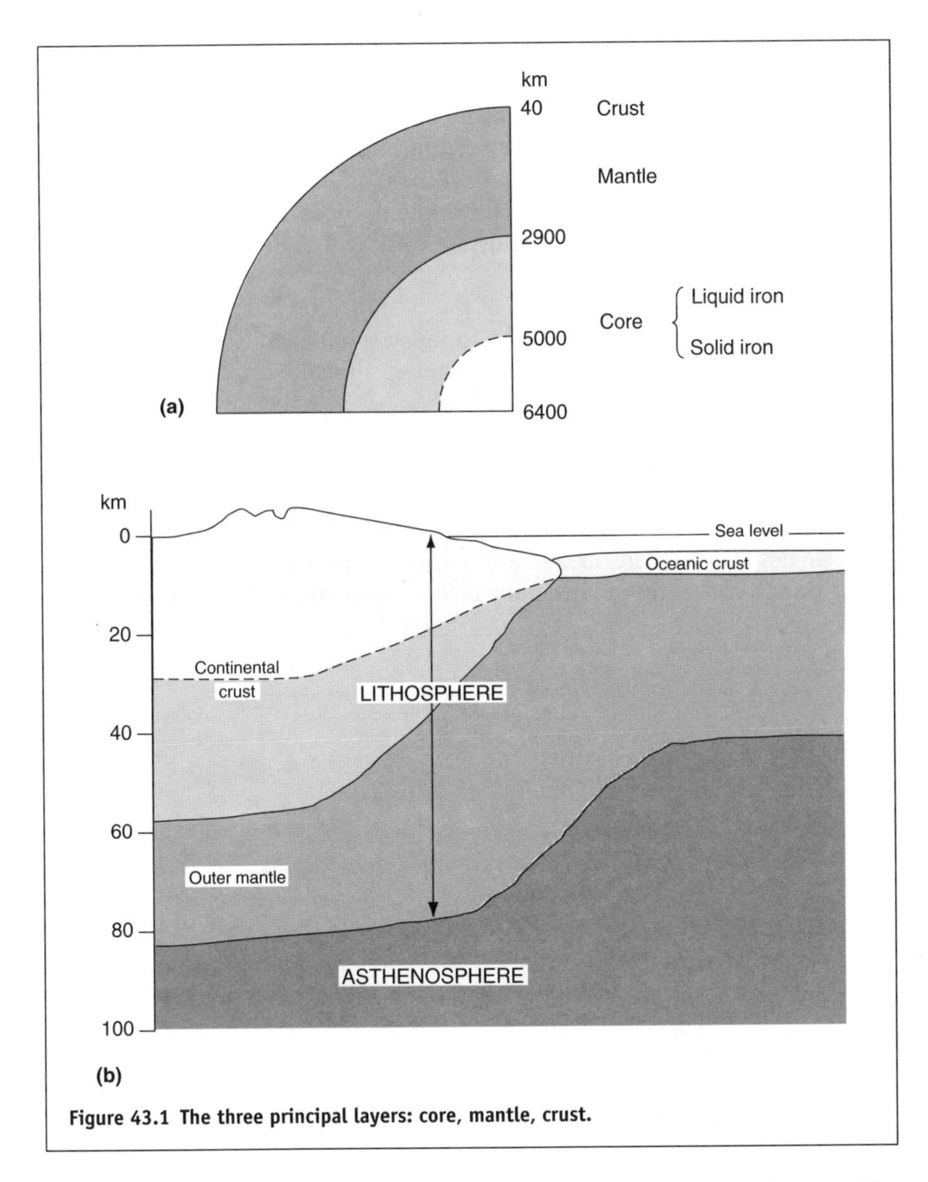

Figure 43.1 The three principal layers: core, mantle, crust.

plates (seven major, several minor), some of which carry continents. The movements of plates relative to each other are responsible for major changes in the configuration of land masses and mountain-building episodes over geological timescales, and are expressed as earthquakes.

A distinction is made between continental crust (thickness typically between 30 and 60 kilometres) and oceanic crust (typical thickness around 6 kilometres). Oceanic crust is composed mainly of mafic rock, primarily basalt and gabbro. (The name mafic derives from magnesium and the F in Fe, the chemical symbol for iron.) The upper part of continental crust is principally composed of felsic rock, usually granite. (The name felsic comes from the mineral feldspar.) The upper part of the crust merges, usually gradually, into a zone of mafic rock. Despite its name this rock has a very different origin from the oceanic crust, which is exuded at the mid-oceanic ridges. The lowest layer of the lithosphere is composed of ultramafic rocks, so called because of the high magnesium and iron content. This layer is beneath the crust and thus forms part of the mantle (Figure 43.1). The lithosphere floats on the plastic, partially molten rocks of the asthenosphere. The lower boundary of the asthenosphere is some 300 kilometres beneath the surface. The rest of the Earth's volume beneath the asthenosphere is called the mesosphere. Knowledge of the composition of the Earth's interior below the zone in which rocks can be sampled is obtained from the behaviour of **seismic waves**, i.e. those associated with earthquake activity.

43.3 The oceans

A scheme for zoning the ocean basins is shown in Figure 43.2a. Note that the slope angles shown here are greatly exaggerated. The continental shelves, which slope very gently, are generally below 180 metres and they extend from

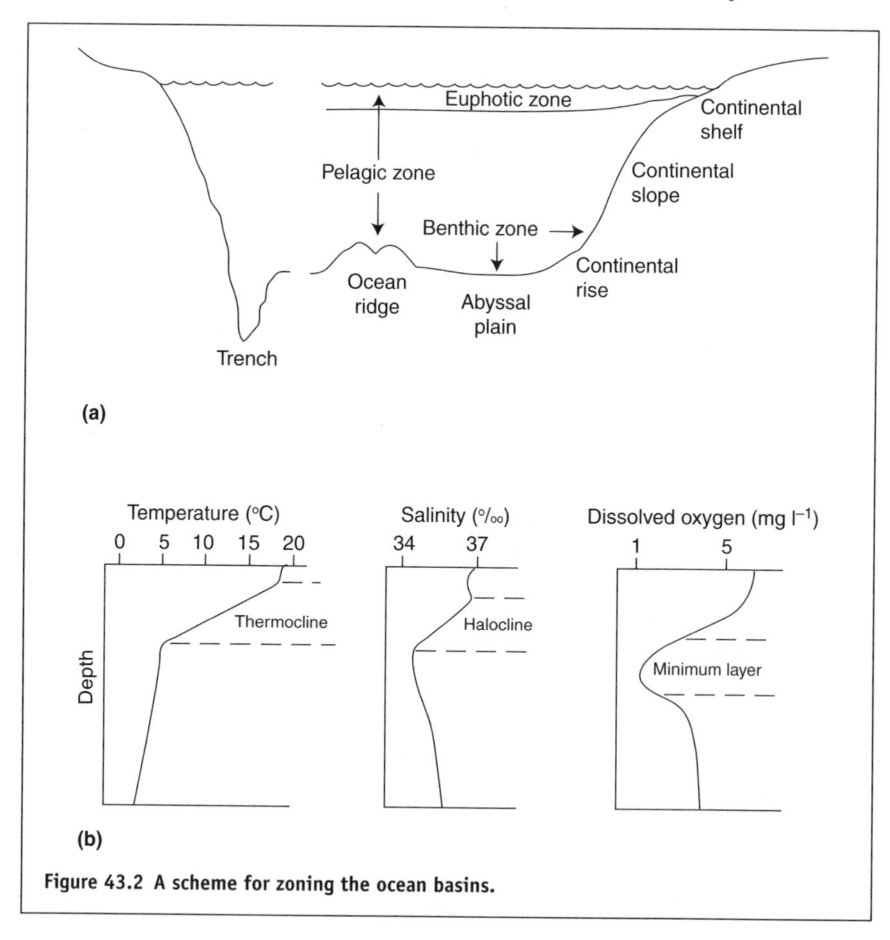

Figure 43.2 A scheme for zoning the ocean basins.

a few kilometres to over 300 kilometres from the coast. The inclination of the continental slope is slightly greater, at around 4 degrees from the horizontal. The depth of continental slope can exceed 3000 metres. The upper boundary of the continental rise is marked by a shallower angle than the slope. The continental rise may extend for several hundred kilometres. The ocean floor consists largely of abyssal plains, which are virtually horizontal and typically between 4.5 and 5.5 kilometres in depth, but this zone also contains canyons, hills and rises. The mid-ocean ridge is a continuous submarine mountain chain extending for some 65 000 kilometres through the ocean basins (but not always midway between two land masses as in the mid-Atlantic ridge). Along the central rift new oceanic crust is exuded. The trenches are the deepest parts of the oceans, extending in some cases to over 10 kilometres below the surface. These linear features mark the locations where oceanic plate material is consumed beneath less dense (and hence more buoyant) continental lithosphere.

The layer of water in which there is sufficient light for photosynthesis is called the **euphotic** (or **photic**) zone. Its depth varies spatially and temporally. In exceptionally clear water it can extend to 200 metres, but is generally very much shallower. The main body of open water is referred to as the pelagic zone while the benthic zone refers to the zone of the sea- (and lake-) floor.

Figure 43.2b shows typical vertical profiles for temperature, salinity and oxygen in the oceans. Near the surface there is a layer, up to 500 metres thick but usually much less, in which temperature declines gradually. In the thermocline below, which is typically between 500 and 1000 metres thick, the temperature gradient is relatively steep. Below the thermocline, temperature drops gradually. The details of the temperature/depth relationship vary considerably with latitude and, outside the tropics, with season. The temperature values shown are for illustration only. Changes in salinity with depth, although comparatively small, have a similar pattern to temperature. Dissolved oxygen is critical for practically all life-forms, and is therefore a key indicator of water quality.

The salinity of seawater is the weight of dissolved salts to seawater. It is conventionally expressed in parts per thousand, which is shown by the symbol ‰. This gives conveniently large values to work with. The average value is usually given as 35 parts per thousand (i.e. 35 grams of salt in every 1000 g of water). Near the surface, values vary, generally between 33 and 38 parts per thousand due to freshwater inflows, evaporation and precipitation. (In inland seas, salinity values can be much higher than in the oceans.) The proportions of the various salinity components are more-or-less constant in oceanic water. These are, in parts per thousand: sodium chloride 23, magnesium chloride 5, sodium sulphate 4, calcium chloride 1, potassium chloride 0.8. Chloride (19 parts per thousand) is the dominant anion, while sodium (10.5 parts per thousand) is the dominant cation.

43.4 The atmosphere

The atmosphere is composed of a mixture of gases, water vapour and tiny dust particles. Below about 100 kilometres, in a zone sometimes called the

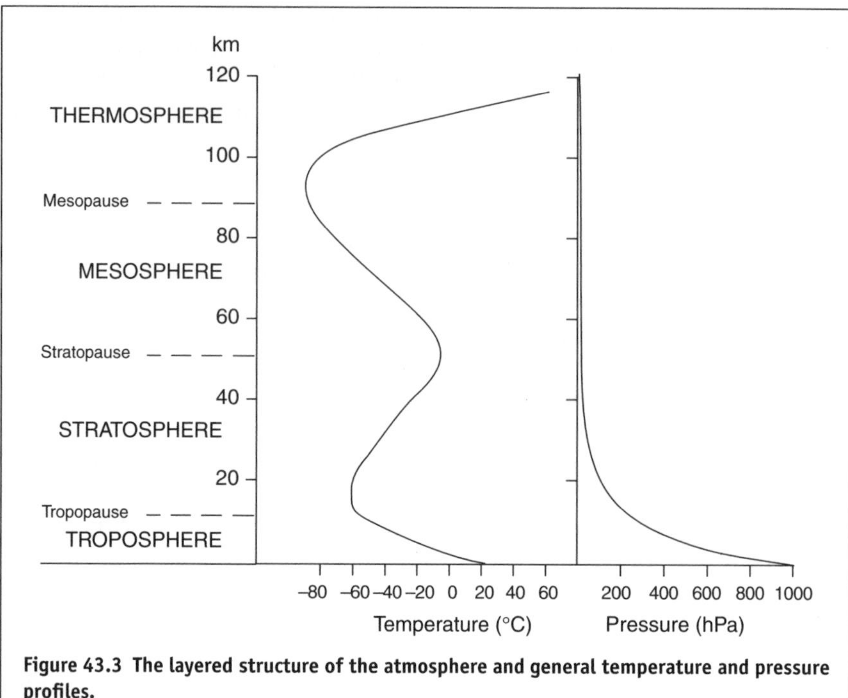

Figure 43.3 The layered structure of the atmosphere and general temperature and pressure profiles.

homosphere (Figure 43.3) the air is so well mixed that the contributions of the major gases are more-or-less constant. By far the most abundant gases in this zone are molecular nitrogen (78 per cent by volume) and molecular oxygen (21 per cent by volume). Argon accounts for just over 0.09 per cent. Other gaseous constituents make a tiny contribution to the total volume but some are crucially important because of their effect on radiation. Because percentage values are so small, amounts are usually expressed in parts per million (ppm), or even parts per billion. The approximate abundances of these gases are as follows: carbon dioxide (350 ppm), neon (1.8 ppm), helium (5.3 ppm), krypton (0.12 ppm), xenon (0.09 ppm), hydrogen (0.05 ppm), methane (0.02 ppm), nitrous oxide (0.05 ppm). The atmosphere also contains gases whose concentrations, although always tiny, vary. These include ozone, oxides of sulphur and nitrogen, and ammonia. Ozone, which is so important because it absorbs UV radiation, is concentrated in the upper stratosphere. Water vapour, whose concentration varies considerably spatially and temporally, plays a key role in the Earth's radiation balance because it absorbs heat. It is also the source of water for clouds and rain. Above about 100 kilometres, mixing processes are less effective and a layering of gases occurs owing to diffusion processes. This zone is therefore sometimes called the heterosphere. There is no outer 'edge' to the upper atmosphere, rather a very diffuse boundary over which matter becomes progressively scarce: for convenience some authorities use 500 kilometres as a convenient reference, some use 1000 kilometres.

Figure 43.3 shows the various layers of the atmosphere and general temperature and pressure profiles. Note the correspondence between the temperature profile and the layers in the atmosphere. In the troposphere, temperature tends to decrease with increasing altitude, although temporary inversions occur near the ground. The processes which give rise to the weather at the surface take place within the troposphere, which extends to about 8 kilometres or so. Above the tropopause, in the lower stratosphere the temperature is relatively uniform. However, temperature rises in the upper stratosphere owing to the absorption of UV radiation by ozone. Temperature declines progressively in the mesosphere, but above the mesopause (about 90 kilometres) a reversal occurs and temperature again rises. This is due to the absorption of radiation of shorter wavelengths (X-rays, gamma rays and some ultraviolet rays).

Gravity tends to pull all matter within the atmosphere towards the Earth's surface, in effect forcing the gas molecules closer together. The density of the atmosphere (i.e. mass of all atmospheric constituents per unit volume) thus declines from the surface outwards. This decline is rapid: at sea level, density is around $1200\,g\,m^{-3}$, at 7 kilometres it is $600\,g\,m^{-3}$, and at 18 kilometres only $120\,g\,m^{-3}$. At 100 kilometres, in the lower mesosphere, it is just $0.001\,g\,m^{-3}$. The concentration of matter continues to decline with altitude throughout the thermosphere and there is a gradual transition to the virtual vacuum conditions of 'space'.

Because the atmosphere has mass it exerts a force on any object within it. This force is known as the atmospheric pressure, and it is one of the key variables measured in meteorological observation. Pressure is simply force per unit area. The atmospheric pressure on a surface at any altitude represents the mass of all atmospheric constituents in a vertical column above the surface. With increasing altitude the length of such a vertical column decreases. The general relationship between altitude and atmospheric pressure is shown in Figure 43.3. At around 5.5 kilometres, atmospheric pressure is about half that at sea level, and at about 16 kilometres it is only one-tenth of the sea-level value. If the logarithms of the pressure values were to be plotted against altitude the relationship between pressure and altitude would approximate a straight line.

Recommended reading

Scientific dictionaries

CLUGSTON, M.J. (1998) *The New Penguin Dictionary of Science*. Penguin Books, London.

ISAACS, A., DAINTITH, J. and MARTIN, E. (1997) *Concise Colour Science Dictionary*. Oxford University Press, Oxford.

ISAACS, A., DAINTITH, J. and MARTIN, E. (1999) *A Dictionary of Science*. Oxford University Press, Oxford.

MACDONALD, A.M. (ed.) (1977) *Chambers Twentieth Century Dictionary*. Chambers, Edinburgh.

Any of the above would be a reliable and helpful companion.

Reference manuals for the practitioner

GREER, A. and HANCOX, D.J. (1989) *Tables, Data and Formulae for Engineers and Mathematicians*. Stanley Thornes, Cheltenham.

STARK, J. and WALLACE, H.G. (1982) *Chemistry Data Book* (2nd edition). John Murray, London.

TALLARIDA, R.J. (1999) *Pocket Book of Integrals and Mathematical Formulas*. Chapman and Hall/CRC, Boca Raton, Florida.

Basic sciences

ATKINS, P. (1995) *The Periodic Kingdom: A Journey into the Land of the Elements*. Weidenfeld & Nicolson, London.

A very enjoyable and informative introduction to the chemical elements and the periodic table.

BALL, B. (1999) *H_2O: A Biography of Water.* Weidenfeld & Nicolson, London.
Well-written and informative account of this key ubiquitous substance.

BRADBURY, I. (1998) *The Biosphere* (2nd edition). John Wiley, Chichester.
Accessible introduction to living processes and the functioning of ecological systems written primarily for non-biologists.

CHOWN, M. (1998) *The Magic Furnace.* Jonathan Cape, London.
Excellent account of the origin and discovery of the elements and how the particulate nature of the atom was discovered.

COX, P. (1995) *The Elements on Earth: Inorganic Chemistry in the Environment.* Oxford University Press, Oxford.
A readable introduction to the chemical elements in the environment: extremely useful.

FEYNMAN, R. (1998) *Six Easy Pieces: The Fundamentals of Physics Explained.* Penguin Books, London.
An elegant introductory account of key physical phenomena by one of the outstanding physicists of the modern era.

HALIDAY, D., RESNICK, R. and WALKER, J. (1993) *Fundamentals of Physics* (4th edition). John Wiley, New York.
Solid introduction to the basics of the subject.

HALL, N. (ed.) (1998) *The Age of the Molecule.* Royal Society of Chemistry, London.
Richly illustrated publication about the place of chemical substances in our everyday lives but with important chemical lessons about the basics.

HALL, N. (ed.) (2000) *The New Chemistry.* Cambridge University Press, Cambridge.
Several distinguished contributors have produced this authoritative but accessible overview of the chemical sciences. Well received and highly recommended.

MUNCASTER, R. (1981) *A-level Physics.* Stanley Thornes, Cheltenham.
Reliable coverage of pre-university-level physics in England and Wales.

ROSE, S. and MILEUSNIC, R. (1999) *The Chemistry of Life* (3rd edition). Penguin Books, London.
Very good – and inexpensive – introduction to biochemistry for the non-specialist.

STRATHERN, P. (2000) *Mendeleyev's Dream: The Quest for the Elements.* Hamish Hamilton, London.
An engaging history of chemistry using the development of the periodic table as vehicle.

GORDON, A., GRACE, W., SCHWERDFETGER, P. and BYRON-SCOTT, R. (1998) *Dynamic Meteorology: A Basic Course*. Arnold, London.
Accessible but rigorous introduction to the subject.

HARRISON, R.M. and DE MORA, S.J. (1996) *Introductory Chemistry for the Environmental Sciences*. Cambridge University Press, Cambridge.
Good text for those with some chemical background. Very good section on analytical methods.

HARVEY, J.G. (1976) *Atmosphere and Ocean: Our Fluid Environment*. Artemis, London.
Despite its age, this remains a very useful text.

MANAHAN, S.E. (1999) *Environmental Chemistry* (7th edition). CRC/Lewis, Boca Raton, Florida.
Comprehensive and accessible introduction to all aspects of the subject – highly recommended.

O'NEILL, P. (1998) *Environmental Chemistry* (3rd edition). Blackie Academic, London.
Succinct introduction to the subject for those familiar with basic chemistry.

STUMM, W. and MORGAN, J. (1995) *Aquatic Chemistry* (3rd edition). John Wiley, New York.
Comprehensive standard reference, but not for the beginner.

Index